NEW DEVELOPMENTS IN MUTATION RESEARCH

NEW DEVELOPMENTS IN MUTATION RESEARCH

CHARLES L. VALON
EDITOR

Nova Science Publishers, Inc.
New York

For permission to use material from this book please contact us:
Telephone 631-231-7269; Fax 631-231-8175
Web Site: http://www.novapublishers.com

Library of Congress Cataloging-in-Publication Data
New developments in mutation research / Charles L. Valon, editor.
p. cm.
Includes index.
ISBN 1-59454-664-9
1. Mutation (Biology) I. Valon, Charles L.
[DNLM: 1. Mutation--genetics. QU 500 N532 2006]
QH460.N49 2006
576.5'49--dc22 2005031837

Published by Nova Science Publishers, Inc. ✣New York

CONTENTS

PREFACE

Mutation refers to any change in the DNA of a cell. Mutations may be caused by mistakes during cell division, or they may be caused by exposure to DNA-damaging agents in the environment. Mutations can be harmful, beneficial, or have no effect. If they occur in cells that make eggs or sperm, they can be inherited; if mutations occur in other types of cells, they are not inherited. Certain mutations may lead to cancer or other diseases. This new book gathers together and presents the latest research in this field.

Chapter one states that hereditary nonpolyposis colorectal cancer (HNPCC) is one of the most common hereditary colon cancer syndromes. The causative genes of HNPCC are DNA mismatch repair genes, and inactivation of these genes triggers HNPCC tumors.

Germline mutations of the hMSH2, hMLH1 and hMSH6 genes were detected in Japanese HNPCC families. These mutations included single-base substitutions and frameshifts, both resulting in truncated proteins, a mutation at the splice donor site of exon 5, and a 2kb genomic deletion encompassing exon 5. The coexistence of germline and somatic mutations of DNA mismatch repair genes was observed in colorectal and extracolorectal cancers. The majority of somatic mutations were frameshift, one was a mutation at the splice donor site of exon 5, and two were loss of the normal allele. All HNPCC cancers exhibited high microsatellite instability. Sixty-four % of HNPCC colorectal cancers included somatic mutations of either the APC or ß-catenin gene, whereas only 13% of the cancers had p53 mutations and none of the cancer cases showed the BRAF mutation. The loss of heterozygosity at tumor suppressor regions (chromosomes 5q, 8p, 17p, 18q and 22q) was not observed. Also detected were frequent somatic frameshift mutations of growth-related genes with coding repeats in HNPCC colorectal cancers. Mutation frequencies at these repeats were 77% at TGFßRII$(A)_{10}$, 55% at hMSH3$(A)_8$, 52% at CASP5$(A)_{10}$, 48% at BAX$(G)_8$, 48% at RIZ$(A)_8$ and $(A)_9$, 45% at RAD50$(A)_9$, 45% at MBD4$(A)_{10}$, 36% at hMSH6$(C)_8$, 23% at BLM$(A)_9$, 19% at IGFIIR$(G)_8$ and 19% at PTEN$(A)_6$.

The present data have the following implications: Inactivation of DNA mismatch repair genes through germline and somatic mutations causes genetic instability resulting in somatic mutations in various growth-related genes. Disruption of the WNT signaling pathway by mutations of the APC or ß-catenin gene and disruption of the TGF-ß signaling pathway by alteration of the TGFßRII gene seem to largely contribute to the development of HNPCC colorectal cancer via adenomas. Moreover, inactivation of apoptosis-related genes, such as the BAX, CASP5 and RIZ genes, may also contribute to the development, and inactivation of the IGFIIR and PTEN genes play a role in the progression of HNPCC colorectal cancer.

Chapter two informs us that polycyclic aromatic hydrocarbons (PAH) are widespread environmental contaminants, proven to be potent carcinogens in rodent bioassays and implicated in the etiology of several human cancers. PAHs must be metabolized to chemically reactive electrophiles to initiate this carcinogenic process by covalent interactions with DNA. This chapter discusses mutations induced by the metabolites of two environmental carcinogenic pollutants, the bay and fjord region-containing PAH, benzo[*a*]pyrene and dibenzo[*a,l*]pyrene, respectively. These PAH-DNA lesions can cause distortions in the DNA molecule and influence enzyme activity, DNA binding, or repair of bulky DNA adducts. Here, considered and discussed is the induction of potentially carcinogenic mutations, which is a biological consequence of the formation of PAH-DNA adducts. The PAH-DNA adduct-induced mutational frequency and spectra has been established through various test systems. This review compares in the mammalian *HPRT* gene model mutation test system, and the types of mutations induced by carcinogenic diol-epoxide metabolites of BP and DBP. Further examination included the influence of PAHs in the environment on mutation induction, specifically PAH-induced mutations in tumor suppressor genes and proto-oncogenes. In this chapter, also highlighted was the point that replication on PAH-DNA adducts could produce mutations leading to cancer.

In chapter three tandem repeats are DNA nucleotide sequences widespread throughout the genomes of various organisms, including humans and the mouse. Most of the tandem repeats are stably transmitted to successive generations; however, some exhibit genetic instability, under certain conditions or without any obvious reasons, during the transmission to the next generation. Tandem repeats that exhibit such instability are often referred to as "hyper-variable" tandem repeats.

The genetic instability of tandem repeats has been utilized in detecting the hazardous effect of environmental mutagens, including chemical compounds and ionizing radiation. For instance, *Pc-1* (*Ms6-hm*) is one of the hyper-variable tandem repeats in the mouse genome, and its hypersensitivity is very useful for the detection of the biological effects caused by environmental mutagens. *Pc-1* comprises (GGCAGG)n and exhibits a genetic instability in its copy number during trans-generational transmission. Although no disease has been identified in relation to the genomic instability of the *Pc-1* tandem repeat thus far, its hyper-mutability has been widely utilized as a sensitive detector able to monitor the hazardous effects of environmental mutagens. The instability of the *Pc-1* tandem repeat is also affected by the genetic background of mice. For instance, functional mutations which occur in DNA-dependent protein kinase catalytic subunit (DNA-PKcs) are able to increase the genetic instability of *Pc-1* tandem repeat.

Triplet repeats are the specific type of tandem repeats composed of trinucleotide repeats in the human genome. While most of triplet repeats are stable during trans-generational transmission, some exhibit genetic instability. Genetic instability of certain trinucleotide repeats has been identified as one of the causes of genetic disease in human populations. The first human disorder revealed to be caused by CCG trinucleotide repeat expansion was the so-called fragile X syndrome. This new type of mutation, caused by repeat expansion (and/or contraction), is categorized as a dynamic mutation. Repeat expansion has explained many characteristic aspects of the syndrome, such as the dominant inheritance and anticipation (i.e., juvenile onset or severer manifestation of the disease phenotype), which are observed in the affected pedigrees. Cells derived from the patients affected by the fragile X syndrome also exhibit a specific chromosomal fragile sites at Xq27.3, thus the identification of CCG repeat expansion has facilitated research into both the clinical manifestations of the syndrome and cytogenetic expression of the chromosomal fragile sites at the molecular level. CAG repeat expansion has also been identified in many neurodegenerative disorders characterized by an autosomal dominant pattern of inheritance and anticipation, such as Huntington's disease and the spinocerebellar ataxias. Recently, new repeat expansions, a tetranucleotide and a pentanucleotide repeat expansion, have been identified in type2 myotonic dystrophy (DM2) and type10 spinocerebellar ataxia (SCA10), respectively.

Dynamic mutations occurred at the polyalanine tract in the coding regions of the transcription factor genes cause the malformation syndromes. Most of the dynamic mutations causing polyalanine diseases seem to have arisen in the past, and these repeats to have been since transmitted stably to the present. However, this being the case, it is nevertheless likely that the original dynamic mutation occurred through a similar molecular mechanism observed in the other repeat diseases caused by genetically unstable repeats. Non-lethal symptoms of polyalanine diseases may have allowed such mutations to be inherited while avoiding strong pressure in the general population, and subsequent introduction of a point mutation interrupting the perfect repeat may have stabilized the repeats. Significant roles for polyalanine expansion in morphological evolution of vertebrates have been suggested.

The molecular mechanism for dynamic mutation has yet to be elucidated, and it is of great interest to understand both the mechanism itself and the dynamism and stability of the mammalian genome.

In chapter four transgenic mouse and rat mutation assays have provided a tool for analyzing *in vivo* mutation in any tissue from which sufficient amounts of DNA can be recovered. Thus these assays provide a method for directly comparing cancer incidence with mutant frequency. In these systems, chromosomally-integrated transgenes are used as the targets for mutation. Following recovery of the transgenes, the target DNA can be assayed *in vitro* to identify and quantify mutations produced *in vivo*. In this article, the progress in the development, validation, and use of transgenic *in vivo* mutation models was reviewed.

Chapter five shows that reactive oxygen species (ROS) are constantly being generated in all living organisms. Their origin may be endogenous, such as those arising from normal aerobic metabolism, or exogenous, produced as a consequence of exposure to environmental chemicals or irradiation. ROS play an essential function in performing inter- and intracellular signaling but, when they are in excess, their effects are deleterious as they modify cellular macromolecules, accumulation of which has been associated with numerous diseases. In DNA, ROS induce different types of base- and sugar-derived DNA lesions, of which the guanine modification, 7,8-dihydro-8-oxo-2′-deoxyguanosine (8-OH-dG) is best studied and is also most often used as a biomarker of oxidative DNA damage. If not repaired, these lesions can have a number of deleterious consequences, including induction of base pair mutations, rearrangements, deletions, insertions and sequence amplification. In addition to point mutations, ROS can produce gross chromosomal alterations, which may lead to the inactivation of tumor suppressor genes or expression of mutated proto-oncogenes during tumor promotion and progression.

DNA repair mechanisms are extremely important for maintaining DNA integrity. Numerous pathways are involved in the repair of oxidative DNA damage, however, it is still not known which pathways and proteins are responsible for removing specific types of oxidative DNA lesions.

There is evidence that many carcinogens and/or their metabolites, apart from inducing DNA damage by direct bonding to DNA, also induce ROS formation and oxidative DNA damage. However, when detecting mutagenic effects it is generally difficult to distinguish the contribution of direct from that of ROS mediated DNA damage. Compounds of arsenic, cadmium, chromium (VI), cobalt and nickel are all classified as human carcinogens. They have all been shown to induce intracellular ROS formation and oxidative DNA damage. They are also known to interfere with DNA repair mechanisms. Therefore the carcinogenicity of metals can, in part, be explained by inducing oxidative DNA damage and by decreasing the repair capacity towards DNA lesions, which may in turn increase the risk of tumor formation. Carcinogenesis is a multistep process requiring accumulation of changes through tumor initiation, promotion and progression, which include DNA mutation, activation of oncogenes and inactivation of tumor suppressor genes. Oxidative mechanisms have been shown to be involved in all these steps. The data from human studies obtained so far support the experimentally based indication of oxidative DNA damage as an important mutagenic and apparently carcinogenic factor. However, proof of a causal relationship in humans is still lacking. In the future the use of biomarkers may provide this evidence and allow further investigation of the qualitative and quantitative importance of oxidative DNA modification and carcinogenesis in humans, and also suggest possible preventive measures.

In chapter six, Pierisin-1 is a potent apoptosis-inducing protein found in a pupal extract of the cabbage white butterfly, *Pieris rapae*. Pierisin-1 catalyzes mono (ADP-ribosyl)ation of the 2'-deoxyguanosine (dG) residue and produces a bulky adduct, N^2-(ADP-ribos-1-yl)-2'-deoxyguanosine N^2-ADPR-dG in DNA, which leads to the apoptosis of tumor cells. N^2-ADPR-dG generated G to C and T transversion mutations in mammalian cells when assayed with the endogenous *HPRT* gene and the *supF* gene on a shuttle vector as the marker gene. Mutations are often generated during bypassing of damaged nucleotides in DNA, as explored in the translesion DNA synthesis across N^2-ADPR-dG conducted by mammalian Y-family DNA polymerases *in vitro*. A site-specifically modified oligodeoxynucleotide containing a single N^2-ADPR-dG was used as a template for ^{32}P-labelled-primer extension reactions catalyzed by DNA polymerase (Pol) η, ι and κ. The results showed that Pol κ bypassed through N^2-ADPR-dG by inserting dCMP preferentially opposite the lesion. Pol κ also extended DNA synthesis from a mismatched primer terminus where dG, dA or dT had been placed opposite the lesion. Pol ι incorporated dCMP and dTMP opposite N^2-ADPR-dG, but could not extend the DNA strand past the lesion. Pol η inserted dCMP and extended the DNA strand poorly past the lesion. These results could not clarify completely which polymerase generates G to C and T transversions as observed *in vivo* mutation analyses, but indicated that Pol κ could extend DNA synthesis from the mismatch terminus once a misincorporation of non-complementary nucleotide occurred. Therefore, Pol κ may play an important role in both the error-free and error-prone bypass of N^2-ADPR-dG.

DNA double-strand breaks (DSB) in chapter seven are considered to be one of the most serious forms of DNA damage, because if left unrepaired, they can cause cell death and, if misrepaired, they can lead to genomic instability and, ultimately, the development of cancer in multicellular organisms. Efficient DSB repair is therefore critical to preserve genomic stability and cell viability. There are two major pathways for the repair of DSB in a cell: homologous recombination (HR) and non-homologous end-joining (NHEJ).

In mammalian cells, DSB are repaired primarily through the NHEJ pathway. The essential mammalian NHEJ factors are the DNA-dependent protein kinase (DNA-PK), consisting of the catalytic subunit (DNA-PKcs) and the two DNA end-binding components (the KU70 and KU80 proteins forming a stable complex), the DNA ligase IV/XRCC4 complex and the ARTEMIS protein. In contrast to mammalian cells, the budding yeast *Saccharomyces cerevisiae* repairs DSB mainly by HR, although NHEJ pathway is also of notable relevance. This organism possesses the KU70, KU80, DNA ligase IV and XRCC4 homologues (YKu70, YKu80, Lig4 and Lif1, respectively), while it lacks clear functional homologues of either ARTEMIS or DNA-PKcs. However, *S. cerevisiae* cells possess an additional regulatory NHEJ factor, the Nej1 protein, which has not yet been found in any other organisms. Furthermore, there is evidence in the budding yeast for a requirement for the Mre11/Rad50/Xrs2 complex for NHEJ, which does not appear to be the case in mammals.

DSB by NHEJ had previously been thought to be restricted to the eukarya. However, functionally homologous end-bridging and ligation activities have now been identified in prokarya and recent papers have demonstrated that bacterial Ku and ligase proteins form a *bona fide* NHEJ system that encodes all the recognition, processing and ligation activities required for DSB repair. These findings provide new evolutionary insights into the basic biochemistry of this repair pathway, and suggest that the basic DSB repair process is evolutionary much older than previously thought.

Here reviewed is the current knowledge on yeast and bacterial NHEJ. The authors investigate the possible origins of this NHEJ repair system, examining the insights it has given us into the biochemistry underlying the DSB repair process and how it may also play a role in stationary phase mutagenesis leading to genome diversification under conditions of stress.

Studies on mutagenic agents in chapter eight have mostly focused on potential effects in humans. Despite the presence of mutagens in most of the pollution sources, the impact of mutagens on natural populations is still poorly understood. Mutagens in the environment may represent a long-term risk for ecosystems. Although exposure of aquatic organisms to genotoxic contaminants can pose a risk to human health via the food chain, there is also an ecological risk that may lead to heritable mutations and loss in the total genetic diversity. Mutagens can also directly affect the reproductive potential of populations by alterations in the fecundity and offspring viability caused by germ cell mutations. Among the biomarkers of germ cell effects, dominant lethal events have been used as convenient indicators of genetic damage since they are easily detectable. The so-called dominant lethal test with rodents has generally been considered one of the most important methods for the estimation of the mutagenic effects in humans. Freshwater mollusks of the genus *Biomphalaria* are good indicators for biomonitoring studies. The genus shows wide geographical distribution, has low dispersion and is easily collected. Other aspects make this system a good model for laboratory studies. The snails are easy to breed, need little space, can reproduce throughout the year under controlled conditions and have a short life-span: an egg-to-egg monitoring can be done in two months in *B. glabrata*. The dominant lethal assay was established in *B.glabrata* to evaluate mutagenic effects on germ cells by using a similar approach to that used in the test in rodents. Albinism was used as genetic marker to adapt the test for *Biomphalaria*, since the animal is a simultaneous hermaphrodite and reproduces by self- and cross-fertilization. Adult wild-type snails are exposed for ten to 30 days to the test agent in aqueous solutions. The range of concentrations is defined in preliminary experiments. After the end of the exposure, animals are crossed with albino snails at different intervals to evaluate the effects in the germ cells at the different stages of spermatogenesis. The analysis of lethal effects in the heterozygous offspring indicates the induction of mutations on germ cells of the exposed wild-type snails. Dominant lethal test in B. glabrata can be an useful tool to predict ecological changes in biomonitoring studies. Germ cell effects are biomarkers of great value in determining the implications for ecosystems of exposure to pollution since they reflect changes that affect the overall reproductive capability.

Chapter nine illustrates that the dendrimer is the common name of the polymers which are characterized by a specific structure, although their chemical composition can be very different. Dendrimers display several characteristics which distinguish them from linear polymers. All dendrimers are synthesized by attaching branched monomers to the central core molecule. Monomers are attached radially, layer by layer. As a result of such a synthesis, dendrimers possess plenty of terminal, reactive groups on their surface and empty internal cavities. Both of these features can be utilized. Soon after the first synthesis in mid-1980's, dendrimers have become interesting and promising molecules for pharmaceutical aims. First, they turned out to be efficient transfection agents. They are able to transport bigger amount of the genetic material than liposomes and do not cause a strong immunological response as viral vectors do. Second, dendrimers can play an important role as drug-delivery-systems. Drug molecules can be attached to the end groups or encapsulated in the dendrimer's interior.

To continue the studies on the dendrimers' role in medicine, it is important to know more about their toxicity and biocompatibility. The last ten years brought many new facts about dendrimers impact on biological systems. Both, general cytotoxic studies, as well as detailed research devoted to particular cellular compounds, as DNA or lipid bilayer, were carried out.

In this review the development of last ten years studies how dendrimers affect different biological systems is reported. The emphasis is put on basic studies on interactions between dendrimers and DNA or a lipid bilayer to explain dendrimers' high efficiency as transfection agents.

Chapter ten states that diabetes is a growing threat to public health worldwide. It is clasified into two main types. Type 1 diabetes is characterized by loss of insulin production due to autoimmune destruction of the pancreatic islet β cells. Type 2 diabetes is characterized by a deficiency in the capacity of insulin secretion and insulin resistance in peripheral tissues. Genetic factors seem to play an important role in the pathogenesis of diabetes. Several genes, which are involved in the metabolic process including the synthesis and secretion of insulin in pancreatic β cells and the insulin action on various target cells, have been proposed to be associated with or responsible for the pathogenesis of the disease. Some mutations in genes encoding for insulin, insulin receptor, glucokinase, glycogen synthase and suppressor of cytokine signalling 3 detected diabetic patients. Also mitochondrial DNA (mtDNA) has been suggested to be one of the genetic factors for diabetic development. mtDNA was shown to be required for the regulation of glucose-stimulated insulin secretion in a pancreatic β cell line of the mouse. Further studies which show a somatic mutation in mtDNA from peripheral blood cells in diabetic patients, have provided strong evidence to support the notion that mitochondrial function is crucial for the secretion of insulin by pancreatic cells. Not only mutational but also quantiative changes in mtDNA have been observed in diabetic patients. In diabetes, both hyperglycaemia and hyperinsulinemia can increase oxidative stress by produces ROS through the polyol pathway, glyco-oxidation, impairment of antioxidant system and activation of endothelial NAD(P)H oxidase. Accordingly, diabetes has been suggested to enhance DNA damage by ROS, which in turn, can cause diabetic complications. It has also been reported that point mutations and deletions of mtDNA accumulate due to intensive exposure of mtDNA to reactive oxygen species (ROS). On the other hand, methylation of chromosomal DNA has genetically important regulatory effects on gene expression , especially when involving cpG-islands located in the promoter regions of many genes. There is considerable indirect evidence that aberrant DNA methylation plays a role in the development of type 2 diabetes. Recently, defective repair of DNA alkylation damage by the DNA repair enzyme, O^6-methylguanine DNA methyltransferase, have shown in streptozotocin-induced experimental diabetes as well in the patients with type 1 and type 2 diabetes in our laboratory.

The purpose of the present chapter is to discuss the mechanisms and role of DNA oxidation and alkylation in the pathogenesis of diabetes.

In: New Developments in Mutation Research
Editor: Charles L. Valon, pp. 1-17

ISBN 1-59454-664-9

Chapter 1

GERMLINE AND SOMATIC MUTATIONS IN COLORECTAL CANCERS FROM PATIENTS WITH HEREDITARY NONPOLYPOSIS COLORECTAL CANCER

Michiko Miyaki[1,]* Tatsuro Yamaguchi[1,2] and Takeo Mori[2]
[1]Hereditary Tumor Research Project, [2]Department of Surgery,
Tokyo Metropolitan Komagome Hospital, Tokyo, Japan

ABSTRACT

Hereditary nonpolyposis colorectal cancer (HNPCC) is one of the most common hereditary colon cancer syndromes. The causative genes of HNPCC are DNA mismatch repair genes, and inactivation of these genes triggers HNPCC tumors.

We detected germline mutations of the hMSH2, hMLH1 and hMSH6 genes in Japanese HNPCC families. These mutations included single-base substitutions and frameshifts, both resulting in truncated proteins, a mutation at the splice donor site of exon 5, and a 2kb genomic deletion encompassing exon 5. The coexistence of germline and somatic mutations of DNA mismatch repair genes was observed in colorectal and extracolorectal cancers. The majority of somatic mutations were frameshift, one was a mutation at the splice donor site of exon 5, and two were loss of the normal allele. All HNPCC cancers exhibited high microsatellite instability. Sixty-four % of HNPCC colorectal cancers included somatic mutations of either the APC or ß-catenin gene, whereas only 13% of the cancers had p53 mutations and none of the cancer cases showed the BRAF mutation. The loss of heterozygosity at tumor suppressor regions (chromosomes 5q, 8p, 17p, 18q and 22q) was not observed. We also detected frequent somatic frameshift mutations of growth-related genes with coding repeats in HNPCC colorectal cancers. Mutation frequencies at these repeats were 77% at TGFßRII$(A)_{10}$, 55% at hMSH3$(A)_8$, 52% at CASP5$(A)_{10}$, 48% at BAX$(G)_8$, 48% at RIZ$(A)_8$ and $(A)_9$,

* *Correspondence:* Michiko Miyaki; Hereditary Tumor Research Project, Tokyo Metropolitan Komagome Hospital, 3-18-22 Honkomagome, Bunkyo-ku, Tokyo 113-8677, Japan; Tel: +81-3-3823-2101 (ext. 4425), Fax: +81-3-3823-5433, E-mail: mmiyaki@opal.famille.ne.jp

45% at RAD50$(A)_9$, 45% at MBD4$(A)_{10}$, 36% at hMSH6$(C)_8$, 23% at BLM$(A)_9$, 19% at IGFIIR$(G)_8$ and 19% at PTEN$(A)_6$.

The present data have the following implications: Inactivation of DNA mismatch repair genes through germline and somatic mutations causes genetic instability resulting in somatic mutations in various growth-related genes. Disruption of the WNT signaling pathway by mutations of the APC or ß-catenin gene and disruption of the TGF-ß signaling pathway by alteration of the TGFßRII gene seem to largely contribute to the development of HNPCC colorectal cancer via adenomas. Moreover, inactivation of apoptosis-related genes, such as the BAX, CASP5 and RIZ genes, may also contribute to the development, and inactivation of the IGFIIR and PTEN genes play a role in the progression of HNPCC colorectal cancer.

INTRODUCTION

Colorectal cancer comprises hereditary and non-hereditary types. Hereditary colorectal cancer with an autosomal dominant mode of inheritance includes familial adenomatous polyposis (FAP), with numerous colonic polyps, and hereditary nonpolyposis colorectal cancer (HNPCC), without multiple colonic polyps. HNPCC cancer develops at an early age (mean of approximately 45 years) and accounts for 3%-6% of all colorectal cancers [1]. HNPCC also develops into extracolonic cancer, such as cancer of the endometrium, ovaries, stomach, small intestine, and hepatobiliary and urinary tracts. HNPCC results from germline mutations of DNA mismatch repair genes, including hMSH2 [2, 3], hMLH1 [4, 5], hMSH6 [6, 7], hPMS1 [8] and hPMS2 [8]. Detection of germline mutations of mismatch repair genes is important in identifying HNPCC patients and in the follow up of patients with both colorectal and extracolorectal cancer, and is also significant in identifying non-carriers in HNPCC families.

Inactivation of these mismatch repair genes through germline and somatic mutations leads to mutations at microsatellite regions (microsatellite instability or replication error) [9, 10] and frameshift mutations of repeated sequences in coding regions of various tumor-related genes [11], which triggers the development of HNPCC tumors. These properties in HNPCC patients are distinct from those in FAP and the majority of sporadic cancer patients. In the latter two cases, colorectal cancers develop via an adenoma-carcinoma sequence [12, 13] by mutations of tumor suppressor genes, such as APC, p53 and Smad4, and oncogenes such as K-ras. To clarify the mechanisms of HNPCC colorectal carcinogenesis, it is necessary to determine the extent of the contribution of frameshift mutations in growth-related genes, as well as the mutations in tumor suppressor genes and oncogenes. In this study we analyzed germline and somatic mutations of DNA mismatch repair genes, as well as mutations in various tumor-related genes in HNPCC cancer.

Materials and Methods

Clinical Characterization of HNPCC Families

Pedigrees of independent Japanese families that include multiple patients with colorectal cancer were clinically analyzed. HNPCC families that fit the Amsterdam criteria II [14] were selected for DNA analysis. The criteria are: (a) at least three relatives must have a cancer associated with HNPCC (colorectal, endometrial, stomach, ovary, ureter or renal-pelvis, brain, small-bowel, hepatobiliary tract, or skin), (b) one must be a first-degree relative of the other two, (c) at least two successive generations must be affected, and (d) at least one of the relatives with cancer associated with HNPCC should have received the diagnosis before the age of 50 years.

Samples and DNA Preparation

Tumor tissues, corresponding normal tissues and peripheral blood were obtained after full informed consent, and analysis of these samples was approved by the Komagome Hospital Review Committee. DNA was prepared from each sample using SDS-proteinase K and phenol-chloroform. RNA was extracted using the guanidine thiocyanate method.

Analysis of Germline Mutations of DNA Mismatch Repair Genes

Analysis of germline mutations of the hMSH2, hMLH1 and hMSH6 genes by PCR-SSCP was performed as follows: The PCR reaction mixture (6 μl) contained 300 ng of genomic DNA from normal tissue, 0.2 μM of each primer, 25 μM each of four dNTPs, 1 X PCR buffer, Taq polymerase and [α-^{32}P]dCTP. PCR conditions were: 5 min at 97 °C once; 1 min at 94 °C, 1 min at 58 °C and 1 min at 72 °C for 35 cycles; and 10 min at 72 °C once. Primers used to amplify the 16 coding exons of the hMSH2 gene were the same as those previously reported [15, 16], and primers used to amplify the 19 exons of the hMLH1 gene were the same as those previously described [4, 5] and primers for the 10 exons of the hMSH6 gene were designed based on the reported sequence [17]. PCR products were diluted with formamide-dye solution, heated for 5 min at 80 °C, and applied to 5% polyacrylamide gel containing 5% glycerol. The gel was then exposed to X-ray film. Aberrant single-strand DNA fragments were extracted with water from the corresponding band on SSCP gel. The extracted DNA fragments were amplified through asymmetrical PCR, purified using a purification kit, and then sequenced with dideoxy chain-termination reaction using T7 Sequenase V 2.0.

Analysis of germline genomic deletions of the hMSH2 gene by Southern blot hybridization was performed as follows: Probes for Southern blot hybridization were prepared by RT-PCR using total RNA from human normal tissue and a primer set (primer for probe A involving exons 1-5, for probe B involving exons 4-11 or for probe C involving exons 9-16 of the hMSH2 gene) under the conditions previously described [18]. RT-PCR products were subcloned in plasmids, and then confirmed by direct sequencing. Purified plasmids containing hMSH2 cDNA were labeled with Dig (Roche), and used for probes. For

Southern blot hybridization, 5 μg genomic DNA from normal tissue of the patients was digested by 25 units of a restriction enzyme (*Eco*RI, *Hind*III or *Nsi*I) for 20 hr at 37 °C, electrophoresed in agarose gel for 20 hr at 20 V, and transferred to nylon membrane. The membrane was pre-hybridized for 4 hr at 42 °C, and hybridized with Dig-labeled cDNA probe for 20 hr at 42 °C. After hybridization the membrane was washed at 65 °C, and exposed to X-ray film. Confirmation of a deleted exon of the hMSH2 gene was performed by RT-PCR-sequencing using total RNA from normal tissue of the patients.

Analysis of Somatic Mutations of DNA Mismatch Repair Genes

DNA samples of tumors were analyzed for somatic mutations of the hMSH2, hMLH1 ans hMSH6 genes by PCR-SSCP-sequencing under the same conditions as those for the analysis of germline mutations.

Analysis of Mutations at Repeated Sequences

DNA samples from tumors and corresponding normal tissues were amplified by PCR. The reaction mixture (6 μl) contained 300 ng of genomic DNA, 0.2 μM of ^{32}P-labeled forward primer, 0.2 μM of unlabeled reverse primer, 25 μM concentrations of each of four dNTPs, 1 X PCR buffer, and Taq polymerase. PCR was performed under the same conditions as those previously described [19]: 5 min at 97 °C once; 1 min at 94 °C, 1 min at 55 °C and 1 min at 72 °C for 35 cycles; and 10 min at 72 °C once. PCR products were diluted with formamide-dye solution and heated for 5 min at 80 °C, then applied to 6% polyacrylamide gel containing 7 M urea. The gel was then exposed to X-ray film. Primer sets used for the analysis of microsatellite instability were BAT25, BAT26, D2S123, D5S346 and D17S250 [20]. Primers used for repeated sequences in the coding region of tumor-related genes were the same as those previously reported: TGFßRII [11], IGFIIR [21], BAX [22], hMSH3 [23], hMSH6 [23], CASP5 [24], PTEN [25], MBD4 [26], RIZ [27], BLM [28] and RAD50 [29].

Analysis of Mutations of Tumor Suppressor Genes and Oncogenes

DNA samples of tumors were analyzed for mutations of the APC, ß-catenin, p53, K-ras and BRAF genes by PCR-SSCP-sequencing under the same conditions as those previously described [30-33]. Primers used were the same as those previously reported: APC [34], p53 [31], ß-catenin [35], K-ras [19] and BRAF [36].

Analysis of Loss of Heterozygosity

To analyze LOH in tumors, Southern blot hybridization was performed as described previously [37].

Results

Germline Mutations of DNA Mismatch Repair Genes

The PCR-SSCP-sequencing method detected 6 hMSH2, 3 hMLH1 and 1 hMSH6 germline mutations in 10 independent Japanese HNPCC families (Table 1). The six hMSH2 mutations included 3 frameshifts and 2 single-base substitutions, both resulting in truncated proteins, and a mutation at the splice donor site of exon 5, resulting in deletion of exon 5. The three hMLH1 mutations included a frameshift mutation and a single-base substitution, resulting in truncated and extended proteins, and a deletion of one amino acid. The mutation of the hMSH6 gene was a frameshift mutation resulting in a truncated protein.

Table 1. Germline mutations of hMSH2, hMLH1 and hMSH6 genes in Japanese HNPCC families

HNPCC family	Gene	Exon	Codon	DNA change	Protein change
HNP1	hMSH2	3	136	TTT to TT	Truncation at codon 173-174
HNP4	hMSH2	4	227-229	AGAAAAAAA to AGAAAAAAAA	Truncation at codon 230-231
HNP2	hMSH2	7	419	CAG to TAG	Truncation at codon 419
HNP3	hMSH2	14	811	TTA to TGA	Truncation at codon 811
HNP5	hMSH2	15	877-878	AGAGAG to AGAG	Truncation at codon 880-881
HNP7	hMSH2	5	265-314	G/gta to G/gtt	Deletion of exon 5
HNP15	hMSH2	5	2 kb	Genomic deletion	Deletion of exon 5
HNP19	hMLH1	11	343	TAC to TAA	Truncation at codon 343
HNP8	hMLH1	16	616-618	AAGAAGAAG to AAGAAG	Deletion of a lysine
HNP20	hMLH1	19	712	TGG to TG	Extension of COOH terminus
HNP13	hMSH6	4	534	AAC to AA	Truncation at codon 570-571

In one HNPCC family a germline mutation of the hMSH2 gene could not be detected by the PCR-SSCP method, although this family fulfills the Amsterdam criteria and tumors from two affected individuals showed microsatellite instability, somatic mutations of the hMSH2 gene and a high frequency of alterations at repeated sequences of tumor-related genes. These characteristics of tumors and the presence of somatic hMSH2 mutations in tumors in this family were suggestive of the existence of a germline mutation in the same gene. The Southern blot hybridization method was then applied to detect this germline mutation. Genomic DNA from normal mucosa of an affected individual was digested by the restriction enzymes *Eco*RI, *Hind*III and *Nsi*I, respectively. These genomic digests were hybridized to three different hMSH2 cDNA probes, probe A involving exons 1-5, probe B involving exons 4-11 and probe C involving exons 9-16. The *Nsi*I digest of this patient's DNA exhibited an aberrant restriction fragment of 8.6 kb in addition to the normal fragment of 10.6 kb when hybridized with probe B. The same aberrant fragment was also observed in the hybridization with probe A. These patterns suggested the presence of a heterozygous 2 kb genomic deletion encompassing exon 4, 5 or 6, when referred to the reported restriction patterns of the hMSH2 gene [38]. Sequencing of the RT-PCR product of total RNA from the patient revealed the loss of exon 5 sequences. These results confirmed that the patient's germline mutation was a 2 kb genomic deletion encompassing exon 5 [18] (Table 1). An aberrant *Nsi*I fragment was also detected in the genomic DNA of normal tissue from the other patient but was not detected in healthy family members, which indicated that the 2 kb genomic deletion is pathogenic. The frequency of such a large genomic deletion is not so high, since 10 of 11 germline mutations detected in Japanese HNPCC families were frameshift, nonsense and alternative splicing mutations. The percentage of HNPCC families fulfilling the Amsterdam criteria and carrying the hMSH2 genomic deletion/rearrangement has been estimated to be nearly 10 % [39].

Table 2. Tumors produced in HNPCC families with hMSH2 and hMSH6 germline mutations

Tumor	hMSH2	(7 families)	hMSH6	(1 family)
	No. of patient	Mean age at diagnosis (yrs)	No. of patient	Mean age at diagnosis (yrs)
Colorectal cancer	33/37 (89%)	42	1/6 (17%)	52
Endometrial cancer	7/21 (33%) [a]	44	3/4 (75%) [a]	58
Ovarian cancer	4/21 (19%) [a]	37	1/4 (25%) [a]	61
Gastric cancer	5/37 (14%)	52	0	
Urinary tract cancer	1/37 (3%)		0	
Prostate cancer	1/16 (6%) [b]	70	0	
Brain tumor	1/37 (3%)	49	0	
Skin cancer	1/37 (3%)	49	0	
Pancreatic cancer	0		1/6 (17%)	68
Multiple cancer	17/37 (46%)		2/6 (33%)	

a, female patient; b, male patient

With respect to the effect of germline mutations on cancer pathogenesis, differences were observed between hMSH2 and hMSH6 (Table 2). Among patients in HNPCC families with hMSH2 germline mutations, 33 of the 37 patients (89%) were affected by colorectal cancers, with a mean age at diagnosis of 42. Of 21 female patients 7 (33%) had endometrial cancer at about the age of 44. In addition to these two most common types, there were ovarian (19%), gastric (14%), brain (3%), skin (3%), prostate (6%), and urinary tract (3%) cancers. In 17 cases (46%), a second, third, or even fourth primary cancer occurred in each patient [40]. In contrast to HNPCC families with hMSH2 mutations, the family with the hMSH6 mutation predominantly produced endometrial (75%) and ovarian (25%) cancers, with colorectal cancers being 17% [6]. The mean age of cancer formation in the family with the hMSH6 mutation was 58 years old, which is later than the mean age of 44 years for the appearance of cancer in families with hMSH2 mutations. Such clinical aspects suggest that the effect of germline mutations of the hMSH6 gene on cancer pathogenesis may be different from that of the hMSH2 mutation. A predominance of endometrial cancer and delayed cancer onset has also been reported for HNPCC families with hMSH6 mutations [41].

Somatic Mutations of DNA Mismatch Repair Genes

PCR-SSCP-sequencing analysis of genomic DNA from HNPCC tumors revealed somatic mutations of DNA mismatch repair genes. To determine whether these mutations were caused by a somatic mutation or a germline mutation, genomic DNA from corresponding normal tissues was also examined using PCR-SSCP. The coexistence of somatic and germline mutations in the same DNA mismatch repair gene was observed in HNPCC tumors. Inactivation of the hMSH2, hMLH1 and hMSH6 genes through germline plus somatic mutations occurred in 13 tumors, including colorectal, duodenal, gastric and endometrial cancers (Table 3). Somatic mutations included six 1-bp deletions, one 11-bp deletion, three 1-bp insertions, and one mutation at the splice donor site of exon 5. Of these eleven somatic mutations, ten mutations led to stop codons resulting in truncated proteins and one mutation led to a deletion of exon 5. Two other tumors showed loss of the normal allele, which was confirmed by the absence of normal bands in the SSCP pattern showing abnormal bands of germline mutation.

Table 3. Somatic mutations of hMSH2, hMLH1 and hMSH6 genes in HNPCC cancers

Tumor		Somatic mutation			Germline mutation
	Gene	Exon	Codon	DNA change	Gene
HNP1 CoCa	hMSH2	5	311-312	AACCTT to AACTT	hMSH2
HNP3 CoCa	hMSH2			Loss of normal allele	hMSH2
HNP4 CoCa	hMSH2	12	661	AAA to AAAA	hMSH2
HNP5 CoCa1	hMSH2	2	82-83	AAAATG to AAAAATG	hMSH2
HNP5 CoCa2	hMSH2	3	173-176	CTAGGACTGTGT to A	hMSH2
HNP7 CoCa1	hMSH2	13	690-691	CAAATT to CAATT	hMSH2
HNP7 CoCa2	hMSH2	5	311-312	AACCTT to AACTT	hMSH2
HNP15 DuoCa	hMSH2	4	227-229	AGAAAAAAA to AGAAAAAA	hMSH2
HNP15 GaCa	hMSH2	1	21	CGC to CAGC	hMSH2
HNP2 EnCa	hMSH2	5	265-314	G/gta to G/gtt	hMSH2
HNP8 CoCa	hMLH1			Loss of normal allele	hMLH1
HNP13 CoCa	hMSH6	2	128	CGT to CG	hMSH6
HNP13 EnCa	hMSH6	5	1085-1087	ACCCCCCCC to ACCCCCCC	hMSH6

CoCa, colorectal cancer; DuoCa, duodenal cancer; GaCa, gastric cancer; EnCa, endometrial cancer

Microsatellite Instability

The fidelity of DNA replication is maintained by a system that recognizes DNA mismatched sequences, excises and corrects the mismatch. Mismatches are caused by incorrect base pairings and by the slippage of DNA polymerase on the template strand containing repeated sequences. hMSH2 and hMSH6 proteins produce a heterodimeric complex that binds to mismatches, and hMLH1 and hPMS2 bind the hMSH2-hMSH6 plus heteroduplex DNA complex. This complex participates in mismatch repair [42]. Microsatellite regions are repeating sequences that are distributed throughout the human genome, most commonly (A)n/(T)n and (CA)n/(GT)n, located within non-coding DNA sequences. Inactivation of DNA mismatch repair genes causes genomic instability, as shown by a 100-700 fold increase in slipped-strand mismatches of (CA)n repeats [43]. Such genomic instability (replication error or microsatellite instability) is one aspect of the remarkable nature of tumors from HNPCC patients.

We analyzed microsatellite instability in 31 colorectal cancers from 23 HNPCC patients in at least 5 microsatellite regions, such as BAT25, BAT26, D2S123, D5S346 and D17S250, including mononucleotide repeats and dinucleotide repeats. All cancers showed a high frequency (more than 4/5) of microsatellite instability (Figure 1). Such microsatellite instability was also observed in duodenal, gastric and endometrial cancers from HNPCC patients.

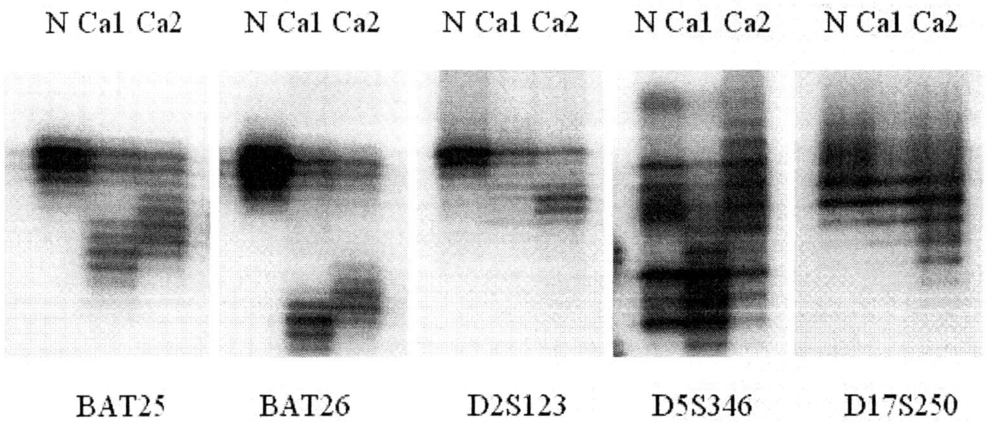

Figure 1. Microsatellite instability in HNPCC colon cancer.
N, normal tissue; Ca, colon cancer. In these two cancers, 5 of 5 regions are altered.

Frameshift Mutations in Coding Repeats

Most microsatellite instability occurs in repeated sequences in non-coding DNA which is assumed to have no effect on protein function. However, there are many genes that have repeated sequences in their coding regions. Replication errors (frameshift mutations) at coding repeats frequently occur in tumors with microsatellite instability. Frameshift mutations of coding repeats of tumor suppressor genes form stop codons leading to the inactivation of proteins, which propagate growth advantages to tumors. To clarify the significance of frameshift mutations in HNPCC colorectal carcinogenesis, we analyzed the frequencies and accumulation profiles of frameshift mutations of 11 genes in 31 HNPCC colorectal cancers.

An example of a frameshift mutation is shown in Figure 2, and the frequencies of mutations of various genes are listed in Table 4. The TGFßRII gene is a component of the TGF-ß signaling pathway, including Smad2, Smad3 and Smad4, and this pathway functions as a growth suppressor of epithelial cells. A frameshift mutation at the $(A)_{10}$ region in the TGFßRII gene forms a truncated protein, and mutations in both alleles cause inactivation of this gene, resulting in a disappearance of the growth suppression activity. Mutations of the TGFßRII gene were frequently (77%) detected in HNPCC colorectal cancers even at the early stage. The BAX gene product, a proapoptotic factor, is directly transactivated by p53. The $(G)_8$ region of the BAX gene was mutated at 29% in the early stage of colorectal cancer and at 65% in the advanced stage. Since the p53 mutation is infrequent in HNPCC colorectal cancer [19], the alteration of the BAX gene probably plays a more important role than p53 gene mutations in both the development and progression of HNPCC cancer. The IGFIIR gene has been proposed as being a growth suppression gene in colorectal cancer cells [21], and the PTEN gene as a tumor suppressor [25]. No mutations of these two genes were detected in the early stage, whereas 35% of advanced cancer had mutations in these genes. This suggests that inactivation of IGFIIR and PTEN genes plays a role in cancer progression in HNPCC colorectal carcinogenesis. CASP5 cleaves pro-caspase-3 resulting in caspase-3, which is involved in apoptotic cell death. Alteration of the $(A)_{10}$ repeat in CASP5 occurred in 52% of the cancers, suggesting that alteration of CASP5 may convey a growth advantage to cancer cells. The $(A)_{10}$ tract of the MBD4 and $(A)_9$ tract of the RAD50, which are DNA repair genes, mutated in 45% of HNPCC cancers, suggesting that genomic instability may be increased in these cancers. Frameshift mutations were detected at $(A)_8$ of the hMSH3 gene in 55% of the cancers, and $(C)_8$ of the hMSH6 gene in 36% of cancers. Alteration of these genes, which are components of the mismatch repair complex [23], may cause more severe replication errors in HNPCC cancers, although these alterations contribute indirectly in tumorigenesis.

Table 4. Frameshift mutations in HNPCC colorectal cancers

Gene	Function	Mutation frequency (%)
TGFßRII $(A)_{10}$	growth factor receptor	77.4
hMSH3 $(A)_8$	mismatch repair	54.8
CASP5 $(A)_{10}$	proapoptotic factor	51.6
BAX $(G)_8$	proapoptotic factor	48.4
RIZ $(A)_8$, $(A)_9$	cell cycle arrest	48.4
RAD50 $(A)_9$	DNA repair	45.2
MBD4 $(A)_{10}$	DNA repair	45.2
hMSH6 $(C)_8$	mismatch repair	35.5
BLM $(A)_9$	DNA repair	22.6
IGFIIR $(G)_8$	growth factor receptor	19.4
PTEN $(A)_6$, $(A)_6$	tumor suppressor	19.4

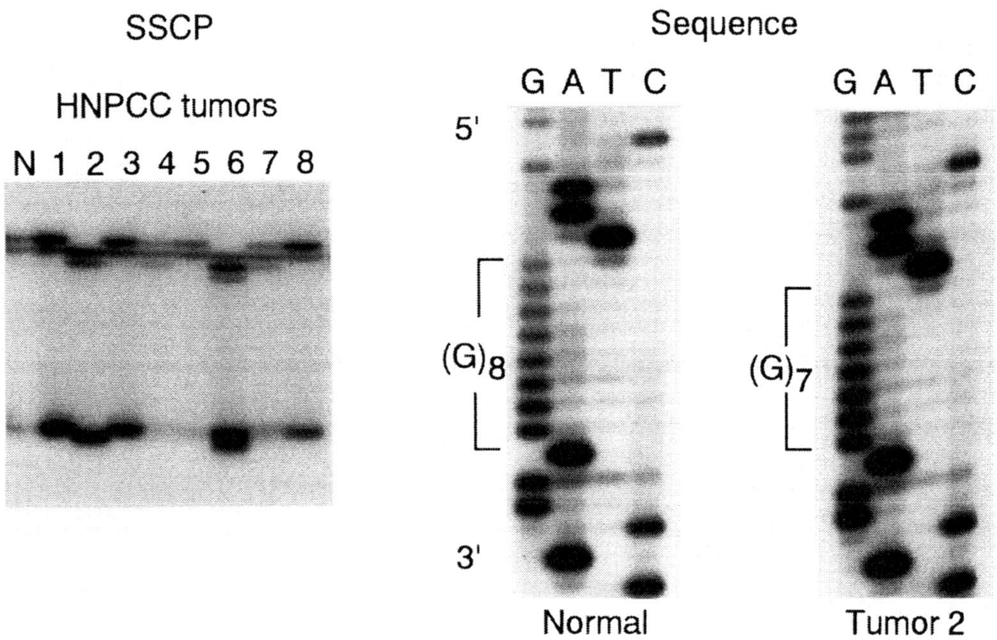

Figure 2 Framashift mutation at $(G)_8$ region of BAX gene in HNPCC colon cancer. N, normal tissue; 1-8, colon cancers. In Tumor 2, $(G)_8$ is changed to $(G)_7$.

Mean numbers of altered genes within the 11 genes were estimated as 3.6 for Dukes A, 5.3 for Dukes B and 5.7 for Dukes C cancers. These data suggest that it is possible to observe the gradual development of HNPCC colorectal cancers by the gradual accumulation of the number of altered genes with repeated sequences.

Somatic Mutations of ß-catenin and APC Genes

We have observed that the frequency of APC gene mutations was low (nearly 20%) in HNPCC tumors compared to the high frequency (more than 70%) in non-HNPCC tumors [19]. It is known that wild-type APC protein forms a complex with ß-catenin and GSK3ß, leading to a degradation of ß-catenin in normal cells, whereas mutant APC protein in tumor cells lacks the ability for a complex formation and ß-catenin accumulates in the nuclei. This causes an interaction of ß-catenin with the nuclear transcription factor (Lef/Tcf) family, resulting in the activation of target genes, including cyclin D1 and c-myc. A similar accumulation of ß-catenin has been demonstrated in colorectal cancer cell lines without APC mutations but with mutations in the regulatory domain (codons 29 to 48) of the ß-catenin gene, suggesting that activation of the ß-catenin-Tcf signaling pathway by mutation in either the APC or ß-catenin gene contributes to colorectal carcinogenesis [44]. To assess the possible contribution of this signaling pathway in HNPCC carcinogenesis, we analyzed ß-catenin mutations in colorectal tumors from HNPCC patients.

Table 5. Mutations of ß-catenin and APC genes in HNPCC colorectal tumors

Tumor	ß-catenin mutation		APC mutation	
	Codon	DNA change	Codon	DNA change
HNP10 CoAd1		-		-
HNP10 CoAd2	32	GAC to GGC		-
HNP14 CoAd		-	1338	CAG to TAG
HNP6 CoCa		-		-
HNP10 CoCa1	45	TCT to TTT		-
HNP1 CoCa1		-		-
HNP3 CoCa		-		-
HNP4 CoCa	34	GGA to GAA		-
HNP5 CoCa1		-		-
HNP5 CoCa2		-	901-902	CAGGAA to CAGAA
			1554-1556	GAAAAAACT to GAAAAAAACT
HNP5 CoCa3		-		-
HNP5 CoCa4		-	874	TCA to TAA
HNP7 CoCa1		-	1466	GGA to GGGA
HNP8 CoCa1	41	ACC to GCC		-
HNP10 CoCa2	45	TCT to TTT		-
HNP12 CoCa	45	TCT to TTT		-
HNP13 CoCa		-		-
HNP14 CoCa2		-	1462-1465	AAGAGAGAGAGT to AAGAGAGAGAGAGT
HNP14 CoCa3		-		-
HNP15 CoCa	45	TCT to CCT		-
	43	GCT deletion		
HNP16 CoCa1	45	TCT to TGT		-
HNP16 CoCa2	32	GAC to TAC		-
HNP16 CoCa3		-	1492	GCC to GC
HNP16 CoCa4	34	GGA to GAA		-
HNP17 CoCa	41	ACC to GCC		-
HNP18 CoCa	37	TCT to TGT		-
HNP19 CoCa1		-		-
HNP19 CoCa2		-		-

CoAd, colorectal adenoma; CoCa, colorectal cancer; -, mutation was not detected; HNP5 CoCa2 had two APC mutations. HNP15 CoCa had two ß-catenin mutations.

Somatic ß-catenin gene mutations were detected in 12 HNPCC colorectal tumors, including 11 cancers and 1 adenoma, in which no APC gene mutations were detected [32] (Table 5). These mutations were single-base substitutions that were located within the NH_2-terminal regulatory domain of ß-catenin (codons 29 to 48): TCT(Ser) to TTT(Phe), CCT(Pro) or TGT(Cys) at codon45; ACC(Thr) to GCC(Ala) at codon 41; TCT(Ser) to TGT(Cys) at codon 37; GGA(Gly) to GAA(Glu) at codon 34; GAC(Asp) to GGC(Gly) or TAC(Tyr) at codon 32; and a 3-bp deletion at codon 43. Such single-base substitutions are assumed to be caused by defects of DNA mismatch repair in HNPCC tumors. Immunohistochemical staining of HNPCC cancers with a ß-catenin mutation showed a high level of ß-catenin protein in both cytoplasm and nuclei, with stronger staining in the nuclei [32], although in normal epithelial cells of the colon ß-catenin protein was localized in the membrane but not in nuclei. ß-Catenin mutations were not detected in HNPCC tumors with APC mutations, indicating an alternative exclusion between ß-catenin mutations and APC mutations. However, the coexistence of mutations of TGFßRII and APC, and TGFßRII and ß-catenin was observed, which suggests that the TGFßRII mutation does not substitute for the activated ß-catenin-Tcf signaling pathway. The activation of the ß-catenin-Tcf signaling pathway, through either ß-catenin mutations (43%) or APC mutations (21%) may significantly contribute to HNPCC colorectal carcinogenesis. Because mutations in either APC or ß-catenin have been known to be associated with adenoma formation, the present observation may imply that more than half (64%) of HNPCC colorectal cancers develop via adenoma.

Somatic Mutations of the p53 gene

p53 gene mutations were detected in 2 of 15 colorectal cancers from HNPCC patients [19]. These included CGT to TGT at codon 110 and CGT to TGT at codon 273. The mutation frequency (13%) was lower than in non-HNPCC cancers (more than 50%), suggesting that colorectal cancer in HNPCC involves a more distinct pathogenic mechanism than in non-HNPCC cancer. Instead of p53 gene mutations, frequent frameshift mutations in the BAX gene appear to contribute to colorectal carcinogenesis in HNPCC patients.

Somatic mutations of K-ras and BRAF Genes

K-ras gene mutations have been demonstrated to generally occur at 30% to 40% in non-HNPCC colorectal cancers [45]. K-ras is a member of the RAS family which participates in the RAS-RAF signaling pathway, which mediates cellular responses to growth signals. BRAF is a member of the RAF family which is regulated by RAS and also mediates growth signals. It has recently been found that activation of RAS-RAF signaling is caused not only by K-ras gene mutations, but also by BRAF gene mutations in various human cancers [36], and a significantly high frequency of BRAF mutations has been observed in mismatch repair-deficient sporadic colorectal cancers compared to mismatch repair-proficient cancers [46]. HNPCC cancers are typical mismatch repair-deficient cancers. To clarify wether HNPCC colorectal cancers exhibit a high frequency of BRAF mutations similar to the mismatch repair-deficient sporadic cancers, we analyzed BRAF mutations in colorectal cancers from HNPCC patients. We also analyzed K-ras mutations in these cancers. We observed a low frequency of K-ras mutations (less than 18%) and no BRAF mutations in HNPCC colorectal cancers [33]. An obvious difference in the BRAF mutation frequency between mismatch repair-deficient sporadic (31%) and HNPCC (0%) cancers suggests a difference in the mechanisms of colorectal carcinogenesis between mismatch repair-deficient sporadic cancers and HNPCC cancers.

Loss of Heterozygosity

Colorectal cancers from FAP patients and the majority of sporadic colorectal cancers develop via an adenoma-carcinoma sequence through the accumulation of inactivations of tumor suppressor genes. High frequencies (40% to 70%) of LOH have been detected in FAP and sporadic colorectal cancers at tumor suppressor regions, including chromosomes 5q, 8p, 17p, 18q and 22q [13]. However, these LOH were not detected in colorectal cancers from HNPCC patients [19]. The absence of chromosomal instability is one aspect of the remarkable nature of HNPCC tumors, which is assumed to be related to the observation that HNPCC-associated tumors have generally diploid or near-diploid DNA.

Conclusion

Hereditary nonpolyposis colorectal cancer (HNPCC) is one of the most commom hereditary colon cancer syndromes with an autosomal dominant inheritance. The causative genes are germline mutations of DNA mismatch repair genes including hMSH2, hMLH1 and hMSH6. HNPCC also develops into extracolonic cancers such as cancer of the endometrium and other extracolonic organs. Inactivation of DNA mismatch repair genes through germline and somatic mutations leads to microsatellite instability, as well as frameshift mutations of repeated sequences in coding regions of various tumor-related genes. Frameshift mutations occurred in TGFßRII, IGFIIR, BAX, CASP5, RIZ, RAD50, MBD4, BLM, hMSH3, hMSH6 and PTEN genes. These genes include members of growth factor receptors, proapoptotic factors, DNA repair genes, components of the mismatch repair system, and tumor suppressor genes. Inactivation of these genes appears to cause the development and progression of HNPCC colorectal cancers. The gradual increase in the number of frameshift mutations during the progression from Dukes A to Dukes B and C cancers suggests the gradual progression of HNPCC colorectal cancers by the accumulation of mutations of genes with coding repeats. Mutations of ß-catenin and APC genes were also high in HNPCC colorectal tumors, which suggests that the majority of HNPCC colorectal cancers develop via adenomas.

The present data revealed that disruption of the WNT signaling pathway by mutations of ß-catenin and APC genes and disruption of the TGF-ß signaling pathway by mutation of the TGFßRII gene largely contribute to the development of HNPCC colorectal cancer. Moreover, it was suggested that mutations of apoptosis-related genes, such as the BAX, CASP5 and RIZ genes, contribute to the development, and inactivation of the IGFIIR and PTEN genes play a role in the progression of HNPCC colorectal cancer.

References

[1] Lynch, H.T. & de la Chapelle, A. (2003). Hereditary colorectal cancer. *New Engl. J. Med., 348*, 919-932.

[2] Fishel, R., Lescoe, M.K., Rao, M.RS., Copeland, N.G., Jenkins, N.A., Garber, J., Kane, M. & Kolodner, R. (1993). The human mutator homolog MSH2 and its association with hereditary nonpolyposis colon cancer. *Cell, 75*, 1028-1038.

[3] Leach, F.S., Nicolaides, N.C., Papadopoulos, N., Liu. B., Jen, J., Parsons, R., Peltomaki, P., Sistonen, P., Aaltonen, L.A., Nystrom-Lahti, M., Guan, X.-Y., Zang, J., Meltzer, P.S., Yu, J.-W., Kao, F.-T., Chen, D.J., Cerosaletti, K.M., Fournier, R.E.K., Todd, S., Lewis, T., Leach, R.J., Naylor, S.L., Weissenbach, J., Mecklin, J.-P., Jarvinen, H., Petersen, G.M., Hamilton, S.R., Green, J., Jass. J., Watson, P., Lynch, H.T., Trent, J.M., de la Chapelle, A., Kinzler, K.W. & Vogelstein, B. (1993). Mutations of a mutS homolog in hereditary nonpolyposis colorectal cancer. *Cell*, *75*, 1215-1225.

[4] Bronner, C.E., Baker, S.M., Morrison, P.T., Warren, G., Smith, L.G., Lescoe, M.K., Kane, M., Earabino, C., Lipfold, J., Lindblom, A., Tannergard, P., Bollag, R.J., Godwin, A.R., Ward, D.C., Nordenskjold, M., Fishel, R., Kolodner. R & Liskay, R.M. (1994). Mutation in the DNA mismatch repair gene homologue hMLH1 is associated with hereditary non-polyposis colon cancer. *Nature*, *368*, 258-261.

[5] Papadopoulos, N., Nicolaides, N.C., Wei, Y.-F., Ruben, S.M., Carter, K.C., Rosen, C.A., Haseltine, W., Fleischmann, R.D., Fraser, C.M., Adams, M.K., Venter, J.C., Hamilton, S.R., Petersen, G.M., Watson, P., Lynch, H.T., Peltomaki, P., Mecklin, J.-P., de la Chapelle, A., Kinzler, K.W. & Vogelstein, B. (1994). Mutation of a mutL homolog in hereditary colon cancer. *Science*, *263*, 1625-1629.

[6] Miyaki, M., Konishi, M., Tanaka, K., Kikuchi-Yanoshita, R., Muraoka, M., Yasuno, M., Igari, T., Koike, M., Chiba, M. & Mori, T. (1997). Germline mutation of MSH6 as the cause of hereditary nonpolyposis colorectal cancer. *Nature Genet.*, *17*, 271-272.

[7] Akiyama, Y., Sato, H., Yamada, T., Nagasaki, H., Tsuchiya, A., Abe, R. & Yuasa, Y. Germ-line mutation of the hMSH6/GTBP gene in an atypical hereditary nonpolyposis colorectal cancer kindred. (1997). *Cancer Res.*, *57*, 3920-3923.

[8] Nicolaides, N.C., Papadopoulos, N., Liu, B., Wei, Y.-F., Carter, K.C., Ruben, S.M., Rosen, C.A., Haseltine, W.A., Fleischmann, R.D., Fraser, C.M., Adams, M.D., Venter, J.C., Dunlop, M.G., Hamilton, S.R., Petersen, G.M., de la Chapelle, A., Vogelstein, B. & Kinzler, K.W. (1994). Mutation of two PMS homologues in hereditary nonpolyposis colon cancer. *Nature*, *371*, 75-80.

[9] Aaltonen, L.A., Peltomaki, P., Leach, F.S., Sistonen, P., Pylkkanen, L., Mecklin, J.-P., Jarvinen, H., Powell, S.M., Jen, J., Hamilton, S.R., Petersen, G.M., Kinzler, K.W., Vogelstein, B. & de la Chapelle, A. (1993). Clues to the pathogenesis of familial colorectal cancer. *Science*, *260*, 812-816.

[10] Ionov, Y., Peinads, M.A., Malkosyan, S., Shibata, D. & Perucho, M. (1993). Ubiquitous somatic mutations in simple repeated sequences reveal a new mechanisms for colonic carcinogenesis. *Nature*, *363*, 558-561.

[11] Markowitz, S., Wang, J., Myeroff, L., Parsons, R., Sun, L., Lutterbaugh, J., Fan, R.S., Zborowska, E., Kinzler, K.W., Vogelstein, B., Brattain, M. & Willson, J.K.W. (1995). Inactivation of the type II TGF-ß receptor in colon cancer cells with microsatellite instability. *Science*, *268*, 1336-1338.

[12] Fearon, E.R. & Vogelstein, B. (1990). A genetic model for colorectal tumorigenesis. *Cell*, *61*, 759-767.

[13] Miyaki, M., Tanaka, K., Kikuch-Yanoshita, R., Muraoka, M.& Konishi, M. (1995). Familial polyposis: recent advances. *Crit. Rev. Oncol. Hematol.*, *19*, 1-31.

[14] Vasen, H.F.A., Watson, P., Mecklin, J.-P. & Lynch, H.T. (1999). New clinical criteria for hereditary nonpolyposis colorectal cancer (HNPCC, Lynch syndrome) proposed by the International Collaborative Group on HNPCC. *Gastroenterology*, *116*, 1453-1456.

[15] Liu, B., Parsons, R.E., Hamilton, S.R., Petersen, G.M., Lynch, H.T., Watson, P., Markowitz, S., Willson, J.K.V., Green, J., de la Chapelle, A., Kinzler, K.W. & Vogelstein, B. (1994). HMSH2 mutations in hereditary nonpolyposis colorectal cancer kindreds. *Cancer Res.*, *54*, 4590-4594.

[16] Kolodner, R.D., Hall, N.R., Lipford, J., Kane, M.F., Rao, M.R.S., Morson, P., Wirth, L., Finan, P.J., Burn, J., Chapman, P., Earabino, C., Merchant, E. & Bishop, D.T. (1994). Structure of the human MSH2 locus and analysis of two Muir-Torre kinders for msh2 mutations. *Genomics*, *24*, 516-526.

[17] Acharya, S., Wilson, T., Gradia, S., Kane, M.F., Guerrette, S., Marsischky, G.T., Kolodner, R. & Fishel, R. (1996). HMSH2 forms specific mispair-binding complexes with hMSH3 and hMSH6. *Proc. Natl. Acad. Sci. USA*, *93*, 13629-13634. Genbank [U73732-7].

[18] Miyaki, M., Iijima, T., Yamaguchi, T., Shirahama, S., Ito, T., Yasuno, M. & Mori, T. (2004). Novel germline hMSH2 genomic deletion and somatic hMSH2 mutations in a hereditary nonpolyposis colorectal cancer family. *Mutation Res.*, *548*, 19-25.

[19] Konishi, M., Kikuchi-Yanoshita, R., Tanaka, K., Muraoka, M., Onda, A., Okumura, Y., Kishi, N., Iwama, T., Mori, T., Koike, M., Ushio, K., Chiba, M., Nomizu, S., Konishi, F., Utsunomiya, J. & Miyaki, M. (1996). Molecular nature of colon tumors in hereditary nonpolyposis colon cancer, familial polyposis, and sporadic colon cancer. *Gastroenterology*, *111*, 307-317.

[20] Boland, C.R., Tibodeau, S.N., Hamilton, S.R., Sidransky, D., Eshleman, J.R., Burt, R.W., Meltzer, S.J., Rodoriguez-Bigas, M.A., Fodde, R., Ranzani, G.N., & Srirastava, S. (1998). A National Cancer Institute workshop on microsatellite instability for cancer detection and familial predisposition: Development of international criteria for the determinations of microsatellite instability in colorectal cancer. *Cancer Res.*, *58*, 5248-5257.

[21] Souza, R.F., Appel, R., Yin, J., Wang, S., Smolinski, K.N., Abraham, J.M., Zou, T.T., Shi, Y.Q., Leu, J., Cottrell, J., Cymes, K., Bider, K., Simms, L., Leggett, B., Lynch, P.M., Frazier, M., Powell, S.M., Harpaz, N., Sugimura, H., Young, J. & Meltzer, S.J. (1996). Microsatellite instability in the insulin-like growth factor II receptor gene in gastrointestinal tumors. *Nature Genet.*, *14*, 255-257.

[22] Rampino, N., Yamamoto, H., Ionov, Y., Li, Y., Sawai, H., Reed, J.C. & Perucho, M. (1997). Somatic frameshift mutations in the BAX gene in colon cancers of the microsatellite mutator phenotype. *Science*, *275*, 967-969.

[23] Malkhosyan, S., Rampino, N., Yamamoto, H. & Perucho, M. (1996). Frameshift mutator mutations. *Nature*, *382*, 499-500.

[24] Schwartz, S. Jr, Yamamoto, H., Navarro, M., Maestro, M., Reventos, J. & Perucho, M. (1999). Frameshift mutations at mononucleotide repeats in Caspase-5 and other target genes in endometrial and gastrointestinal cancer of the microsatellite mutator phenotype. *Cancer Res.*, *59*, 2995-3002.

[25] Guanti, G., Resta, N., Simone, C., Cariola, F., Demma, I., Fiorente, P. & Gentile, M. (2000). Involvement of PTEN mutations in the genetic pathway of colorectal carcinogenesis. *Hum. Mol. Genet.*, *9*, 283-287.

[26] Riccio, A., Aaltonen, L.A., Godwin, A.K., Loukola, A., Percesepe, A., Salovaara, R., Masciullo, V., Genuardi, M., Papavatou-Petsotas, M., Bassi, D.E., Ruggeri, B.A., Klein-Szanto, A.J., Testa, J.R., Neri, G. & Bellacosa, A. (1999). The DNA repair gene

MBD4 (MED1) is mutated in human carcinomas with microsatellite instability. *Nature Genet.*, *23*, 266-268.

[27] Chadwick, R.B., Jiang, G.L., Bennington, G.A., Yuan, B., Johnson, C.K., Stevens, M.W., Niemann, T.H., Peltomaki, P., Huang, S. & de la Chapelle, A. (2000). Candidate tumor suppressor RIZ is frequently involved in colorectal carcinogenesis. *Proc. Natl. Acad. Sci. USA*, *97*, 2662-2667.

[28] Calin, G., Herlea, V., Barbanti-Brodano, G & Negrimi, M. (1998). The coding region of the Bloom syndrome BLM gene and of the CBL proto-oncogene is mutated in genetically unstable sporadic gastrointestinal tumors. *Cancer Res.*, *58*, 3777-3781.

[29] Kim, N.J., Choi, Y.R., Beak, M.J., Kim, Y.H., Kang, H., Kim, N.K., Min, J.S. & Kim, H. (2001). Frameshift mutations at coding mononucleotide repeats of the hRAD50 gene in gastrointestinal carcinomas with microsatellite instability. *Cancer Res.*, *61*, 36-38.

[30] Miyaki, M., Konishi, M., Kikuchi-Yanoshita, R., Enomoto, M., Igeri, T., Tanaka, K., Muraoka, M., Takahashi, H., Amada, Y., Fukayama, M., Maeda, Y., Iwama, T., Mishima, Y., Mori, T. & Koike, M. (1994). Characteristics of somatic mutation of the adenomatous polyposis coli gene in colorectal tumors. *Cancer Res.*, *54*, 3011-3020.

[31] Kikuch-Yanoshita, R., Konishi, M., Ito, S., Seki, M., Tanaka, K., Maeda, Y., Iino, H., Fukayama, M., Koike, M., Mori, T., Sakuraba, H., Fukunari, M., Iwama, T. & Miyaki, M. (1992). Genetic changes of both p53 alleles associated with the conversion from colorectal adenoma to early carcinoma in familial adenomatous polyposis and non-familial adenomatous polyposis patients. *Cancer Res.*, *52*, 3965-3971.

[32] Miyaki, M., Iijima, T., Kimura, J., Yasuno, M., Mori, T., Hayashi, Y., Koike, M., Shitara, N., Iwama, & Kuroki, T. (1999). Frequent mutations of ß-catenin and APC genes in primary colorectal tumors from patients with hereditary nonpolyposis colorectal cancer. *Cancer Res.*, *59*, 4506-4509.

[33] Miyaki, M., Iijima, T., Yamaguchi, T., Kadofuku, T., Funata, N. & Mori, T. (2004). Both BRAF and KRAS mutations are rare in colorectal carcinomas from patients with hereditary nonpolyposis colorectal cancer. *Cancer Lett.*, *211*, 105-109.

[34] Groden, J., Thliveris, A., Samowitz, W., Carlson, M., Gelbert, L., Albertsen, H., Joslyn, G., Stevens, J., Spirio, L., Robertson, M., Sargeant, L., Krapcho, K., Walff, E., Burt, R., Hunghes, J.P., Warrington, J., McPherson, J., Wasmuth, J., LePaslier, D., Abderrahim, H., Cohen, D., Leppert, M. & White, R. (1991). Identification and characterization of the familial adenomatous polyposis coli gene. *Cell*, *66*, 589-600.

[35] Voeller, H.J., Truica, C.I. & Gelmann, E.P. (1998). ß-Catenin mutation in human prostate cancer. *Cancer Res.*, *58*, 2520-2523.

[36] Davis, H., Bignell, G.R., Cox, C., Stephens, P., Edkins, S., Clegg, S., Teague, J., Woffendin, H., Garnett, M.J., Bottomley, W., Davis, N., Dicks, E., Ewing, R., Floyd, Y., Gray, K., Hall, S., Hawes, R., Hughes, J., Kosmidou, V., Menzies, A., Mould, C., Parker, A., Stevens, C., Watt, S., Hooper, S., Wilson, R., Jayatilake, H., Gusterson, B.A., Cooper, C., Shipley, J., Hargrave, D., Pritchard-Jones, K., Maltland, N., Chenevix-Trench, G., Riggs, G.J., Bigner, D.D., Palmleri, G., Cossu, A., Flanagan, A., Nicholson, A., Ho, J.W.C., Leung, S.Y., Weber, B.L., Selgler, H.F., Darrow, T., Paterson, H., Marals, R., Marshall, C.J., Wooster, R., Stratton, M. & Futreal, P.A. (2002). Mutations of the BRAF gene in human cancer. *Nature*, *417*, 949-954.

[37] Miyaki, M., Seki, M., Okamoto, M., Yamanaka, A., Maeda, Y., Tanaka, K., Kikuchi, R., Iwama, T., Ikeuchi, T., Tonomura, A., Nakamura, Y., White, R., Miki, Y., Utsunomiya,

J. & Koike, M. (1990). Genetic changes and histopathological types of colorectal tumors from patients with familial adenomatous polyposis. *Cancer Res.*, *50*, 7166-7173.

[38] Wijnen, J., van der Klift, H., Vasen, H., Meera Khan, P. & Fodde, R. (1998). MSH2 genomic deletions are a frequent cause of HNPCC. *Nature Genet.*, *20*, 326-328

[39] Chanbonnier, F., Olschwang, S., Wang, Q., Boisson, C., Martin, C., Buisine, M.-P., Puisieux, A. & Frebourg, T. (2002). MSH2 in contrast to MLH1 and MSH6 is frequently inactivated by exonic and promoter rearrangements in hereditary nonpolyposis colorectal cancer. *Cancer Res.*, *62*, 848-853.

[40] Miyaki, M., Konishi, M., Muraoka, M., Kikuchi-Yanoshita, R., Tanaka, K., Iwama, T., Mori, T., Koike, M., Ushio, K., Chiba, M., Nomizu, S. & Utsunomiya, J. (1995). Germ line mutations of hMSH2 and hMLH1 genes in Japanese families with hereditary nonpolyposis colorectal cancer (HNPCC): usefulness of DNA analysis for screening and diagnosis of HNPCC patients. *J. Mol. Med.*, *73*, 515-520.

[41] Wijnen, J., de Leeuw, W., Vasen, H., van der Klift, H., Moller, P., Stormorken, A., Meijers-Heijboer, H., Lindhout, D., Menko, F., Vossen, S., Moslein, G., Tops, C., Brocker-Vriends, A., Wu, Y., Hofstra, R., Sijmons, R., Cornelisse, C., Morreau, H. & Fodde, R. (1999). Familial endometrial cancer in female carriers of MSH6 germline mutations. *Nature Genet.*, *23*, 142-144.

[42] Rhyu, M.S. (1996). Molecular mechanisms underlying hereditary nonpolyposis colorectal carcinoma. *J. Natl. Cancer Inst.*, *88*, 240-251.

[43] Strand, M., Prolla, T.A., Liskay, R.M. & Petes, T.D. (1993). Destabilization of tracts of single repetitive DNA in yeast by mutation affecting DNA mismatch repair. *Nature*, *365*, 274-276.

[44] Morin, P.J., Sparks, A.B., Korinek, V., Baker, N., Clevers, H., Vogelstein, B. & Kinzler, K.W. (1997). Activation of ß-catenin-Tcf signaling in colon cancer by mutation in ß-catenin or APC. *Science*, *275*, 1787-1790.

[45] Bos, J.L., Fearon, E.R., Hamilton, S.R., Verlaan-de Vries, M., van Boom, J.H., van der Eb, A.J. & Vogelstein, B. (1987). Prevalence of ras gene mutations in human colorectal cancers. *Nature*, *327*, 293-297.

[46] Rajagopalan, H., Bardelli, A., Lengauer, C., Kinzler, K.W., Vogelstein, B. & Veculescu, V.E. (2002). RAF/RAS oncogenes and mismatch-repair status. *Nature*, *418*, 934.

In: New Developments in Mutation Research
Editor: Charles L. Valon, pp. 19-40
ISBN 1-59454-664-9

Chapter 2

MUTATIONS INDUCED BY POLYCYCLIC AROMATIC HYDROCARBONS

Brinda Mahadevan,[*] Lauren A. Courter and William M. Baird
Oregon State University, Corvallis, OR

ABSTRACT

Polycyclic aromatic hydrocarbons (PAH) are widespread environmental contaminants, proven to be potent carcinogens in rodent bioassays and implicated in the etiology of several human cancers. PAHs must be metabolized to chemically reactive electrophiles to initiate this carcinogenic process by covalent interactions with DNA. This chapter discusses mutations induced by the metabolites of two environmental carcinogenic pollutants, the bay and fjord region-containing PAH, benzo[*a*]pyrene and dibenzo[*a,l*]pyrene, respectively. These PAH-DNA lesions can cause distortions in the DNA molecule and influence enzyme activity, DNA binding, or repair of bulky DNA adducts. Here we consider and discuss the induction of potentially carcinogenic mutations, which is a biological consequence of the formation of PAH-DNA adducts. The PAH-DNA adduct-induced mutational frequency and spectra has been established through various test systems. This review compares in the mammalian *HPRT* gene model mutation test system, and the types of mutations induced by carcinogenic diol-epoxide metabolites of BP and DBP. We further examined the influence of PAHs in the environment on mutation induction, specifically PAH-induced mutations in tumor suppressor genes and proto-oncogenes. In this chapter, we also highlight the point that replication on PAH-DNA adducts could produce mutations leading to cancer.

Keywords: mutation, DNA adducts, mutagenesis, DNA repair, polycyclic aromatic hydrocarbons

---

[*] *Correspondence*: Brinda Mahadevan; Department of Environmental and Molecular Toxicology, Oregon State University, Agricultural and Life Sciences 1007, Corvallis, OR 97331-7302; Phone: 541-737-5413; Fax: 541-737-0497; Email: mahadevb@onid.orst.edu

Abbreviations

PAH	Polycyclic Aromatic Hydrocarbon;
DE	dihydrodiol-epoxide
BP	Benzo[*a*]pyrene;
DBP	Dibenzo[*a,l*]pyrene;
HPRT	Hypoxanthine Guanine Phosphoribosyl Transferase;
CYP	Cytochrome P450;
IARC	International Agency for Research on Cancer;
CHO	Chinese Hamster Ovary V79 cells;
dA	Deoxyadenosine;
dG	Deoxyguanosine;
PAHDE	PAH-dihydrodiol-epoxide;
BPDE	Benzo[*a*]pyrene-dihydrodiol-epoxide;
DBPDE	Dibenzo[*a,l*]pyrene-dihydrodiol-epoxide;
B[*c*]Ph	benzo[*c*]phenanthrene;
B[*c*]PhDE	Benzo[*c*]phenanthrene-dihydrodiol-epoxide;
B[*c*]C	Benzo[*c*]chrysene;
DMBA	Dimethylbenz[*a*]anthracene;
B[*g*]C	Benzo[*g*]chrysene;
NER	Nucleotide Excision Repair;
AP	Apurinic/Apyrimidinic;
NAAAF	*N*-Acetoxy-2-(acetylamino)fluorine;
N-OH-AF	*N*-hydroxy-2-aminofluorene.

Introduction

Polycyclic aromatic hydrocarbons (PAHs) are a class of ubiquitous environmental pollutants generated from incomplete combustion processes. While harmful chemicals present in the environment cause various diseases, complex environmental mixtures containing PAHs (e.g. tobacco smoke, coke production, coal gasification) have been characterized by the International Agency for Research on Cancer (IARC) as carcinogenic to humans [1]. Many biochemical and toxicological studies have evaluated the genotoxicity and mutagenicity of PAHs to contribute information necessary to assess human cancer risk to PAH exposure.

PAH-induced carcinogenesis involves a number of steps, including: (i) metabolism and the enzymatic activation of the PAH into reactive metabolites (ii) the covalent binding of the PAH metabolites to DNA and (iii) the induction of mutations that serve to initiate the transformation process as a result of PAH-DNA adducts (as reviewed in [2]). Most mutagenic and tumorigenic PAHs can be categorized into two broad classes based on their electronic and structural contributions [3], the bay and fjord region containing compounds. It is at these regions where reactive oxygen is inserted into an aromatic double bond by cytochrome P450s (CYP). Epoxide hydrolase then hydrolyzes the oxide to form a *trans*-dihydrodiol. The same ring of the PAH can then undergo epoxidation to form an electrophilic, reactive dihydrodiol-epoxide (DE) metabolite capable of binding to specific residues on cellular DNA forming PAH-DNA adducts [4-6]. This is illustrated in Figure 1 with benzo[*a*]pyrene (BP) and dibenzo[*a,l*]pyrene (DBP) as examples and will be further discussed in this chapter.

Covalent PAH-DNA adducts are often fixed as mutations if left to error-prone excision repair [7], misrepair, or replication errors during early S-phase [8, 9]. Most importantly, if these fixed mutations occur in proto-oncogenes and/or tumor suppressor genes, it could lead to neoplastic growth and tumor formation through the process of promotion and progression [8, 10-16]. The processes of replication, repair and recombination are performed by multiprotein assemblies that share many common features. The enzymes active in replicating DNA during S phase are needed to synthesize new stretches of DNA during various types of recombination and repair, while repair of all types must occur in tandem with replication and transcription. The p53 protein, often referred to as the "guardian of the genome" provides assistance in managing the cell cycle for the repair of any common lesions that may occur in DNA processing.

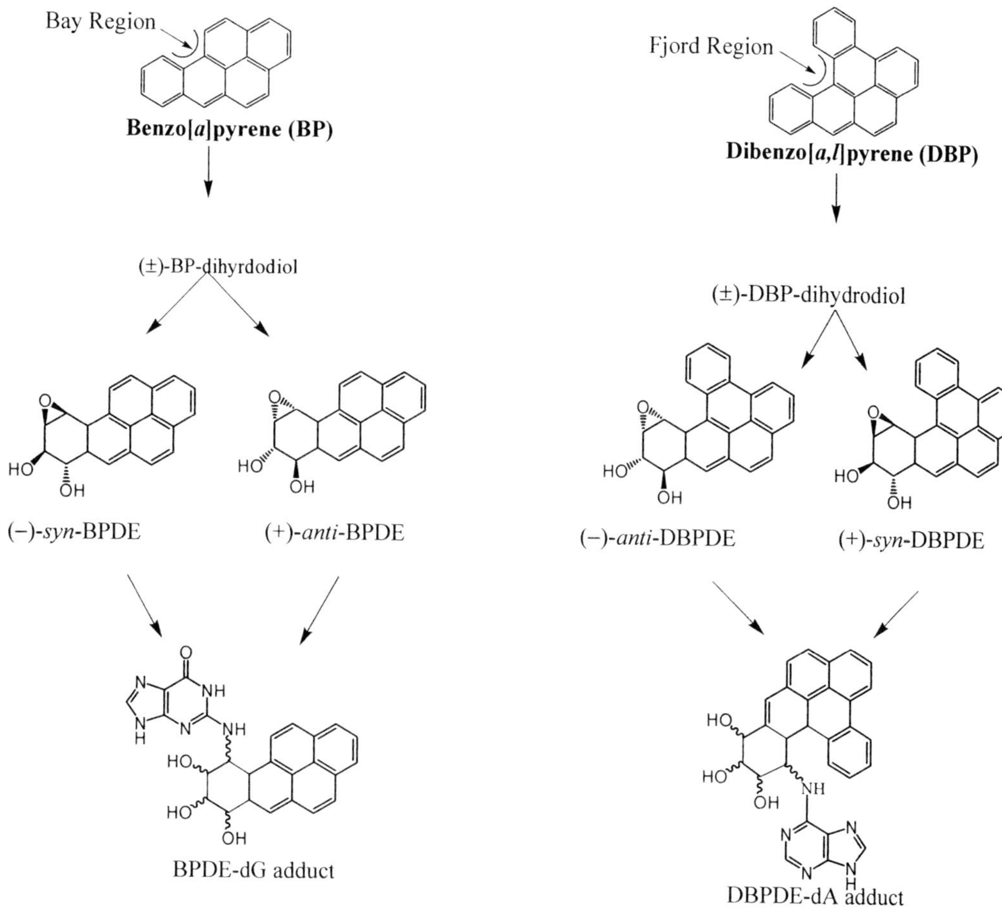

Figure 1. Metabolic activation pathways of the bay and fjord region containing PAH, BP and DBP. CYP enzymes mainly catalyze the formation of dihydrodiol epoxides, the (+)-*anti*-BPDE and (−)-*syn*-BPDE of BP; the (−)-*anti*-DBPDE and (+)-*syn*-DBPDE of DBP. These electrophilic derivatives of BP and DBP react with purine bases of DNA predominantly resulting in dG and dA adducts respectively.

In this chapter, we discuss PAH-DNA adducts and their relationship to mutation induction, specifically in tumor suppressor genes and proto-oncogenes, and the influence of PAHs in the environment and mutation induction.

Relationship between PAH-DNA adducts and Mutations

A common mechanism by which PAHs initiate the process of malignant transformation is through the mutations induced by DNA damage in the target organ. In most cases this damage is in the form of a chemical modification of nucleotides in DNA, in which the PAH or its metabolites are covalently bound to form a stably modified nucleotide, otherwise known as a DNA adduct [17]. When a replication error occurs at a critical site in a gene essential for cell cycle control such as *p53*, or in proto-oncogenes, it could lead to neoplastic growth and tumor formation through the process of promotion and progression [8, 10-16]. Brookes and Lawley [18] first discovered the relationship between the interactions of carcinogenic PAHs with DNA and their carcinogenic potency. However, inconsistencies still remain today in understanding the relationship of the level of DNA adducts formed by a PAH and its carcinogenic potency. Although, it is well established that strong correlations exist between PAH-DNA adducts and mutations in specific human genes, including cancer related genes. These mutations are thought to be initiated when a DNA polymerase encounters a carcinogen-DNA adduct or a resultant apurinic/apyrimidinic (AP) site during replication. For example, the ultimate metabolite of DBP, which is DBPDE has been shown to react with the purine bases of DNA causing stable and depurinating DNA adduct damage. This PAH-induced stably adducted DNA is thought to undergo error free NER [19, 20], whereas AP sites generated from depurinating adducts undergo error-prone repair to form tumor-initiating mutations [21].

Studies indicate that the formation of DNA adducts is dependant on the dose of the PAH. Earlier work on DNA adducts in Chinese hamster ovary cells [22] and mutagenicity determined in the CHO (Chinese hamster ovary cells)/hypoxanthine phosphoribosyl transferase assay (*HPRT)* demonstrated a linear relationship between the numbers of BPDE-DNA adducts and mutagenicity over the concentration of BP-diol assayed [23]. Also, a study with the ultimate carcinogenic metabolite of DBP demonstrated a dose dependent increase in both the total level of DNA adducts as well as the frequency of DNA mutations within the *HPRT* gene [24]. There was clear indication that the mean number of base substituitions per mutant clone increased as the treatment dose increased [24]. An example of this linear relationship between DNA adducts and mutation induction in an occupational exposure setting was also reported by Perera et al [25] where the mutations at the *HPRT* in workers in an iron foundry with exposure to PAHs were evaluated. The *HPRT* mutant frequencies demonstrated an upward trend with increasing exposure dose and a high correlation with PAH-DNA adducts. Even at relatively low exposures to PAH, *HPRT* mutations increased in parallel with PAH-DNA adducts [25].

Studies with (+)-*anti*-dihydrodiol-epoxide of BP [(+)-*anti*-BPDE] and (−)-*anti*-dihydrodiol-epoxide of DBP [(−)-*anti*-DBPDE] demonstrate that the sequence context affects binding to oligodeoxynucleotides [26-28]. Also, the conformation of a PAH-DNA adduct can be modulated by the sequence context, where different conformations can lead to different mutational outcomes [29, 30]. Bay region PAHs, such as BPDE, react with nucleophilic sites on the bases of DNA, including guanine N^2 [31-33], and adenine N^6 [34, 35] at a lower frequency. They are known to predominantly induce G-to-T transversions at high frequencies when tested in standard mutagenesis assays, while DBPDE predominately binds to deoxyadenosine (dA) residues in DNA and deoxyguanosine (dG) residues at a lower frequency [36, 37].

The types of DNA mutations induced by PAH dihydrodiol-epoxides can depend on sequence context and metabolite stereochemistry [38], where the latter is attributed as the dominant factor in determining mutational spectra. In an effort to gain insights into the molecular mechanisms involved in the recognition and excision of bulky lesions, Geacintov et al [39] investigated a series of site specifically modified oligonucleotides containing single, well defined PAH dihydrodiol-epoxide- dA adducts. They observed that although the covalent adduct derived from the bay region PAH (BP) are removed by human nucleotide excision repair (NER) enzymes *in vitro*, the stereochemically analogous N^6-dA adducts derived from the topologically different fjord region PAH benzo[*c*]phenanthrene (B[*c*]Ph) were resistant to repair [39]. They suggest that the evasion of DNA repair may play a role in the highly tumorigenic fjord region PAH dihydrodiol-epoxides [39].

Fjord region containing PAH, such as B[*c*]Ph, B[*g*]C, benzo[*c*]chrysene (B[*c*]C), and DBP, predominantly bind to dA residues in DNA while binding to dG residues at a lower frequency [36, 37]. As a result, fjord region containing PAH predominantly induce A-to-T transversions, along with G-to-T transversions at a lower frequency. This has been demonstrated in the *supF* gene of the pSP189 shuttle vector [40], the *HPRT* gene [24, 41], and in Big-Blue transgenic mouse cells [42]. Fjord region PAH (DBP) are characterized as the strongest tumor initiator in mouse skin and rat mammary gland among the carcinogenic PAHs tested thus far [43-45]. This is in part due to the repulsive interactions between the two opposing hydrogen bonds present at the fjord region [3, 46]. As a result, the molecule is rendered out of plane where their dihydrodiol-epoxides can bind more extensively to DNA compared to the metabolically activated bay region PAH, BPDE [39, 47]. The fjord region dihydrodiol-epoxide conformation also provides the molecule robustness against NER [48]. Furthermore, DBP and B[*c*]Ph predominately bind to N^6 of dA and dG, producing more depurinating adducts compared to bay region containing PAH, such as BP and dimethylbenz[*a*]anthracene (DMBA) , leading to erroneous excision repair, and a higher incidence in tumor initiation [7, 49]. Recent observations by Yoon et al [42] on DBPDE induced mutations although revealed A-to-T transversions in addition to G-to-T, the distribution of mutational hotspots differed at several positions. They reported that the major mutation hotspots were at sites of strong PAH-DNA adduct formation, but not all of the strong adduct formation sites were mutational hotspots. Also, adducts produced by both DBPDE isomers were repaired at a rather slow rate, and repair was not greatly sequence dependent [42] as previously noted for BPDE-DNA adducts [26-28].

The studies discussed in the section below focus on the DNA binding and carcinogenic potential of the environmental PAHs. We review the various types of mutations induced by the very potent BPDE and DBPDE in the *HPRT* gene in mammalian cells as the test system. We also outline their relationship between doses of the carcinogen and induction of mutation. A summary of the mutational profiles of the carcinogens discussed, grouped in a dose range for easy compilation, are compiled in Table 1.

Table 1. Effect of varying doses of PAHDE on base substitutions occurring at AT or GC base pairs

PAH	5nM	10-20nM	40-100nM	0.1-1.25μM	1-3μM
(+)-*anti*-BPDE	ND	36 (AT)[‡]	24 (AT)[‡]	6 (AT)[‡]	ND
	ND	64 (GC)[‡]	76 (GC)[‡]	94 (GC)[‡]	ND
(–)-*syn*-BPDE	ND	ND	ND	21 (AT)[*]	14 (AT)[*]
	ND	ND	ND	79 (GC)[*]	86 (GC)[*]
(–)-*anti*-	40 (AT)[§]	38 (AT)[§]	61 (AT)[¥]	ND	ND
DBPDE	60 (GC)[§]	62 (GC)[§]	39 (GC)[¥]	ND	ND
(+)-*syn*-DBPDE	ND	ND	64 (AT)[¥]	ND	ND
	ND	ND	36 (GC)[¥]	ND	ND
(–)-B[*c*]PhDE	ND	ND	ND	89 (AT)[†]	69 (AT)[†]
	ND	ND	ND	11 (GC)[†]	31 (GC)[†]
(+)-B[*c*]PhDE	ND	ND	ND	50 (AT)[†]	49 (AT)[†]
	ND	ND	ND	50 (GC)[†]	51 (GC)[†]

Note: This table represents data compiled from research on Chinese hamster V79 cells and the base substitutions in the *HPRT* gene (until otherwise mentioned) on exposure to varying doses (grouped into a range for comparison) of PAHDE. The numbers indicate the percentage of base substituitions with the transversions noted in parentheses. Data compiled in this table are from research performed by several authors as indicated as follows: ‡ taken from Conney et al 2001 [132]; * taken from Wei et al 1994 [54]; §, taken from Mahadevan et al 2003 [24]; ¥, taken from Yoon et al 2004 [42], where transversions were detected in the *cII* transgene of Big-Blue mouse embryo cells; † taken from Wei et al 1996 [41]. ND indicates data not determined.

Mutations Induced by BP and DBP

In the early years of PAH research, major focus was on BP and DMBA as examples of the unsubstituted and methyl substituted PAH with differing carcinogenic potencies. After learning about the carcinogenic potency and environmental relevance of DBP, determination of the pathways of metabolic activation of DBP and establishment of how it interacts mainly with dA in DNA was questioned. DMBA is not found as an environmental contaminant perhaps because of its very great susceptibility to destruction by photooxidation. The emphasis of the research therefore shifted towards studies on BP and DBP. These PAHs are ideal for the study of chemical carcinogenesis due to the unique differences between their stereochemistry, DNA adduct conformation, and the different mutations they induce. In addition, they represent the two broad classes of PAH carcinogens, the bay and fjord region containing compounds. It was demonstrated in earlier studies that out of the optical isomers of the metabolically activated bay-region 7,8-diol-9,10 epoxides of BP, only (+)-*anti*-BPDE was both a potent mutagen in CHO cells and a potent carcinogen in mice [4, 50, 51]. Wei et al [52] analyzed the kinds of mutations in the coding region of the *HPRT* gene obtained by treating CHO cells with two doses of (+)-*anti*-BPDE, where the low noncytotoxic dose of (+)-*anti*-BPDE had a higher proportion of base substitution mutations at AT base pairs than those treated with a higher cytotoxic dose. This observation was further studied using a high cytotoxic dose, an intermediate noncytotoxic dose, and a low noncytotoxic dose of (+)-*anti*-

BPDE [53]. The examination of *HPRT* mutants indicated a dose-dependent difference in the profile of mutations induced by (+)-*anti*-BPDE [53]. The proportion of GC→TA transversions decreased and the proportion of AT→CG transversions increased when the dose of (+)-*anti*-BPDE was decreased. Also, dose-dependent differences in hotspots for base substitutions and exon deletions were also observed [53]. This was followed up by another study where the mutation profile was evaluated and compared for both (–)-*syn*-BPDE (a weakly carcinogenic or inactive enantiomer) and (+)-*anti*-BPDE [54]. The mutation profiles revealed that (+)-*anti*-BPDE was more mutagenic than (–)-*syn*-BPDE. While (+)-*anti*-BPDE caused a significantly higher proportion of mutations at AT base pairs as the dose decreased and showed high selectivity for GC→TA transversions, this was not the case for (–)-*syn*-BPDE [54]. The hotspots for base substitutions were also different with (+)-*anti*-BPDE and (–)-syn-BPDE [54]. Therefore the results revealed not only dose-dependent differences in the profile of mutations but also variation in mutation induction depending on the enantiomer used.

Studies of the carcinogenic activity of the enantiomers of the fjord region containing PAH B[*c*]PhDE demonstrated that although all bind extensively to dA, there were differences in carcinogenic potency . The mutational spectra of B[*c*]PhDE and DBP have been shown previously to be very similar [40-42]. Wei et al [41] evaluated the possibility of dose-dependent differences in the mutational profile of (–)-(1R,2S,3S,4R)-3,4-dihyroxy-1,2-epoxy-1,2,3,4-tetrahydrobenzo[*c*]phenanthrene [(–)-B[*c*]PhDE] (a potent fjord region dihydrodiol-epoxide derived from B[*c*]Ph) that covalently binds to DNA predominantly at dA residues [55, 56] causing significant mutations at AT base pairs [40, 57]. Results demonstrated that although dose dependent differences were observed for the mutagenic profile of the highly carcinogenic (–)-B[*c*]PhDE, that was not the case with the less active (+)-B[*c*]PhDE [41].

Recently, Mahadevan et al [24] using (–)-*anti*-DBPDE, reported that the ultimate carcinogenic metabolite of DBP at 1 nM resulted in mutant frequencies comparable to a relatively nontoxic dose of about 10-20 nM (+)-*anti*-BPDE [58]. At these low doses both reactive intermediates of BPDE mainly induced base substitutions. However, in accordance with their preferential base target sites within DNA, (+)-*anti*-BPDE induced mainly substitutions at dG [58], whereas (–)-*anti*-DBPDE displayed dA substituitions [24]. Very recently Yoon et al [42] have performed a detailed comparison of the mutagenicity of the two isomeric fjord region diol-epoxides from DBP that stero-selectively formed in cells in culture and *in vivo*. Their work clearly reveal that the more carcinogenic enantiomer (–)-*anti*-DBPDE was slightly more mutagenic than (+)-*syn*-DBPDE, although the mutational spectra were very similar. The mutation spectra of the DBPDE enantiomers were similar to those of (–)-B[*c*]PhDE [41]. The majority of the induced mutations were transversions, with A-to-T transversions being the major type for both stereoisomers, while G-to-T transversions were also significantly induced [42].

The studies on PAHDE discussed above demonstrate that the ultimately DNA-reactive metabolites of PAHs, induce a variety of mutations with base substitutions always the predominant kind of mutation irrespective of concentrations used. However, among base substitutions, the main nucleotides targeted by PAHDE vary with treatment dose (see Table 1). Chen and Thilly [59] discuss their findings on the effect of exposure conditions of BP in human cells in culture, with regard to the variation in the mutational spectra. They concluded that the more difficult protocols of long-term and low-concentration mutation studies could be justified as conditions necessary, although perhaps not sufficient to approximate human *in vivo* mutational pathways. Therefore, a comparison study using various PAHDE at low, relatively noncytotoxic, environmentally relevant concentration would help to determine the persistence of base substitutions at the nucleotides. It may also allow the prediction of the types of mutations induced in tumor suppressor genes or proto-oncogenes within neoplastic tissue during the process of carcinogenesis.

PAH Induced Mutations in Tumor Suppressor Genes and Proto-Oncogenes

The PAH-DNA adduct induced mutational frequency and spectra was established through various test systems such as the *his* alleles of *Salmonella* [60], the *lacI* gene in *E. coli* [61], and the *HPRT* gene in mammalian cells [52, 53, 62]. In addition, several *in vivo* models have been utilized, including rodents [24, 63-66] and transgenic zebrafish [67, 68]. PAH-DNA adduct induced mutations *in vivo* are often point mutations, targeted towards codons within *p53*. These mutational hotspots are often at 5'-CpG-3' (CpG) sequences in codons 157, 158, 248, 249, and 273 of *p53* [15, 16, 69]. Specifically, the aforementioned hotspot codons are major binding sites for BPDE and other PAH-diol-epoxides [15, 16, 70, 71]. PAHs have been reported to target mutational hotspots in the *p53* gene in bronchial epithelial cells [71], and lung cancer tumors which often exhibit G-to-T transversions. It is this particular mutation profile within lung tumors that is causally linked to PAH carcinogens within tobacco smoke.

In addition to *p53*, metabolically activated PAH often target oncogenic sequences within the *ras* family of proto-oncogenes, including Ha-*ras*, Ki-*ras*, and N-*ras* (codons 12, 13, and 61) [72]. Mutation of *ras* at these particular codons converts it into an oncogene, capable of transforming cells into a malignant phenotype. Generally, PAHs that form DNA adducts at dA primarily induce mutations at codon 61, whereas those PAHs that form DNA adducts at dG primarily induce mutations at codons 12 and 13. Those PAHs that induce adducts at both bases induce both types of mutations (as reviewed by [10]).

DNA sequence analyses of PAH-initiated tumor tissues have revealed various mutations. Within the Ki-*ras* gene, codon 12 predominantly yielded G-to-T transversions when initiated with certain PAHs [10, 73-75]. PAHs have also shown to induce CGC mutations within codon 13 (74, 76-81), and A-toT transversions at codon 61 [75, 79-81]. DNA sequence analyses of PAH-induced Ha-*ras* mutations within tumor tissue have shown G to T transversions at codon 13 [74], and A-to-T transversions at codon 61 [75, 82, 83]. For N-*ras*, PAH mainly induced A-to-T mutations at codon 61 [84]. The research discussed above on various carcinogen-induced fixed mutations that occur in tumor suppressor genes and or/ proto-oncogenes emphasize the importance and consequence of such mutations. These mutations, may lead to progenitor cells lacking in cell cycle control, which could initiate clonal expansion and result in tumor formation. In the section below, we specifically address the mutations induced by two PAHs, BP and DBP in the tumor suppressor gene *p53* and the *ras* family of proto-oncogenes.

BP AND DBP INDUCED *P53* AND *RAS* MUTATIONS

Mutations in *p53*, specifically G-to-T transversions, in smoking-associated lung cancer is linked to the presence of BP in cigarette smoke [15]. It is the metabolically activated form, BPDE, which preferentially binds to dG residues in methylated CpG sites within the *p53* gene. According to the human *p53* tumor mutation database, six dominant mutation hotspots associated with lung cancer have been identified, five of which (codons 157, 158, 245, and 273) are guanines at methylated CpG sites [85, 86]. Bulky chemical carcinogens preferentially form DNA adducts at methylated CpG sites; however, mutational frequencies at methylated CpG sites ($C^{m}pG^{BPDE}$) were found to be much lower compared to CpG^{BPDE} sites. Therefore, it was hypothesized that the excision of BPDE-dG adducts by nucleotide excision repair (NER) is stimulated by CpG methylation [87-90]; however, this remains to be elucidated.

The predominant (+) and (−)-*trans*-dG-N^2-BPDE adducts positioned within these codons of *p53* has shown to cause G-to-T transversions with fewer G-to-A transitions [87], where the (+)-*trans*-dG-N^2-BPDE adducts were more mutagenic than the (−)-*trans*-enantiomers [38, 87, 91]. In a study using the bronchial epithelial cell line, BEAS-2B cells, a high frequency of G-to-A and C-to-T transitions at codons 249 and 250 were found upon BPDE treatment, respectively [92]. Not only were these hotspots major bindings sites for BPDE, other PAH dihydrodiol-epoxides, including the metabolically activated fjord region containing PAH, B[*c*]PhDE, was also shown to preferentially bind there [71].

The mutagenic potential of (+)- and (−)-*trans*-dG-N^2-BPDE within codons 60 and 61 of H-*ras* has been found to predominantly produce G-to-T transversions [38, 93]. In mouse skin tumors, it has been reported that low and high doses of BPDE cause G-to-A and G-to-T mutations at codon 12; G-to-C and G-to-T mutations at codon 13; and, A-to-T, and A-to-G mutations at codon 61 within the Ha-*ras* gene. Moreover, it was the G-to-T transversion at nucleotide 38, codon 13 which indicated high mutational frequency, rendering a Gly-to-Val amino acid change [93].

For the Ki-*ras* proto-oncogene, the most frequent mutation in human cancers was found at codon 12, however, the bulky chemical carcinogens which initiated mutations targeted codons 13 and 61 [94, 95] at a lower frequency. Among the Ki-*ras* oncogene mutations, codon 12 was thought to be the most frequently mutated because the adducts formed at that particular sequence were poorly repaired [96]. Since the first guanine of codon 14 of Ki-*ras* is within a CpG site and is highly methylated in human genomic DNA, DNA-adduct formation of BPDE was more predominant at codon 14 compared to codon 12 [97]. The type and frequency of mutations induced at codon 14 by BPDE is characteristic, where other bulky carcinogens such as *N*-Acetoxy-2-(acetylamino)fluorene (NAAAF), 4-aminobiphenyl and *N*-hydroxy-2-aminofluorene (N-OH-AF) do not cause the same mutational profile at that codon [97]. This indicates that mutations at codon 14 in Ki-*ras* are carcinogen dependent. It is important however to note that mutations at codon 14 of the Ki-*ras* gene are very rare in human cancers which may be accounted for the efficiency of DNA repair at this site. Less frequent BPDE-DNA adduct formation was observed at codons 13 while none were detected at codon 61 of the Ki-*ras* gene [96].

The N-*ras* gene was found to be a target of BPDE-DNA adduct formation, specifically at codon 12 [96]. However, neither codons 12, 13, nor 61 were commonly found in human lung cancer and human cancers in general [96].

The fjord region containing PAH, DBP is found in cigarette smoke [98] and in other environmental mixtures [99]. DBPDE induces mutations during early post-initiation stage; however, these mutations increase over time, specifically during clonal expansion of pre-papillomas to tumors [49]. A study by Yoon et al, used the *cII* transgene in mouse cells, and reported that both (+)-*syn*- and (–)-*anti*-DBPDE (the two important enantiomers of DBP) predominantly induced A-to-T transversions, and a considerable level of G-to-T transversions at methylated CpG sites [42]. These results suggested that fjord region PAHs, such as DBP, could theoretically contribute to the G-to-T mutational signature, present in lung cancers from smokers [42, 85]. Currently, no study has investigated the effects of DBP exposure on the mutational spectra of *p53*. As mentioned earlier, the predominant mutations induced by DBP are A-to-T transversions which are often the kind of mutations observed in the Ha-*ras*, Ki-*ras* and N-*ras* genes [7, 48, 49]. DBP has shown to specifically induce a high frequency of A toT transversions within codon 61 (CAA→CTA) of the SENCAR mouse Ha-*ras* proto-oncogene [49]. The most common Ki-*ras* mutations detected in DBP-induced lung tumors were G-to-T transversions in the first base of codon 12, A to G transitions in the second base of codon 12, and A-to-T transversions in the second or third base of codon 61 [75]. Currently, no spectrum has been determined for DBPDE induced mutations in N-*ras*.

Influence of PAHs in the Environment and Mutation Induction

Humans are environmentally exposed to complex mixtures of PAHs in the atmosphere from combustion sources such as cigarette smoking, vehicle emission, as well as industrially through coal gasification, aluminum production, steel founding, and coal tar production [100-103]. Many individual PAHs are potent carcinogens in rodents and several PAH mixtures have been implicated in the induction of human cancer [100-103]. Recently, it was demonstrated that exposure to particulate air pollution increases the risk of heritable mutations in mice, however the reduction of exposure lowered this risk [104, 105]. Evidence using high-density oligonucleotide array expression analysis has shown that indeed an environmental PAH mixture affects gene expression of cancer associated genes. Exposure of human cells in culture to a standardized urban dust particulate mixture (SRM 1649a) altered gene expression patterns of several genes involved in metabolism, cell cycle, proliferation/apoptosis, DNA repair, tumor suppression, immune response, cell structure/adhesion, transport, receptor/inhibitor and proto-oncogenes (106). Specifically, co-treatment of human cells with BP and SRM 1649a indicated an additive induction of *CYP1A1* and *CYP1B1* genes and protein expression [106].

CYP1A1 is the isoenzyme most active in the metabolic activation of PAHs. Cigarette smoke induces CYP1A1 protein levels and thereby increases the metabolism of BP [107]. A growing number of studies link the mutations in the *CYP1A1* gene to increased susceptibility to DNA damage. For example, the allele carrying the Ile-to-Val mutation was more commonly found in lung parenchyma tissue from lung cancer patients and in white blood cells from PAH-exposed individuals with high levels of DNA adducts derived from (+) *anti*-BPDE [108]. This allele was also associated with increased occurrence of *p53* mutations in tumor tissue [109-111]. Another example to note was the variation in aryl hydrocarbon hydroxylase activity, an enzyme involved in the metabolism of certain PAHs, where at least four common mutations resulted in amino acid exchanges in the human *CYP1B1* gene (coding for another PAH-metabolizing enzyme) [112]. Epidemiological studies have reported increase in mutation induction on exposure to PAHs.

On occupational exposure to PAHs, workers in an iron foundry, coke-oven workers, and bus maintenance workers have revealed the genotoxic effects caused by PAHs even at low-levels of exposure. Perera et al [25] evaluated the mutations at the *HPRT* gene in workers from an iron foundry with exposure to PAHs. The *HPRT* mutant frequencies demonstrated an upward trend with increasing exposure dose and a high correlation with PAH-DNA adducts. Even at relatively low exposures to PAH, *HPRT* mutations increased in parallel with PAH-DNA adducts. Another study [113] suggested a molecular link between somatic gene mutations and PAHs. A study involving mutant T-lymphocytes derived from coke-oven workers were analyzed for point mutations in the *HPRT* gene where deletions/insertions accounted for 22% of the mutations, and base substitutions for 78% [114]. In yet another small study, samples of human T-lymphocyte clones from bus maintenance workers exposed to diesel exhaust in comparison to control individuals, revealed deletions and point mutations including splice site mutations and coding errors [115]. They report that splice mutation was more frequent among the garage workers compared to the controls, suggesting a PAH specific mutation induction [115].

PAHs that were found negative in the *Salmonella* mutagenesis assay, such as naphthalene and anthracene, may be capable of inducing DNA deletions. These PAHs have been tested positive in somatic mutation and recombination test in *Drosophila melanogaster* [116]. Recently, in a study involving extract of smoky coal emissions and the mutational spectra obtained in *Salmonella* TA98 and TA100, indicated the presence of mutagenic PAHs in the mixture [117]. These studies not only supported the role for PAHs in the induction of mutations but also identified the primary cause of the elevated lung cancer seen in the population of non-smoking women using smoky coal for cooking in China. Transplacental studies have demonstrated that maternal exposure to diesel exhaust particulate during gestation results in a significantly elevated frequency of DNA deletions in the mouse fetus [118].

DeMarini et al [119] reported from the first mutation spectra of a complex mixture of a whole urban air sample in comparison to BP, the induction of base-substitution (G-to-T transversions) and frameshift mutations. Peluso et al [120] performed a meta-analysis of 13 DNA-adduct studies on occupational cohorts exposed to air pollution. The association between levels of DNA adducts and air pollution exposure was significant both in heavily exposed industrial workers and in less severely exposed urban workers. In the same study, analysis of seven studies that reported measuring levels of BP-DNA adduct levels in exposed workers significantly correlated with air levels of BP [120]. These examples of the exposure to environmental PAH mixtures urge us to obtain a broader understanding of the complex mixtures of PAHs found in the environment and examine their effects that form the context for carcinogen exposure [121].

CONCLUSION

The accumulated evidence to date confirms the environmental ubiquity of PAHs and their negative association with human health. Several lines of evidence presented in this chapter lead us to believe that the interaction of the PAHs or their electrophilic metabolites with DNA is complex. As discussed earlier in this chapter, the mutational complexity that arises due to the differences in stereochemistry of PAH, adduct conformation, sequence context, DNA repair and replication errors, adds to the urgency of the ongoing studies of these compounds. In this chapter, we have discussed important aspects relevant to understanding the DNA damage caused by PAHs, given that DNA adduct formation followed by mutation induction is the critical early event in PAH-induced carcinogenesis.

Brookes and Lawley [122] pointed out in very early years that there is a discrepancy between the level of DNA binding and the carcinogenicity of BP compared to BPDE. Overall, research done on both BP and DBP suggest that hydrocarbons which form high proportions of dA adducts relative to dG adducts may be more potent carcinogens. However, there is presently not sufficient evidence to prove whether this is a general function of PAH-modified dA or conformation of PAH metabolites bound to dA. The development of very sensitive analytical techniques and the application of these techniques to human monitoring requires that we are able to distinguish factors affecting PAH-DNA adduct formation and its removal. Therefore, future needs require the designing of experiments to address several important factors that influence the mutational complexity of PAHs.

In principle, mutational complexity can be attributed to either a collection of different adducts each inducing a single kind of mutation or a single adduct inducing different kinds of mutations. Seo et al [123] have summarized both the scenarios with examples where structural factors such as adduct structure, DNA sequence context and biological factors (viz, differing DNA polymerases) play a role in the pattern of adduct mutagenesis. The PAH mutational spectrum can also be shaped by the efficiency and specificity of lesion bypass by DNA polymerase and selection, thus giving rise to adduct mutational complexity [123].

The ability of cells to remove PAH-DNA adducts prior to replication of DNA is a major factor in determining whether particular adducts induce mutation [124, 125]. Although the removal of BPDE-DNA adducts has been examined in cell cultures and *in vivo* [126, 127] there is very little information available on the relative rates of removal of different PAH-DNA adducts. Tang et al [22] observed that polymorphisms in DNA repair enzyme XPD were associated with increased levels of PAH-DNA adducts in a case-control study of breast cancer. This suggested that diminished XPD mediated DNA repair capacity may be associated with increased DNA damage in tumor tissue. In a recent pilot molecular epidemiological study in breast cancer patients, reduced DNA repair of BPDE-induced adducts and common XPD polymorphisms were observed, suggesting that women with reduced DNA repair capacity may be at increased risk of developing breast cancer [128]. The above mentioned studies clearly suggest that PAH-DNA adducts and the mutations they render can serve as useful biomarkers of PAH exposure and predict human risk.

Future studies are required to allow us to answer the following questions: (a) Are dA and dG adducts repaired at similar or different rates? (b) Does the stereochemical configuration of the adduct (i.e. from the (+) or (–) enantiomer of the *syn*- and *anti*-diol-epoxide attached by *cis* or *trans* addition by the base) affect its relative rate of repair? (c) Does the carcinogenic potency of complex mixtures of PAHs present in the environment, affect the fidelity of DNA repair of PAH-DNA adducts?

As a concluding remark, studies on the processes of DNA repair on a molecular and structural level are important to assess the carcinogenic potencies of stereoisomeric PAH dihydrodiol-epoxide-DNA lesions, as they play an indirect role [90, 129-131] in cellular mutagenesis. Futhermore, there is a future need to determine the potency of these well studied carcinogenic PAH within an environmental complex mixture. This would help in assessing human risk on exposure to environmental PAH mixtures, by providing a representation of the carcinogenic PAH.

ACKNOWLEDGEMENTS

The authors wish to acknowledge Dan Albershardt for assistance in the preparation of this chapter and compilation of data for the table. We also thank Eric Brooks and Tuan Nguyen for procuring and organizing pertinent references. Support for this research was provided by National Cancer Institute Grant CA28825 to WMB. Ms. Courter was partially supported by a training grant from the National Institute of Environmental Health Sciences, T32 ES007060.

REFERENCES

[1] 1983. IARC, Polynuclear aromatic compounds, Part 1. In *Chemical, environmental, and experimental data, IARC Monogr. Eval. Carcinog. Risks Hum. Suppl*

[2] Baird W, Ralston S. 1997. Carcinogenic polycyclic aromatic hydrocarbons. In *Comprehensive Toxicology*, ed. C McQueen, A Gandolfi, pp. 171-200. Cambridge: Cambridge University Press

[3] Harvey RG. 1991. *Polycyclic Aromatic Hydrocarbons: Chemistry and Carcinogenicity*. Cambridge: Cambridge University Press

[4] Wood AW, Chang RL, Levin W, Yagi H, Thakker DR, Jerina DM, Conney AH. 1977. Differences in mutagenicity of the optical enantiomers of the diastereomeric benzo[a]pyrene 7,8-diol-9,10-epoxides. *Biochem Biophys Res Commun* 77: 1389-96

[5] Thakker DR, Levin W, Yagi H, Tada M, Ryan DE, Thomas PE, Conney AH, Jerina DM. 1982. Stereoselective metabolism of the (+)- and (-)-enantiomers of trans-3,4-dihydroxy-3,4-dihydrobenz[a]anthracene by rat liver microsomes and by a purified and reconstituted cytochrome P-450 system. *J Biol Chem* 257: 5103-10

[6] Puga A, Nebert DW. 1990. Evolution of the P450 gene superfamily and regulation of the murine Cyp1a1 gene. *Biochem Soc Trans* 18: 7-10

[7] Chakravarti D, Mailander PC, Cavalieri EL, Rogan EG. 2000. Evidence that error-prone DNA repair converts dibenzo[a,l]pyrene-induced depurinating lesions into mutations: formation, clonal proliferation and regression of initiated cells carrying H-ras oncogene mutations in early preneoplasia. *Mutat Res* 456: 17-32

[8] Harris CC. 1985. Future directions in the use of DNA adducts as internal dosimeters for monitoring human exposure to environmental mutagens and carcinogens. *Environ Health Perspect* 62: 185-91

[9] Perera FP. 1988. The significance of DNA and protein adducts in human biomonitoring studies. *Mutat Res* 205: 255-69

[10] Ross JA, Nesnow S. 1999. Polycyclic aromatic hydrocarbons: correlations between DNA adducts and ras oncogene mutations. *Mutat Res* 424: 155-66

[11] Bishop JM. 1987. The molecular genetics of cancer. *Science* 235: 305-11

[12] Dittrich KA, Krugh TR. 1991. Mapping of (+/-)-anti-benzo[a]pyrene diol epoxide adducts to human c-Ha-ras1 protooncogene. *Chem Res Toxicol* 4: 277-81

[13] DiGiovanni J, Beltran L, Rupp A, Harvey RG, Gill RD. 1993. Further analysis of c-Ha-ras mutations in papillomas initiated by several polycyclic aromatic hydrocarbons and papillomas from uninitiated, promoter-treated skin in SENCAR mice. *Mol Carcinog* 8: 272-9

[14] Chen JX, Pao A, Zheng Y, Ye X, Kisleyou AS, Morris R, Slaga TJ, Harvey RG, Tang MS. 1996. Sequence preference of 7,12-dimethylbenz[a]anthracene-syn-diol epoxide-DNA binding in the mouse H-ras gene detected by UvrABC nucleases. *Biochemistry* 35: 9594-602

[15] Denissenko MF, Pao A, Tang M, Pfeifer GP. 1996. Preferential formation of benzo[a]pyrene adducts at lung cancer mutational hotspots in P53. *Science* 274: 430-2

[16] Denissenko MF, Chen JX, Tang MS, Pfeifer GP. 1997. Cytosine methylation determines hot spots of DNA damage in the human P53 gene. *Proc Natl Acad Sci U S A* 94: 3893-8

[17] Phillips DH. 2002. The formation of DNA adducts. *The Cancer Handbook,* In Allison, M.R. (ed), Macmilan, London.: 293-306

[18] Brookes P, Lawley PD. 1964. Reaction of Some Mutagenic and Carcinogenic Compounds with Nucleic Acids. *J Cell Physiol* 64: SUPPL 1:111-27

[19] Watanabe M, Maher VM, McCormick JJ. 1985. Excision repair of UV- or benzo[a]pyrene diol epoxide-induced lesions in xeroderma pigmentosum variant cells is 'error free'. *Mutat Res* 146: 285-94

[20] Choi DJ, Roth RB, Liu T, Geacintov NE, Scicchitano DA. 1996. Incorrect base insertion and prematurely terminated transcripts during T7 RNA polymerase transcription elongation past benzo[a]pyrenediol epoxide-modified DNA. *J Mol Biol* 264: 213-9

[21] Chakravarti D, Pelling JC, Cavalieri EL, Rogan EG. 1995. Relating aromatic hydrocarbon-induced DNA adducts and c-H-ras mutations in mouse skin papillomas: the role of apurinic sites. *Proc Natl Acad Sci U S A* 92: 10422-6

[22] Tang D, Cho S, Rundle A, Chen S, Phillips D, Zhou J, Hsu Y, Schnabel F, Estabrook A, Perera FP. 2002. Polymorphisms in the DNA repair enzyme XPD are associated with increased levels of PAH-DNA adducts in a case-control study of breast cancer. *Breast Cancer Res Treat* 75: 159-66

[23] Recio L, Shugart LR, Hsie AW. 1987. The relationship between benzo(a)pyrene diol-epoxide-DNA adducts and mutagenicity in the CHO/HPGRT assay. *Fundam Appl Toxicol* 8: 243-52

[24] Mahadevan B, Dashwood WM, Luch A, Pecaj A, Doehmer J, Seidel A, Pereira C, Baird WM. 2003. Mutations induced by (-)-anti-11R,12S-dihydrodiol 13S,14R-epoxide

of dibenzo[a,l]pyrene in the coding region of the hypoxanthine phosphoribosyltransferase (Hprt) gene in Chinese hamster V79 cells. *Environ Mol Mutagen* 41: 131-9
[25] Perera FP, Tang DL, O'Neill JP, Bigbee WL, Albertini RJ, Santella R, Ottman R, Tsai WY, Dickey C, Mooney LA, et al. 1993. HPRT and glycophorin A mutations in foundry workers: relationship to PAH exposure and to PAH-DNA adducts. *Carcinogenesis* 14: 969-73
[26] Schwartz JJ, Lau HH, Baird WM. 1994. Base sequence selectivity in the binding of 7(R),8(S)-dihydroxy- 9(S),10(R)-epoxy-7,8,9,10-tetrahydrobenzo[a]pyrene to oligodeoxyribonucleotide duplexes. *Chem Res Toxicol* 7: 29-40.
[27] Page JE, Zajc B, Oh-hara T, Lakshman MK, Sayer JM, Jerina DM, Dipple A. 1998. Sequence context profoundly influences the mutagenic potency of trans-opened benzo[a]pyrene 7,8-diol 9,10-epoxide-purine nucleoside adducts in site-specific mutation studies. *Biochemistry* 37: 9127-37
[28] Musafija-Jeknic T., Luch A., Seidel A., Johns C., Pereira C. and Baird W. M. 2005. Effect of Nucleotide Sequence on the Binding of (-)-anti-Dibenzo[a,l]pyrene-11,12-diol 13,14-epoxide to Short Oligodeoxyribonucleotides. *Polycyclic Aromatic Compounds* 25: 103-111.
[29] Li Z, Tamura PJ, Wilkinson AS, Harris CM, Harris TM, Stone MP. 2001. Intercalation of the (1R,2S,3R,4S)-N6-[1-(1,2,3,4-tetrahydro-2,3,4-trihydroxybenz[a]anthracenyl)]-2'-deoxyadenosyl adduct in the N-ras codon 61 sequence: DNA sequence effects. *Biochemistry* 40: 6743-55
[30] McNees AG, O'Donnell M, Horton PH, Kim HY, Kim SJ, Harris CM, Harris TM, Lloyd RS. 1997. Lack of correlation between in vitro and in vivo replication of precisely defined benz-a-anthracene adducted DNAs. *J Biol Chem* 272: 33211-9
[31] Dipple A, Moschel RC, Bigger CAH. 1984. In *Chemical Carcinogens*, ed. CE Searle, pp. 41-163. Washington D.C.: American Cancer Society
[32] Geacintov NE. 1985. In *Polycyclic Hydrocarbons and Carcinogenesis*, ed. RG Harvey, pp. 107-24. Washington D.C.: American Chemical Society
[33] Thakker DR, Yagi H, Levin W, Wood AW, Conney AH, Jerina DM. 1985. In *Bioactivation of Foreign Compounds*, ed. MW Anders, pp. 177-242. New York: Academic Press
[34] Jennette KW, Jeffrey AM, Blobstein SH, Beland FA, Harvey RG, Weinstein IB. 1977. Nucleoside adducts from the in vitro reaction of benzo[a]pyrene-7,8-dihydrodiol 9,10-oxide or benzo[a]pyrene 4,5-oxide with nucleic acids. *Biochemistry* 16: 932-8
[35] Osborne MR, Jacobs S, Harvey RG, Brookes P. 1981. Minor products from the reaction of (+) and (-) benzo[a]-pyrene-anti-diolepoxide with DNA. *Carcinogenesis* 2: 553-8
[36] Ralston SL, Lau HH, Seidel A, Luch A, Platt KL, Baird WM. 1994. The potent carcinogen dibenzo[a,l]pyrene is metabolically activated to fjord-region 11,12-diol 13,14-epoxides in human mammary carcinoma MCF-7 cell cultures. *Cancer Res* 54: 887-90
[37] Ralston SL, Seidel A, Luch A, Platt KL, Baird WM. 1995. Stereoselective activation of dibenzo[a,l]pyrene to (-)-anti (11R,12S,13S,14R)- and (+)-syn(11S,12R,13S,14R)-11,12-diol-13,14-epoxides which bind extensively to deoxyadenosine residues of DNA in the human mammary carcinoma cell line MCF-7. *Carcinogenesis* 16: 2899-907

[38] Moriya M, Spiegel S, Fernandes A, Amin S, Liu T, Geacintov N, Grollman AP. 1996. Fidelity of translesional synthesis past benzo[a]pyrene diol epoxide-2'-deoxyguanosine DNA adducts: marked effects of host cell, sequence context, and chirality. *Biochemistry* 35: 16646-51

[39] Geacintov NE, Broyde S, Buterin T, Naegeli H, Wu M, Yan S, Patel DJ. 2002. Thermodynamic and structural factors in the removal of bulky DNA adducts by the nucleotide excision repair machinery. *Biopolymers* 65: 202-10

[40] Bigger CA, Strandberg J, Yagi H, Jerina DM, Dipple A. 1989. Mutagenic specificity of a potent carcinogen, benzo[c]phenanthrene (4R,3S)-dihydrodiol (2S,1R)-epoxide, which reacts with adenine and guanine in DNA. *Proc Natl Acad Sci U S A* 86: 2291-5

[41] Wei SJ, Chang RL, Cui XX, Merkler KA, Wong CQ, Yagi H, Jerina DM, Conney AH. 1996. Dose-dependent differences in the mutational profiles of (-)-(1R,2S,3S,4R)-3,4-dihydroxy-1, 2-epoxy-1,2,3,4-tetrahydrobenzo[c]phenanthrene and its less carcinogenic enantiomer. *Cancer Res* 56: 3695-703

[42] Yoon JH, Besaratinia A, Feng Z, Tang MS, Amin S, Luch A, Pfeifer GP. 2004. DNA damage, repair, and mutation induction by (+)-Syn and (-)-anti-dibenzo[a,l]pyrene-11,12-diol-13,14-epoxides in mouse cells. *Cancer Res* 64: 7321-8

[43] Higginbotham S, RamaKrishna NV, Johansson SL, Rogan EG, Cavalieri EL. 1993. Tumor-initiating activity and carcinogenicity of dibenzo[a,l]pyrene versus 7,12-dimethylbenz[a]anthracene and benzo[a]pyrene at low doses in mouse skin. *Carcinogenesis* 14: 875-8

[44] Cavalieri EL, Higginbotham S, RamaKrishna NV, Devanesan PD, Todorovic R, Rogan EG, Salmasi S. 1991. Comparative dose-response tumorigenicity studies of dibenzo[alpha,l]pyrene versus 7,12-dimethylbenz[alpha]anthracene, benzo[alpha]-pyrene and two dibenzo[alpha,l]pyrene dihydrodiols in mouse skin and rat mammary gland. *Carcinogenesis* 12: 1939-44

[45] Cavalieri EL, Higginbotham S, Rogan EG. 1994. Dibenzopyrene: the most potent carcinogenic aromatic hydrocarbon. *Polycyclic Aromatic Compounds* 6: 177-83

[46] Singer B, Grunberger D. 1983. *Molecular Biology of Mutagens and Carcinogens*. New York: Plenum Press

[47] Wu M, Yan S, Patel DJ, Geacintov NE, Broyde S. 2002. Relating repair susceptibility of carcinogen-damaged DNA with structural distortion and thermodynamic stability. *Nucleic Acids Res* 30: 3422-32

[48] Buterin T, Hess MT, Luneva N, Geacintov NE, Amin S, Kroth H, Seidel A, Naegeli H. 2000. Unrepaired fjord region polycyclic aromatic hydrocarbon-DNA adducts in ras codon 61 mutational hot spots. *Cancer Res* 60: 1849-56

[49] Chakravarti D, Mailander P, Franzen J, Higginbotham S, Cavalieri EL, Rogan EG. 1998. Detection of dibenzo[a,l]pyrene-induced H-ras codon 61 mutant genes in preneoplastic SENCAR mouse skin using a new PCR-RFLP method. *Oncogene* 16: 3203-10

[50] Buening MK, Wislocki PG, Levin W, Yagi H, Thakker DR, Akagi H, Koreeda M, Jerina DM, Conney AH. 1978. Tumorigenicity of the optical enantiomers of the diastereomeric benzo[a]pyrene 7,8-diol-9,10-epoxides in newborn mice: exceptional activity of (+)-7beta,8alpha-dihydroxy-9alpha,10alpha-epoxy-7,8,9,10-tetrahydro-benzo[a]pyrene. *Proc Natl Acad Sci U S A* 75: 5358-61

[51] Slaga TJ, Bracken WJ, Gleason G, Levin W, Yagi H, Jerina DM, Conney AH. 1979. Marked differences in the skin tumor-initiating activities of the optical enantiomers of the diastereomeric benzo(a)pyrene 7,8-diol-9,10-epoxides. *Cancer Res* 39: 67-71

[52] Wei SJ, Chang RL, Wong CQ, Bhachech N, Cui XX, Hennig E, Yagi H, Sayer JM, Jerina DM, Preston BD, et al. 1991. Dose-dependent differences in the profile of mutations induced by an ultimate carcinogen from benzo[a]pyrene. *Proc Natl Acad Sci U S A* 88: 11227-30

[53] Wei SJ, Chang RL, Bhachech N, Cui XX, Merkler KA, Wong CQ, Hennig E, Yagi H, Jerina DM, Conney AH. 1993. Dose-dependent differences in the profile of mutations induced by (+)-7R,8S-dihydroxy-9S,10R-epoxy-7,8,9,10-tetrahydrobenzo(a)pyrene in the coding region of the hypoxanthine (guanine) phosphoribosyltransferase gene in Chinese hamster V-79 cells. *Cancer Res* 53: 3294-301

[54] Wei SJ, Chang RL, Hennig E, Cui XX, Merkler KA, Wong CQ, Yagi H, Jerina DM, Conney AH. 1994. Mutagenic selectivity at the HPRT locus in V-79 cells: comparison of mutations caused by bay-region benzo[a]pyrene 7,8-diol-9,-10-epoxide enantiomers with high and low carcinogenic activity. *Carcinogenesis* 15: 1729-35

[55] Dipple A, Pigott MA, Agarwal SK, Yagi H, Sayer JM, Jerina DM. 1987. Optically active benzo[c]phenanthrene diol epoxides bind extensively to adenine in DNA. *Nature* 327: 535-6

[56] Jerina DM, Sayer JM, Agarwal SK, Yagi H, Levin W, Wood AW, Conney AH, Pruess-Schwartz D, Baird WM, Pigott MA, et al. 1986. Reactivity and tumorigenicity of bay-region diol epoxides derived from polycyclic aromatic hydrocarbons. *Adv Exp Med Biol* 197: 11-30

[57] Bigger CA, St John J, Yagi H, Jerina DM, Dipple A. 1992. Mutagenic specificities of four stereoisomeric benzo[c]phenanthrene dihydrodiol epoxides. *Proc Natl Acad Sci U S A* 89: 368-72

[58] Schiltz M, Cui XX, Lu YP, Yagi H, Jerina DM, Zdzienicka MZ, Chang RL, Conney AH, Wei SJ. 1999. Characterization of the mutational profile of (+)-7R,8S-dihydroxy-9S, 10R-epoxy-7,8,9,10-tetrahydrobenzo[a]pyrene at the hypoxanthine (guanine) phosphoribosyltransferase gene in repair-deficient Chinese hamster V-H1 cells. *Carcinogenesis* 20: 2279-86

[59] Chen J, Thilly WG. 1996. Mutational spectra vary with exposure conditions: benzo[a]pyrene in human cells. *Mutat Res* 357: 209-17

[60] DeMarini DM, Shelton ML, Levine JG. 1995. Mutation spectrum of cigarette smoke condensate in Salmonella: comparison to mutations in smoking-associated tumors. *Carcinogenesis* 16: 2535-42

[61] Eisenstadt E, Warren AJ, Porter J, Atkins D, Miller JH. 1982. Carcinogenic epoxides of benzo[a]pyrene and cyclopenta[cd]pyrene induce base substitutions via specific transversions. *Proc Natl Acad Sci U S A* 79: 1945-9

[62] Chen RH, Maher VM, McCormick JJ. 1990. Effect of excision repair by diploid human fibroblasts on the kinds and locations of mutations induced by (+/-)-7 beta,8 alpha-dihydroxy-9 alpha,10 alpha-epoxy-7,8,9,10- tetrahydrobenzo[a]pyrene in the coding region of the HPRT gene. *Proc Natl Acad Sci U S A* 87: 8680-4

[63] Kohler SW, Provost GS, Fieck A, Kretz PL, Bullock WO, Sorge JA, Putman DL, Short JM. 1991. Spectra of spontaneous and mutagen-induced mutations in the lacI gene in transgenic mice. *Proc Natl Acad Sci U S A* 88: 7958-62

[64] Gondo Y, Shioyama Y, Nakao K, Katsuki M. 1996. A novel positive detection system of in vivo mutations in rpsL (strA) transgenic mice. *Mutat Res* 360: 1-14

[65] Gossen JA, de Leeuw WJ, Tan CH, Zwarthoff EC, Berends F, Lohman PH, Knook DL, Vijg J. 1989. Efficient rescue of integrated shuttle vectors from transgenic mice: a model for studying mutations in vivo. *Proc Natl Acad Sci U S A* 86: 7971-5

[66] Nohmi T, Katoh M, Suzuki H, Matsui M, Yamada M, Watanabe M, Suzuki M, Horiya N, Ueda O, Shibuya T, Ikeda H, Sofuni T. 1996. A new transgenic mouse mutagenesis test system using Spi- and 6-thioguanine selections. *Environ Mol Mutagen* 28: 465-70

[67] Amanuma K, Takeda H, Amanuma H, Aoki Y. 2000. Transgenic zebrafish for detecting mutations caused by compounds in aquatic environments. *Nat Biotechnol* 18: 62-5

[68] Winn RN, Norris MB, Brayer KJ, Torres C, Muller SL. 2000. Detection of mutations in transgenic fish carrying a bacteriophage lambda cII transgene target. *Proc Natl Acad Sci U S A* 97: 12655-60

[69] Weisenberger DJ, Romano LJ. 1999. Cytosine methylation in a CpG sequence leads to enhanced reactivity with Benzo[a]pyrene diol epoxide that correlates with a conformational change. *J Biol Chem* 274: 23948-55

[70] Chen JX, Zheng Y, West M, Tang MS. 1998. Carcinogens preferentially bind at methylated CpG in the p53 mutational hot spots. *Cancer Res* 58: 2070-5

[71] Smith LE, Denissenko MF, Bennett WP, Li H, Amin S, Tang M, Pfeifer GP. 2000. Targeting of lung cancer mutational hotspots by polycyclic aromatic hydrocarbons. *J Natl Cancer Inst* 92: 803-11

[72] Chakravarti D, Cavalieri EL, Rogan EG. 1998. Linear amplification mapping of polycyclic aromatic hydrocarbon-reactive sequences in H-ras gene. *DNA Cell Biol* 17: 529-39

[73] Mass MJ, Abu-Shakra A, Roop BC, Nelson G, Galati AJ, Stoner GD, Nesnow S, Ross JA. 1996. Benzo[b]fluoranthene: tumorigenicity in strain A/J mouse lungs, DNA adducts and mutations in the Ki-ras oncogene. *Carcinogenesis* 17: 1701-4

[74] Mass MJ, Jeffers AJ, Ross JA, Nelson G, Galati AJ, Stoner GD, Nesnow S. 1993. Ki-ras oncogene mutations in tumors and DNA adducts formed by benz[j]aceanthrylene and benzo[a]pyrene in the lungs of strain A/J mice. *Mol Carcinog* 8: 186-92

[75] Prahalad AK, Ross JA, Nelson GB, Roop BC, King LC, Nesnow S, Mass MJ. 1997. Dibenzo[a,l]pyrene-induced DNA adduction, tumorigenicity, and Ki-ras oncogene mutations in strain A/J mouse lung. *Carcinogenesis* 18: 1955-63

[76] Xia Q, Yi P, Zhan DJ, Von Tungeln LS, Hart RW, Heflich RH, Fu PP. 1998. Liver tumors induced in B6C3F1 mice by 7-chlorobenz[a]anthracene and 7-bromobenz[a]anthracene contain K-ras protooncogene mutations. *Cancer Lett* 123: 21-5

[77] Wessner LL, Fan M, Schaeffer DO, McEntee MF, Miller MS. 1996. Mouse lung tumors exhibit specific Ki-ras mutations following transplacental exposure to 3-methylcholanthrene. *Carcinogenesis* 17: 1519-26

[78] You L, Wang D, Galati AJ, Ross JA, Mass MJ, Nelson GB, Wilson KH, Amin S, Stoner JC, Nesnow S, et al. 1994. Tumor multiplicity, DNA adducts and K-ras mutation pattern of 5-methylchrysene in strain A/J mouse lung. *Carcinogenesis* 15: 2613-8

[79] Loktionov A, Hollstein M, Martel N, Galendo D, Cabral JR, Tomatis L, Yamasaki H. 1990. Tissue-specific activating mutations of Ha- and Ki-ras oncogenes in skin, lung,

and liver tumors induced in mice following transplacental exposure to DMBA. *Mol Carcinog* 3: 134-40

[80] Manam S, Storer RD, Prahalada S, Leander KR, Kraynak AR, Hammermeister CL, Joslyn DJ, Ledwith BJ, van Zwieten MJ, Bradley MO, et al. 1992. Activation of the Ki-ras gene in spontaneous and chemically induced lung tumors in CD-1 mice. *Mol Carcinog* 6: 68-75

[81] Qing WG, Conti CJ, LaBate M, Johnston D, Slaga TJ, MacLeod MC. 1997. Induction of mammary cancer and lymphoma by multiple, low oral doses of 7,12-dimethylbenz[a]anthracene in SENCAR mice. *Carcinogenesis* 18: 553-9

[82] Tang MS, Vulimiri SV, Viaje A, Chen JX, Bilolikar DS, Morris RJ, Harvey RG, Slaga TJ, DiGiovanni J. 2000. Both (+/-)syn- and (+/-)anti-7,12-dimethylbenz[a]anthracene-3,4-diol-1,2-epoxides initiate tumors in mouse skin that possess -CAA- to -CTA- mutations at Codon 61 of c-H-ras. *Cancer Res* 60: 5688-95

[83] Gill RD, Beltran L, Nettikumara AN, Harvey RG, Kootstra A, DiGiovanni J. 1992. Analysis of point mutations in murine c-Ha-ras of skin tumors initiated with dibenz[a,j]anthracene and derivatives. *Mol Carcinog* 6: 53-9

[84] Osaka M, Matsuo S, Koh T, Liang P, Kinoshita H, Maeda S, Sugiyama T. 1995. N-ras mutation in 7,12-dimethylbenz[a]anthracene (DMBA)-induced erythroleukemia in Long-Evans rats. *Cancer Lett* 91: 25-31

[85] Pfeifer GP, Denissenko MF, Olivier M, Tretyakova N, Hecht SS, Hainaut P. 2002. Tobacco smoke carcinogens, DNA damage and p53 mutations in smoking-associated cancers. *Oncogene* 21: 7435-51

[86] Olivier M, Eeles R, Hollstein M, Khan MA, Harris CC, Hainaut P. 2002. The IARC TP53 database: new online mutation analysis and recommendations to users. *Hum Mutat* 19: 607-14

[87] Dong H, Bonala RR, Suzuki N, Johnson F, Grollman AP, Shibutani S. 2004. Mutagenic potential of benzo[a]pyrene-derived DNA adducts positioned in codon 273 of the human P53 gene. *Biochemistry* 43: 15922-8

[88] Muheim R, Buterin T, Colgate KC, Kolbanovsij A, Geacintov NE, Naegeli H. 2003. Modulation of human nucleotide excision repair by 5-methylcytosines. *Biochemistry* 42: 3247-54

[89] Huang X, Colgate KC, Kolbanovskiy A, Amin S, Geacintov NE. 2002. Conformational changes of a benzo[a]pyrene diol epoxide-N(2)-dG adduct induced by a 5'-flanking 5-methyl-substituted cytosine in a (Me)CG double-stranded oligonucleotide sequence context. *Chem Res Toxicol* 15: 438-44

[90] Denissenko MF, Pao A, Pfeifer GP, Tang M. 1998. Slow repair of bulky DNA adducts along the nontranscribed strand of the human p53 gene may explain the strand bias of transversion mutations in cancers. *Oncogene* 16: 1241-7

[91] Fernandes A, Liu T, Amin S, Geacintov NE, Grollman AP, Moriya M. 1998. Mutagenic potential of stereoisomeric bay region (+)- and (-)-cis-anti-benzo[a]pyrene diol epoxide-N2-2'-deoxyguanosine adducts in Escherichia coli and simian kidney cells. *Biochemistry* 37: 10164-72

[92] Hussain SP, Amstad P, Raja K, Sawyer M, Hofseth L, Shields PG, Hewer A, Phillips DH, Ryberg D, Haugen A, Harris CC. 2001. Mutability of p53 hotspot codons to benzo(a)pyrene diol epoxide (BPDE) and the frequency of p53 mutations in nontumorous human lung. *Cancer Res* 61: 6350-5

[93] Wei SJ, Chang RL, Merkler KA, Gwynne M, Cui XX, Murthy B, Huang MT, Xie JG, Lu YP, Lou YR, Jerina DM, Conney AH. 1999. Dose-dependent mutation profile in the c-Ha-ras proto-oncogene of skin tumors in mice initiated with benzo[a]pyrene. *Carcinogenesis* 20: 1689-96

[94] Bos JL. 1989. ras oncogenes in human cancer: a review. *Cancer Res* 49: 4682-9

[95] Leon J, Pellicer A. 1993. In *The ras superfamily of GTPase*, ed. JC Lacal, F McCormick, pp. 1-36. Boca Raton, FL: CRC Press

[96] Feng Z, Hu W, Chen JX, Pao A, Li H, Rom W, Hung MC, Tang MS. 2002. Preferential DNA damage and poor repair determine ras gene mutational hotspot in human cancer. *J Natl Cancer Inst* 94: 1527-36

[97] Hu W, Feng Z, Tang MS. 2003. Preferential carcinogen-DNA adduct formation at codons 12 and 14 in the human K-ras gene and their possible mechanisms. *Biochemistry* 42: 10012-23

[98] Snook ME, Severson RF, Arrendale RF, Higman HC, Chortyk OT. 1977. The identification of high molecular weight polynuclear aromatic hydrocarbons in a biologically active fraction of cigarette smoke condensate. *Beitr. Tabakforsch* 9: 79-101

[99] De Raat WK, Kooijman SA, Gielen JW. 1987. Concentrations of polycyclic hydrocarbons in airborne particles in The Netherlands and their correlation with mutagenicity. *Sci Total Environ* 66: 95-114

[100] Grimmer G. 1983. *Environmental Carcinogens: Polycyclic Aromatic Hydrocarbons.* Boca Raton, FL: CRC Press Inc

[101] 1983. Polynuclear aromatic compounds, Part 1, Chemical, environmental and experimental data. *IARC Monogr Eval Carcinog Risk Chem Hum* 32: 1-453

[102] 1984. Polynuclear Aromatic Compounds Part 3, Industrial Exposure in Aluminum Production, Coal Gasification, Coke Production, and Iron and Steel Founding. *IARC Monogr Eval Carcinog Risk Chem Hum*

[103] 1985. Polynuclear aromatic compounds, Part 4, Bitumens, coal-tars and derived products, shale-oils and soots. *IARC Monogr Eval Carcinog Risk Chem Hum* 35: 1-247

[104] Somers CM, McCarry BE, Malek F, Quinn JS. 2004. Reduction of particulate air pollution lowers the risk of heritable mutations in mice. *Science* 304: 1008-10

[105] Samet JM, DeMarini DM, Malling HV. 2004. Biomedicine. Do airborne particles induce heritable mutations? *Science* 304: 971-2

[106] Mahadevan B, Keshava, C, Musafia-Jeknic, T, Pecaj, A, Weston, A, Baird, WM. 2005. Altered gene expression patterns in MCF-7 cells induced by the urban dust particulate complex mixture standard reference material 1649a. *Cancer Res.,* 65: 1251-8

[107] Hecht SS. 1999. Tobacco smoke carcinogens and lung cancer. *J Natl Cancer Inst* 91: 1194-210

[108] Rojas M, Alexandrov K, Cascorbi I, Brockmoller J, Likhachev A, Pozharisski K, Bouvier G, Auburtin G, Mayer L, Kopp-Schneider A, Roots I, Bartsch H. 1998. High benzo[a]pyrene diol-epoxide DNA adduct levels in lung and blood cells from individuals with combined CYP1A1 MspI/Msp-GSTM1*0/*0 genotypes. *Pharmacogenetics* 8: 109-18

[109] Kawajiri K, Eguchi H, Nakachi K, Sekiya T, Yamamoto M. 1996. Association of CYP1A1 germ line polymorphisms with mutations of the p53 gene in lung cancer. *Cancer Res* 56: 72-6

[110] Lazarus P, Sheikh SN, Ren Q, Schantz SP, Stern JC, Richie JP, Jr., Park JY. 1998. p53, but not p16 mutations in oral squamous cell carcinomas are associated with specific CYP1A1 and GSTM1 polymorphic genotypes and patient tobacco use. *Carcinogenesis* 19: 509-14

[111] Przygodzki RM, Bennett WP, Guinee DG, Jr., Khan MA, Freedman A, Shields PG, Travis WD, Jett JR, Tazelaar H, Pairolero P, Trastek V, Liotta LA, Harris CC, Caporaso NE. 1998. p53 mutation spectrum in relation to GSTM1, CYP1A1 and CYP2E1 in surgically treated patients with non-small cell lung cancer. *Pharmacogenetics* 8: 503-11

[112] Stoilov I, Akarsu AN, Alozie I, Child A, Barsoum-Homsy M, Turacli ME, Or M, Lewis RA, Ozdemir N, Brice G, Aktan SG, Chevrette L, Coca-Prados M, Sarfarazi M. 1998. Sequence analysis and homology modeling suggest that primary congenital glaucoma on 2p21 results from mutations disrupting either the hinge region or the conserved core structures of cytochrome P4501B1. *Am J Hum Genet* 62: 573-84

[113] Perera FP, Dickey C, Santella R, O'Neill JP, Albertini RJ, Ottman R, Tsai WY, Mooney LA, Savela K, Hemminki K. 1994. Carcinogen-DNA adducts and gene mutation in foundry workers with low-level exposure to polycyclic aromatic hydrocarbons. *Carcinogenesis* 15: 2905-10

[114] Mognato M, Graziani M, Celotti L. 1999. Analysis of point mutations in HPRT mutant T-lymphocytes derived from coke-oven workers. *Mutat Res* 431: 271-8

[115] Osterholm AM, Falt S, Lambert B, Hou SM. 1995. Classification of mutations at the human hprt-locus in T-lymphocytes of bus maintenance workers by multiplex-PCR and reverse transcriptase-PCR analysis. *Carcinogenesis* 16: 1909-12

[116] Delgado-Rodriguez A, Ortiz-Marttelo R, Graf U, Villalobos-Pietrini R, Gomez-Arroyo S. 1995. Genotoxic activity of environmentally important polycyclic aromatic hydrocarbons and their nitro derivatives in the wing spot test of Drosophila melanogaster. *Mutat Res* 341: 235-47

[117] Granville CA, Hanley NM, Mumford JL, DeMarini DM. 2003. Mutation spectra of smoky coal combustion emissions in Salmonella reflect the TP53 and KRAS mutations in lung tumors from smoky coal-exposed individuals. *Mutat Res* 525: 77-83

[118] Reliene R, Hlavacova A, Mahadevan B, Baird WM, Schiestl RH. 2005. Diesel exhaust particles cause increased levels of DNA deletions after transplacental exposure in mice. *Mutat Res* 570: 245-52

[119] DeMarini DM, Shelton ML, Bell DA. 1994. Mutation spectra in Salmonella of complex mixtures: comparison of urban air to benzo[a]pyrene. *Environ Mol Mutagen* 24: 262-75

[120] Peluso M, Ceppi M, Munnia A, Puntoni R, Parodi S. 2001. Analysis of 13 (32)P-DNA postlabeling studies on occupational cohorts exposed to air pollution. *Am J Epidemiol* 153: 546-58

[121] Baird WM, Hooven LA, Mahadevan B. 2005. Carcinogenic polycyclic aromatic hydrocarbon-DNA adducts and mechanism of action. *Environ Mol Mutagen*

[122] Brookes P, Lawley PD. 1964. Evidence for the binding of polynuclear aromatic hydrocarbons to the nucleic acids of mouse skin: relation between carcinogenic power of hydrocarbons and their binding to deoxyribonucleic acid. *Nature* 202: 781-4

[123] Seo KY, Jelinsky SA, Loechler EL. 2000. Factors that influence the mutagenic patterns of DNA adducts from chemical carcinogens. *Mutat Res* 463: 215-46

[124] Yang LL, Maher VM, McCormick JJ. 1980. Error-free excision of the cytotoxic,mutagenic N2-deoxyguanosine DNA adduct formed in human fibroblasts by (+/-)-7 beta, 8 alpha-dihydroxy-9 alpha, 10 alpha-epoxy-7,8,9,10-tetrahydrobenzo-[a]pyrene. *Proc Natl Acad Sci U S A* 77: 5933-7

[125] Yang LL, Maher VM, McCormick JJ. 1982. Relationship between excision repair and the cytotoxic and mutagenic effect of the 'anti' 7,8-diol-9,10-epoxide of benzo[a]pyrene in human cells. *Mutat Res* 94: 435-47

[126] DiGiovanni J, Decina PC, Prichett WP, Fisher EP, Aalfs KK. 1985. Formation and disappearance of benzo[a]pyrene DNA-adducts in mouse epidermis. *Carcinogenesis* 6: 741-7

[127] Nakayama J, Yuspa SH, Poirier MC. 1984. BaP-DNA adduct formation and removal in mouse epidermis in vivo and in vitro: relationship of DNA binding to initiation of skin carcinogenesis. *Cancer Res* 44: 4087

[128] Shi Q, Wang LE, Bondy ML, Brewster A, Singletary SE, Wei Q. 2004. Reduced DNA repair of benzo[a]pyrene diol epoxide-induced adducts and common XPD polymorphisms in breast cancer patients. *Carcinogenesis* 25: 1695-700

[129] Wei D, Maher VM, McCormick JJ. 1995. Site-specific rates of excision repair of benzo[a]pyrene diol epoxide adducts in the hypoxanthine phosphoribosyltransferase gene of human fibroblasts: correlation with mutation spectra. *Proc Natl Acad Sci U S A* 92: 2204-8

[130] Braithwaite E, Wu X, Wang Z. 1999. Repair of DNA lesions: mechanisms and relative repair efficiencies. *Mutat Res* 424: 207-19

[131] Pfeifer GP, Denissenko MF. 1998. Formation and repair of DNA lesions in the p53 gene: relation to cancer mutations? *Environ Mol Mutagen* 31: 197-205

[132] Conney AH, Chang RL, Cui XX, Schiltz M, Yagi H, Jerina DM, Wei SJ. 2001. Dose-dependent differences in the profile of mutations induced by carcinogenic (R,S,S,R) bay- and fjord-region diol epoxides of polycyclic aromatic hydrocarbons. *Adv Exp Med Biol* 500: 697-707

In: New Developments in Mutation Research
Editor: Charles L. Valon, pp. 41-70
ISBN 1-59454-664-9

Chapter 3

Progress in Mutation Research on the Tandem Repeat in the Genome

Masatake Yamauchi* and Reiko Uchiyama
National Institute of Radiological Sciences, Chiba, Japan

Abstract

Tandem repeats are DNA nucleotide sequences widespread throughout the genomes of various organisms, including humans and the mouse. Most of the tandem repeats are stably transmitted to successive generations; however, some exhibit genetic instability, under certain conditions or without any obvious reasons, during the transmission to the next generation. Tandem repeats that exhibit such instability are often referred to as "hyper-variable" tandem repeats.

The genetic instability of tandem repeats has been utilized in detecting the hazardous effect of environmental mutagens, including chemical compounds and ionizing radiation. For instance, *Pc-1* (*Ms6-hm*) is one of the hyper-variable tandem repeats in the mouse genome, and its hypersensitivity is very useful for the detection of the biological effects caused by environmental mutagens. *Pc-1* comprises (GGCAGG)n and exhibits a genetic instability in its copy number during trans-generational transmission. Although no disease has been identified in relation to the genomic instability of the *Pc-1* tandem repeat thus far, its hyper-mutability has been widely utilized as a sensitive detector able to monitor the hazardous effects of environmental mutagens. The instability of the *Pc-1* tandem repeat is also affected by the genetic background of mice. For instance, functional mutations which occur in DNA-dependent protein kinase catalytic subunit (DNA-PKcs) are able to increase the genetic instability of *Pc-1* tandem repeat.

Triplet repeats are the specific type of tandem repeats composed of trinucleotide repeats in the human genome. While most of triplet repeats are stable during trans-generational transmission, some exhibit genetic instability. Genetic instability of certain trinucleotide repeats has been identified as one of the causes of genetic disease in human

* Laboratory of Genetics, National Institute of Radiological Sciences, 4-9-1 Anagawa, Inage, Chiba 263-8555 Japan; Tel: +81-43-206-4069; Fax: +81-43-206-4603; Email: m_yam@nirs.go.jp, reiko@nirs.go.jp

populations. The first human disorder revealed to be caused by CCG trinucleotide repeat expansion was the so-called fragile X syndrome. This new type of mutation, caused by repeat expansion (and/or contraction), is categorized as a dynamic mutation. Repeat expansion has explained many characteristic aspects of the syndrome, such as the dominant inheritance and anticipation (i.e., juvenile onset or severer manifestation of the disease phenotype), which are observed in the affected pedigrees. Cells derived from the patients affected by the fragile X syndrome also exhibit a specific chromosomal fragile sites at Xq27.3, thus the identification of CCG repeat expansion has facilitated research into both the clinical manifestations of the syndrome and cytogenetic expression of the chromosomal fragile sites at the molecular level. CAG repeat expansion has also been identified in many neurodegenerative disorders characterized by an autosomal dominant pattern of inheritance and anticipation, such as Huntington's disease and the spinocerebellar ataxias. Recently, new repeat expansions, a tetranucleotide and a pentanucleotide repeat expansion, have been identified in type2 myotonic dystrophy (DM2) and type10 spinocerebellar ataxia (SCA10), respectively.

Dynamic mutations occurred at the polyalanine tract in the coding regions of the transcription factor genes cause the malformation syndromes. Most of the dynamic mutations causing polyalanine diseases seem to have arisen in the past, and these repeats to have been since transmitted stably to the present. However, this being the case, it is nevertheless likely that the original dynamic mutation occurred through a similar molecular mechanism observed in the other repeat diseases caused by genetically unstable repeats. Non-lethal symptoms of polyalanine diseases may have allowed such mutations to be inherited while avoiding strong pressure in the general population, and subsequent introduction of a point mutation interrupting the perfect repeat may have stabilized the repeats. Significant roles for polyalanine expansion in morphological evolution of vertebrates have been suggested.

The molecular mechanism for dynamic mutation has yet to be elucidated, and it is of great interest to understand both the mechanism itself and the dynamism and stability of the mammalian genome.

INTRODUCTION

The existence of tandem repeat sequences in the human genome has been recognized for many years, but its biological significance in humans has not. On the other hand, the importance of the repetitive DNA sequences has been noticed in plants in the adaptive response to stressed conditions. [1] Plant genomes undergo rapid genomic changes even during the period of normal growth prior to flowering. [2,3] It has also been noticed that similar molecular events occur repeatedly at the specific sites in the plant genomes.

A predominant part of the human genome consists of repetitive sequences of various types encompassing large segmental duplications, interspersed repeats derived from transposons, and tandem repeats. Nearly half of the human genome derives from transposable elements such as the long terminal repeat (LTR), L1, and the short interspersed nuclear elements (SINE) family. SINEs are highly abundant components of mammalian genomes that are propagated by retrotransposition. SINEs have been recognized as a causal agent of human disease and must also have had a profound influence in shaping eukaryotic genomes in general. [4] The SINE family includes Alu, MIR, and MIR3. For instance, the Alu sequence is specific in the primate genome and estimated to be 1.1 million copies, which account for approximately 10% of the human genome. Transposable elements are abundant in gene sequences and are also present in a significant portion of mature mRNAs, mostly in the untranslated regions. [5]

Tandem repeats are also widespread throughout the genomes of various organisms, including the human and mouse. Tandem repeats include minisatellites (~30bp per repeat unit) and microsatellites (2-5bp per repeat unit). Most of the tandem repeats are found between coding sequences and in the untranslated regions of genes. This does not necessarily mean that tandem repeats are separately located from the coding regions and their regulatory sequences in the genome. Some tandem repeats are located near the coding regions as well as within the coding sequence and promoter regions. Most tandem repeats are stably transmitted to successive generations, however, some of them exhibit genetic instability, under certain conditions or for no obvious reasons, during transmission to the next generation. Tandem repeats that are different between individuals include the hypervariable tandem repeats. The majority of hypervariable tandem repeats in an individual are inherited from his/her parents to the successive generations stably, which results in these repeats being very specific to a given individual. Thus, hypervariable tandem repeats are often used for forensic purposes. In this text, we focus on the genetic instability of tandem repeats, such as the dynamic mutations identified in human genetic diseases as well as in chromosomal fragile sites, and on the outcomes caused by the hypermutability of these repetitive DNA sequences.

Hypervariable Tandem Repeats

Tandem-repetitive minisatellite regions in vertebrate DNA frequently exhibit substantial allelic variation in the number of repeat units. This variation is thought to arise through processes such as unequal crossover or replication slippage. Pc-1 (Ms6-hm) [6] is one of the best examples of hypervariable tandem repeats in the mouse genome. Pc-1 comprises a pattern of (GGCAGG)n and exhibits genetic instability in its copy number during trans-generational transmission. Although no disease has been identified in relation to the genomic instability of the Pc-1 tandem repeat to date, its hyper-mutability has been widely utilized as a sensitive detector for monitoring the hazardous effects of environmental mutagens such as chemical compounds and ionizing radiation. Genetic stability of tandem repeats seems to be controlled by many proteins. For instance, mutations which occur in the human flap endonuclease 1 (h-FEN1) have been shown to have an effect on repeat instability. [7] Furthermore, the functional mutations which occur in the DNA-dependent protein kinase catalytic subunit (DNA-PKcs) are able to increase the genetic instability of the Pc-1 tandem repeat. [8]

Triplet repeats are a specific type of tandem repeats composed of trinucleotide repeats in the human genome. While most triplet repeats are stable during trans-generational transmission, some exhibit genetic instability. Genetic instability of certain trinucleotide repeats has been identified as a cause of genetic disease in human populations. These genetic diseases have been characterized by their non-Mendelian patterns of inheritance. Typically the manifestation of disease is caused by an expansion of trinucleotide repeats, although the subsequent molecular mechanisms may be different from each other.

Table 1. Human genetic diseases caused by repeat expansions

OMIM	human genetic disease	repeat	gene involved	position	encoding	inheritance	location
309550	Fragile X Syndrome		*FMR-1*			dominant	Xq27.3
309548	FRAXE mental retardation	CCG	*FMR-2*	5'-UTR	none		Xq28
147791	Jacobsen syndrome		*CBL2*			sporadic	11q23.3
313200	Spinal-bulbar muscular atrophy (SBMA)		*AR*				Xq11-q12
125370	Dentatorubral-pallidoluysian atrophy (DRPLA)		*Atrophin-1*				12p13.31
143100	Huntington disease (HD)		*IT15*				4p16.3
164400	Spinocerebellar ataxia 1 (SCA1)		*ataxin-1*				6p24-p23
183090	Spinocerebellar ataxia 2 (SCA2)		*ataxin-2*	exon	Poly-Gln		12q24.1
109150	Spinocerebellar ataxia 3 (SCA3)	CAG	*ataxin-3*				14q32.1
183086	Spinocerebellar ataxia 6 (SCA6)		*CACNL1A4*				19p13
164500	Spinocerebellar ataxia 7 (SCA7)		*ataxin-7*			dominant	3p13-p12
607136	Spinocerebellar ataxia 17 (SCA17)		*TBP*				6p27
604326	Spinocerebellar ataxia 12 (SCA12)		*PPP2R2B*	5'-UTR			5q31-q33
160900	Myotonic dystrophy 1 (DM1)		*DMPK*		none		19q13.2-q13.3
608768	Spinocerebellar ataxia 8 (SCA8)	CTG	*SCA8*	3'-UTR			13q21
606438	Huntington disease-like-2 (HDL2)		*JPH3*		poly-Leu		16q23-q24
164300	Oculopharyngeal muscular dystrophy (OPMD)		*PABPN1*				14q11.2-q13
257950						recessive	
18600	Syndactyly type II		*HOXD13*			dominant	2q31-q32
119600	Cleidocranial dysplasia	GCG	*CBFA1*	exon	poly-Ala	recessive	6p21
603073	Holoprosencephaly		*ZIC2*			sporadic	13q32
140000	Hand-foot-genital (uterus) syndrome		*HOXA13*			dominant	7p15-p14.2
110100	Blepharophimosis, ptosis and epicanthus inversus		*FOXL2*				3q23

Table 1. Human genetic diseases caused by repeat expansions (Continued)

OMIM	human genetic disease	repeat	gene involved	position	encoding	inheritance	Location
300123	Mental retardation; X-linked, with isolated growth hormone deficiency		*SOX3*				Xq26.3
308350	Infantile spasm syndrome, X-linked;					recessive	
309510	Partington X-linked mental retardation syndrome	GCG					
300215	Lissencephaly, X-linked, with ambiguous genitalia		*ARX*	exon	poly-Ala	dominant	Xp22.13
300430	Mental retardation, X-linked 36					recessive	
300419	Mental retardation, X-linked 54						
209880	Congenital central hypoventilation syndrome		*PHOX2B*				4p12
177170	Pseudoachondroplasia (PSACH)	GAC	*COMP*			dominant	19p13.1
132400	Multiple epiphyseal dysplasia (MED)				poly-Asp		
229300	Friedreich ataxia (FA)	GAA	*FRDA*			recessive	9q13
602668	Myotonic dystrophy 2 (DM2)	CCTG	*ZNF9*	intron	none	dominant	3q13.3-q24
603516	Spinocerebellar ataxia 10 (SCA10)	ATTCT	*ataxin-10*				22q13

OMIM (Online Mendelian Inheritance in Man): http://www.genome.ad.jp/dbget-bin/www_bfind?omim

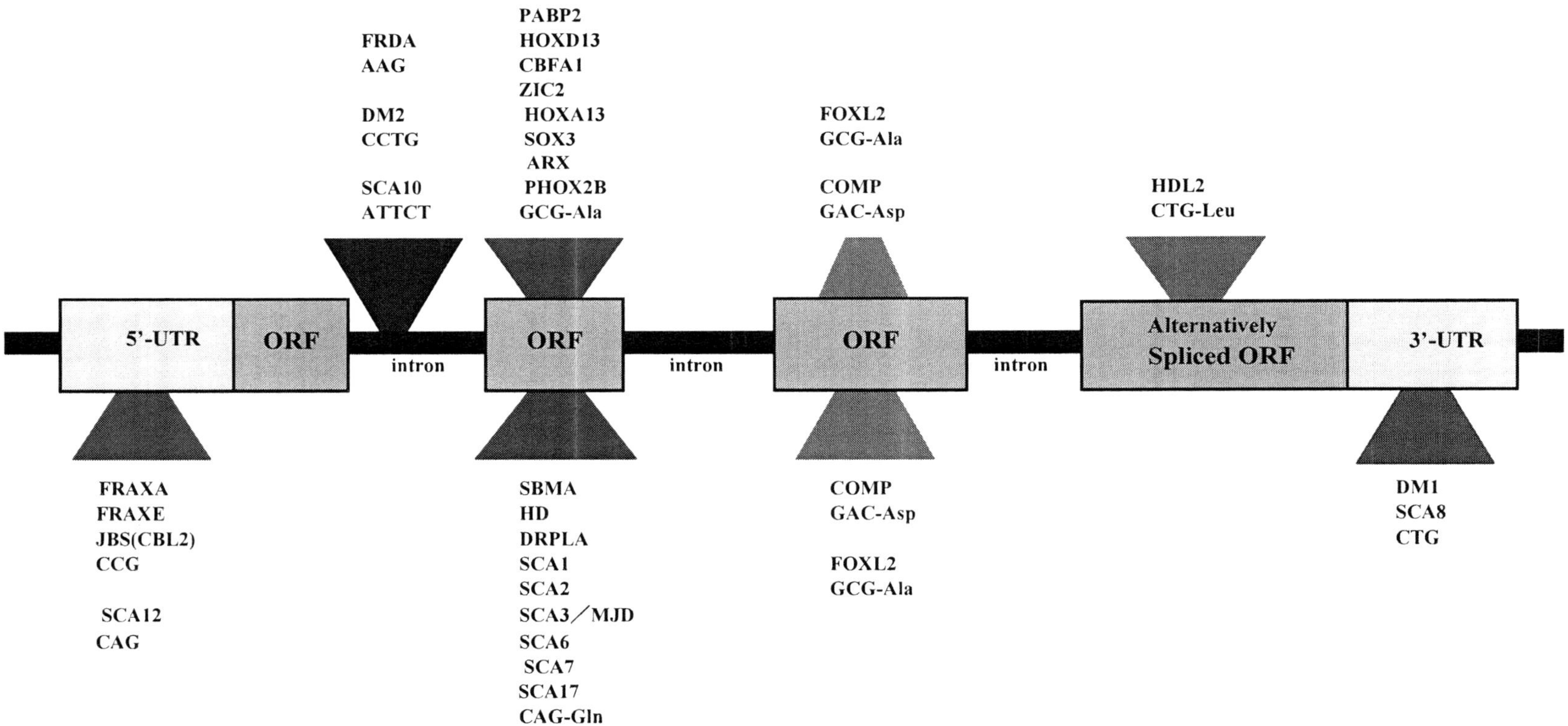

Figure 1. Summary of repeat expansions in disease-responsible genes
Locations of responsible tandem repeats which expansion and/or contraction cause human genetic diseases are summarized. Trapezoidal shape means the type of dynamic mutation, repeat expansion or repeat contraction. Name of the disease or the responsible gene as well as the repeat unit are also presented below or above the trapeziod. Except for pseudoachondroplasia (PSACH) and blepharophimosis, ptosis and epicanthus inversus (BPEI), both caused by the repeat contraction in *COMP* gene and *FOXL2* gene, respectively, all other repeat diseases are caused by the repeat expansion within 5'-UTR (*FRAXA, FRAXE, JBS (CBL2)*, and *SCA12*), open reading frame (ORF) (*SBMA, HD, DRPLA, SCA1, SCA2, SCA3/MJD, SCA6, SCA7, SCA17, PABP2, HOXD13, CBFA1, ZIC2, HOXA13, SOX3, ARX*, and *PHOX2B*), alternatively spliced ORF (*HDL2*), intron (*FRDA, DM2*, and *SCA10*), or 3'-UTR (*DM1* and *SCA8*). PSACH and BPEI are also caused by the repeat expansion.

This new type of mutation, caused by repeat expansion (and/or contraction), is referred to as a dynamic mutation. The molecular mechanisms for dynamic mutation has yet to be elucidated, and it is of great interest to help understand the dynamism and stability of mammalian genome.

The characteristics of repeat diseases are summarized in Table 1 and described in Figure 1.

Fragile X syndromes

Fragile X syndrome [9] is the most common form of familial mental retardation. This syndrome occurs one in 1500-2000 births in most human populations. The syndrome was named after the characteristic fragility of the X chromosome with a break point at q27.3 observed in the lymphocytes derived from mentally retarded male patients. In 1991, expansion of the CCG trinucleotide repeat was reported to be the cause of fragile X syndrome. [10] Molecular cloning of the fragile X CCG repeat was carried out through a so-called positional cloning strategy. This was the very first case that showed that the genetic instability of tandem repeats could be a cause of human genetic disease. The copy number of the expanded fragile X CCG repeat in mentally retarded patients is not statically fixed but rather dynamically changing, therefore, this new type of mutation has been classified as a "dynamic mutation", in contrast to the classical static mutations such as point mutation, insertion, and deletion. Typically, the copy number of the CCG repeats in the somatic cells of mentally retarded fragile X patients are a mosaic, and the dynamic mutations are detected as multiple bands or a smear on Southern blot hybridization analysis. The genetically unstable CCG tandem repeat is located in the upstream region of the responsible gene, *FMR-1*. Expansion of the CCG repeat induces hypermethylation at the promoter region of the *FMR-1* gene, and the transcriptional activity of the *FMR-1* gene is suppressed. Fragile X syndrome has been recognized as not obeying the rules of Mendelian inheritance, and the CCG repeat expansion and subsequent hypermethylation clearly explain the semi-dominant pattern of inheritance. This pattern shows reduced penetrance such as non-symptomatic carrier females, mild mental retardation in some carrier females, and normal transmitting male carriers. In non-symptomatic carrier females, the copy number of the CCG repeat does not exceed the upper limit for hypermethylation, but is significantly more than the normal copy number (referred to as "premutation", 54-150 copies), and the transcript of *FMR-1* gene and the subsequent protein product are normally expressed. In contrast, the copy number of the CCG repeat in carrier females who exhibit mild mental retardation exceeds the limit for hypermethylation (referred to as "full mutation", >150 copies), and the promoter regions of the *FMR-1* gene containing the expanded CCG repeat are methylated, and *FMR-1* gene expression is subsequently repressed. In mentally retarded male patients, the copy number of the fragile X CCG repeat often exceeds one thousand and the neighboring genomic region is heavily methylated. Identification of genetically unstable CCG repeats also explained one of the most puzzling mysteries of the fragile X syndrome, the existence of normal transmitting males. A normal transmitting male does not manifest any symptoms of mental retardation but passes the responsible gene to his daughters. Molecular analysis revealed the unmethylated premutation status of the CCG repeat in normal transmitting males. [11] Before molecular diagnosis became available, fragile X syndrome had been diagnosed by clinical manifestations such as the folate-sensitive fragile site at Xq27.3 (*FRAXA*), mental retardation, and macro-orchism, however, karyotype analysis is highly technical and the physical manifestations are not very consistent, so the diagnosis of fragile X syndrome had been quite limited. Introduction of the molecular diagnosis for the fragile X syndrome was a drastic advancement in understanding of one of the most common familial mental impairment disorders. Molecular diagnosis of fragile X syndrome has been applied not only to individual patients but also to the fetus as a method of prenatal diagnosis. [12,13] Haplotype analyses have been carried out in different populations including Caucasians [14,15] and Japanese, [16] and the involvement of a few founder chromosomes has been suggested in both populations, although different haplotypes were involved in each. Subsequent analyses also suggested that the founder chromosomes of fragile X syndrome are different between Caucasian and Asian. [17,18] When and how the first Fragile X chromosome appeared has yet to be elucidated.

FRAXE is another folate-sensitive fragile site located at Xq28, ~600kb distal to *FRAXA*. *FRAXE* [19] has also been cloned by using a positional cloning technique. [20] Similar CCG tandem repeat expansion and the responsible gene, *FMR2*, for mild mental retardation syndrome, were identified adjacent to *FRAXE*. [21,22] Molecular cloning of the third folate-sensitive fragile site, *FRAXF*, [23] which is located a further 600kb distal to *FRAXE*, revealed the existence of another CCG expansion. [24,25] A novel gene called *FAM11A* [26] was identified and a genetically unstable CCG repeat was located in the 5'-untranslated region (5'-UTR). In an individual carrying full *FRAXF* mutation, transcription of the *FAM11A* gene was silenced by hypermethylation, however, no clinical manifestations have been reported. [27]

Chromosomal Fragile Sites

One of the biological indexes related to the expansion of tandem repeats is the chromosomal fragile site. Chromosomal fragile sites are specific sites on human chromosomes that exhibit physical fragility when karyotype analysis is carried out under certain conditions. Chromosomal fragile sites are loci that show gaps or breaks during the metaphase of cells grown under specific cell culture conditions. Most of the fragile sites in humans are not associated with any diseases, however, there are exceptional cases such as fragile X syndrome, *FRAXE* mental retardation, and Jacobsen syndrome. [28] More than one hundred chromosomal fragile sites have been identified, and common and rare fragile sites distinguished. The common fragile sites are expressed in almost all individuals, whereas the rare fragile sites are not. Rare fragile sites are inherited within a specific pedigree according to Mendelian inheritance. Cytogenetic expression of the majority of common fragile sites are induced in the cells cultured in the presence of aphidicolin, an inhibitor of specific DNA polymerases. [29,30] Despite its importance to prostate cancer development, [31] viral integration such as human papilloma virus 16 (HPV16) in cervical tumors [32] and simian virus 40, [33] and autosomal recessive juvenile Parkinsonism, [34,35,36] no characteristic nucleotide sequences, such as repeat motifs, have not been identified to date in relation to the cytogenetic expression of the common fragile sites. Common fragile sites are broad regions on the human genome. They sometimes span large genomic regions of several hundred kilobases. [37,38] The molecular mechanism for cytogenetic expression of common fragile site has yet to be elucidated. It is also known that common fragile sites are evolutionary conserved, [39,40] suggesting a functional importance.

Folate-sensitive fragile sites are classified as rare fragile sites. Rare fragile sites are inherited within specific pedigrees according to the Mendelian rule. Identification of the CCG repeat expansion in Fragile X syndrome brought us new insight into the workings of the mammalian genome. It showed that dynamic mutations of CCG tandem repeats affect gene expression and the physical durability of the chromosome. The molecular reason for the down regulation of transcriptional activity is explained by hypermethylation in the promoter region of the gene, where the expanded CCG repeat lies. Methylation status can be readily examined by Southern blot hybridization analysis using methylation sensitive restriction enzymes. Expansion of CCG repeat has also been identified at other folate-sensitive fragile sites, such as *FRA10A*, *FRA11B*, and *FRA16A*.. [41,42]

FRA10A [43] is the most recent folate-sensitive fragile site cloned. *FRA10A* lies within the 5'-UTR of a novel gene *FRA10AC1*, which transcript was not detected in a non-symptomatic *FRA10A* carrier. [44] The *FRA10AC1* gene encodes highly conserved nuclear protein of as yet unknown function. The FRA10AC1 protein is ubiquitously expressed.

The *FRA11B* CCG repeat lies within the *CBL2* proto-oncogene. [45] *FRA11B* has been suspected to be a deletion breakpoint related to the manifestation of Jacobsen syndrome. Jacobsen syndrome is a clinically characteristic malformation [46] due to deletion of the terminal band 11q23, where the folate-sensitive fragile site *FRA11B* has been mapped. However, the majority of deletion breakpoints for Jacobsen syndrome have been mapped distal to *FRA11B*, and the cytogenetic expression of *FRA11B* is not likely to be directly related to the manifestation of the syndrome. [47,48]

Fragility at folate-sensitive fragile sites has been accounted for by the nucleosome exclusion model. [49,50] This model may be the most applicable one to explain the molecular mechanism underlying fragile site expression. This molecular explanation is based on the observations that long tracts of CCG trinucleotide repeats in DNA resist assembly into nucleosomes. Typically, competitive nucleosome reconstitution, i.e., the ability of the sample DNA fragment to assemble into the nucleosome, is examined by gel shift assay. Methylation of CpG dinucleotides at or near these fragile sites has been known to enhance the fragile phenotype of the chromosomes. According to the nucleosome exclusion model, the DNA strand at the fragile site is not cut or broken but rather the chromatin structure is decondensed at the site. In fluorescent in situ hybridization analysis, signals of the probe spanning the target fragile site are sometimes observed in the middle of the non-staining gap of the fragile chromosome (our unpublished observations). It has also been shown at the breakpoint clusters in the most unstable common fragile site, *FRA3B*, that they do not contain strong nucleosome positioning elements. [51]

A similar nucleosome exclusion model seems to be applicable to other types of rare fragile sites composed of highly AT-rich minisatellite repeats, *FRA10B* and *FRA16B*. These have been positionally cloned and characterized at the molecular level. In the distamycin A inducible fragile site, *FRA16B*, a highly AT-rich 33-bp [52] or 35-bp [53] repeat has been identified at the fragile sites, and the copy number of the repeat is significantly increased in *FRA16B*-positive individuals. *FRA16B* DNA fragments have been demonstrated to exclude nucleosome assembly only in the presence of distamycin. Increasing the number of AT-rich minisatellite repeats increases the effect of distamycin in the destabilization of the nucleosome formation. [54] Distamycin is a naturally occurring amide-linked oligopyrole antibiotic with affinity for AT-rich regions of the DNA minor groove. [55] The interaction with the minor groove is non-covalent and non-intercalative, and it is based on hydrogen bonding, van der Waals contacts, hydrophobic effects and electrostatic effects of the charged ends with distamycin having one cationic charge. [56] The distamycin molecule possesses a crescent shape that is complementary to the shape of the minor groove. At least four consecutive AT base pairs are involved in the interaction of DNA with one distamycin molecule. The formation of a distamycin–DNA complex is known to be highly dependent on DNA sequence. [57,58] Cytogenetic expression of *FRA16B* has not been shown thus far to be involved in any diseases. [59]

A similar amplified structure caused by dynamic mutation of the AT-rich repeat (~42bp) was identified at *FRA10B*. [60] *FRA10B* is a bromodeoxyuridine-inducible type of chromosomal fragile site, and it is insensitive to distamycin. The applicability of the nucleosome exclusion model needs to be examined at the molecular level for the cytogenetic expression of *FRA10B*.

The identification of AT-rich repeat expansion at *FRA16B* and *FRA10B* showed that dynamic mutation has no evident preference in terms of the GC/AT content to occur.

TRIPLET REPEAT DISEASES

Identification of the dynamic mutation of CCG trinucleotide expansion in Fragile X syndrome facilitated a series of discoveries of other genetically unstable trinucleotide repeats as causes for various types of human genetic diseases. These diseases are recongized as having certain features in common, such as non-Mendelian inheritance, dominant phenotype, and anticipation in the affected pedigrees. Anticipation refers to an increase in disease severity and/or a juvenile onset of disease phenotype in successive generations.

A growing number of human disorders have been associated with expansions of a tract of a single type of amino acid. A series of neurodegenerative disorders caused by a toxic polyglutamine tract have been identified. More recently, another series of malformation disorders caused by an expanded polyalanine tract has been identified. In this section, a brief description of the diseases caused by dynamic mutation is provided.

CAG TRIPLET DISEASES (POLYQ DISEASES)

CAG triplet diseases are famous for their role in dominantly inherited neurodegenerative disorders. The CAG repeat expanded in an affected individual is located in the coding region of a responsible gene, and the extended polyglutamine tract (polyQ) confers a toxic activity. This category of disease includes spinocellebelar ataxia (SCA). SCA8, SCA12, and SCA10 are also categorized to the type SCA with similar clinical manifestations, however, their repeat expansions seem to be in the untranslated regions (SCA8 and SCA12) [61] including one in an intron (SCA10). [62] These three disorders are unlikely to be caused by the toxic effect of polyQ-containing proteins. The repeat unit of SCA10 is not CAG but ATTCT pentanucleotides, therefore, SCA10 is described in another section. The characteristics of the polyQ diseases are summarized in Table 1. Brief descriptions of the phenotypes of the CAG triplet-causing polyQ diseases are as follows;

The first polyQ disease identified was spinal-bulbar muscular atrophy (SMBA). [63] The onset of this disease starts with fasciculations and is followed by muscle weakness and wasting at approximately 40 years of age. Bulbar signs and facial fasciculations are characteristic. Abnormal CAG repeat expansion in the *androgen receptor* (*AR*) [64] gene has been identified in SBMA patients, although the *AR* gene CAG repeat is polymorphic in the general population. The CAG repeat encodes a polyglutamine tract in a portion of the AR protein that is not directly involved in binding to the hormone or DNA. There is no evident correlation between the size of the CAG repeat and the presence of altered *in vitro* androgen binding. However, there is a correlation between CAG repeat length and disease severity. The mildest clinical manifestations are associated with the smallest CAG repeat. The greater the number of CAG repeats, the lower the age of onset of limb muscular weakness and the higher the age-adjusted disability score. The expanded CAG alleles undergo alteration in length when transmitted from parent to offspring similar to the unstable CCG repeats identified in the fragile X syndrome. There is a greater rate of instability in male meiosis than in female meiosis. A case of a man who rapidly progressed from being a healthy firefighter to wheelchair-dependency in less than 1.5 years has been reported. Somatic mosaicism has not been reported in SBMA.

Dentatorubral-pallidoluysian atrophy (DRPLA) [65] is characterized by selective cell loss in particular neuronal pathways. There is a high level of heterozygosity in the length of the polyglutamine tract in the responsible gene, *atrophin 1*. [66] Northern blot analysis revealed a 4.7-kb transcript that is widely expressed in various tissues, including heart, lung, kidney, placenta, skeletal muscle, and brain. There is a good correlation between the size of the CAG repeat expansion and the age of onset. Patients with earlier onset tend to have a phenotype of progressive myoclonic epilepsy and larger expansions.

Huntington disease (HD) [67] is characterized by progressive chorea, rigidity, and dementia frequently associated with seizures, due to progressive neural cell death. The responsible gene, *IT15*, coding huntingtin, with the genetically unstable CAG repeat, is located at 4p16.3. Typically, there is a prodromal phase of mild psychotic and behavioral symptoms which precede frank chorea by up to 10 years.

Spinocerebellar ataxia (SCA) is a heterogeneous group of disorders with a primary feature of progressive cerebellar ataxia. They are classified into three groups. In the autosomal dominant cerebellar ataxias group I (ADCA I), cerebellar ataxia of gait and limbs is invariably associated with supranuclear ophthalmoplegia, pyramidal or extrapyramidal signs, mild dementia, and peripheral neuropathy. In ADCA II, macular and retinal degeneration are added to the features. ADCA III is a pure form of late-onset cerebellar ataxia.

ADCA I includes SCA1, [68] SCA2, [69] and SCA3 (also called as Machado-Joseph disease, MJD). [70] For instance, the symptoms of SCA1 usually begin to manifest around the age of 30. In addition to cerebellar signs, there are upper motor neuron signs and extensor plantar responses. Involuntary choreiform movements may occur. These three disorders are characterized at the molecular level by CAG repeat expansions in responsible genes, *ataxin-1*, [71] *ataxin-2*, [72] and *ataxin-3*, [73] at 6p24-p23, 12q24.1, and 14q32.1, respectively.

Moderate CAG expansions in a calcium ion channel gene, *CACNL1A4*, [74] at 19p13 result in SCA6, [75] which is also associated (to) the ADCA I phenotype.

In contrast to the genetic heterogeneity of ADCA I, a single genotype of SCA7 [76] has been determined by a highly unstable CAG repeat expansion in the *ataxin-7* gene [77] at 3p13-p12. SCA7 is assigned to ADCA II.

ADCA III was found in a large pedigree with distinctive clinical manifestations. Primarily its genetic linkage has been allocated to chromosome 11 (SCA5), however it is also caused by CAG expansions in the *CACNL1A4* gene (also called *CACNA1A* gene) at 19p13, which is responsible for SCA6.

SCA17 [78] is unique among the spinocerebellar ataxias. Clinical manifestation and inheritance of SCA17 is similar to Huntington disease, thus, this disorder is also called Huntington disease-like type 4 (HDL4). The responsible CAG repeat expansion has been identified in the *TATA box-binding protein* (*TBP*) [79] gene. TBP is the DNA-binding subunit of the RNA polymerase II transcription factor D (TFIID). [80]

SCA12 [81] displays a variable phenotype, but its prototypical phenotype resembles the classic spinocerebellar ataxia more closely than any other form of neurodegenerative disorder. The responsible CAG repeat has been identified in the 5'-untranslated region (5'-UTR) of the *PPP2R2B* [82] gene, encoding a brain-specific regulatory subunit of the protein phosphatase, PP2A. The possibility that the CAG repeat may be within an unidentified gene overlapping or adjacent to *PPP2R2B* has not been excluded. An antibody did not detect polyQ expansion in affected family members. Thus, it is believed that the responsible CAG repeat expansion lies within the 5'-UTR. The molecular mechanism for the manifestation of disease symptoms is yet to be elucidated.

POLYALANINE DISEASES (POLYA DISEASES)

Identification of repeat expansions in the fragile X syndromes and the polyQ diseases shed light on the molecular basis for those in which peculiar inheritance patterns and anticipation have been observed. More recently, the expansion of alanine tracts, particularly in transcription factor genes, have been shown to cause both certain types of mental retardation and malformation of the brain, digit, and other organ structures. Although the physiological function of the alanine tract is yet to be elucidated, abnormal expansion of alanine tracts seem to result in either a loss of function or a gain of abnormal function. The majority of disorders caused by polyalanine expansion also show a dominant inheritance pattern. The characteristics of the polyalanine diseases are summarized in Table 1. In this section, we provide brief descriptions of the disorders caused by polyalanine tract expansion.

Oculopharyngeal muscular dystrophy (OPMD) [83] is an autosomal dominant disorder characterized by dysphagia and progressive ptosis of the eyelids. This disorder is late-onset and presents after age 50. *Poly adenine binding protein 2 (PABP2)* [84] gene has been identified as being responsible. Six copies of the GCG tandem repeat encoding a polyalanine tract located at the N-terminus of the protein is expanded to 8-13 copies in affected patients. Interestingly, a homozygous version of seven GCG copies leads to autosomal recessive OPMD. [85] Thus, the (GCG)7 allele is an example of a polymorphism that can act either as a modifier of a dominant phenotype or as a recessive mutation.

Syndactyly is one of the most common congenital hand malformations. The name of the disorder derives from the ancient Greek words 'syn' and 'dactylos', meaning joined or webbed fingers. In Syndactyly type 2, [86] syndactyly of the third and fourth fingers in the hand is observed in association with polydactyly, or multiple copies, of all components or of part of the fourth finger in the web. In the feet of syndactyly type 2 patients, polydactyly of the fifth toe is included in a web of syndactyly of the fourth and fifth toes. The responsible gene for syndactyly type 2 has been identified and it has turned out to be the *homeobox D13* (*HOXD13*) [87] gene. This was the first mutation discovered in a *HOX* gene to cause malformation in humans. The repeat expansion of the alanine tract is characteristic in the *HOXD13* gene of the affected individuals.

Unlike the other disorders caused by genetically unstable trinucleotide repeats, such as the fragile X syndrome and the polyQ diseases, the expanded *HOXD13* tract encoding polyalanines is stable when transmitted from one generation to the next. Genetic stability of the alanine tract in the *HOXD13* gene is achieved by cryptic repeat-coding alanine such as GCG, GCA, GCT, and GCC. Dynamic mutation has been known to occur as a perfect repeat, and cryptic interruptions are an effective inhibitor to stabilize the repeats at dynamic mutation loci. Therefore, some similarities exist between polysyndactyly and dynamic mutations at the protein level, but the mechanisms of mutation appear to be different. The lengthening of tracts of single amino acids leading to protein change-of-function may be a common mechanism of human genetic disease. The introduction and changing of the polyalanine tract in the *HOXD13* gene over evolutionary time may have led to the distinctive limb differences between species ranging from teleost fish through to mammals. The molecular mechanism for this type of repeat expansion, which are currently stable emerged at some time-point in the past, also awaits discovery.

The clinical features of cleidocranial dysplasia (CCD) [88] include persistently open skull sutures with bulging calvaria, hypoplasia or aplasia of the clavicles permitting abnormal facility in apposing the shoulders, a wide pubic symphysis, a short middle phalanx of the fifth fingers, dental anomalies, and often vertebral malformation. These conditions were prevalent in descendants of a Chinese man named Arnold who was a member of the Muslim religion and had 7 wives in South Africa. Of his 356 descendants, 70 manifested the 'Arnold head'. [89] It is estimated that more than 1,000 descendants of the first progenitor now have the disorder. In this case, brachydactyly of the hands and feet is also evidently associated. The *CBFA1* [90] gene was identified as the gene responsible for CCD. The *CBFA1* gene is a member of the runt family that had previously been described as a bone-specific nuclear-matrix-binding transcription factor. An in-frame expansion of a polyalanine stretch was identified in an affected family. Uninterrupted alanine residues lie on the N-terminal side of the runt domain of CBFA1. Since the deletion of the *CBFA1* gene has been identified in other affected families, heterozygous loss of function of CBFA1 appeared to be sufficient to manifest CCD. Therefore, polyalanine expansion in *CBFA1* gene is likely to result in loss-of-function of the protein.

Holoprosencephaly (HPE) [91] is the most common structural anomaly of the human brain and is one of the anomalies seen in patients with deletions and duplications of chromosome 13. The human *ZIC2* gene, a homolog of the Drosophila '*odd-paired*' (*opa*) gene, is located at 13q32 has been identified as the responsible gene. Since heterozygous mutation in ZIC2 is sufficient for the clinical manifestation of HPE, a gene dosage effect of the ZIC2 product is likely to cause the brain malformation seen in 13q deletion patients. Zic2 expression during mouse embryonic development is consistent with a major role in brain and distal limb development. In humans, Northern blot analysis revealed expression of ZIC2 transcript only in the fetal brain.

Hand-foot-genital (uterus) syndrome [92] is also an autosomal dominant disorder. The clinical features include small feet with unusually short great toes and abnormal thumbs. Females with the disorder have duplication of the genital tract, including the longitudinal vaginal septum. Various types of mutations in the homeodomein of *HOXA13* gene [93] result in this syndrome. This was the second mutation discovered with a *HOX* gene being the cause of a malformation in humans. Types of mutation include the nonsense mutations, as well as an expansion of an N-terminal polyalanine tract that causes in-frame addition of 6-9 extra alanines, both resulting in a similar phenotype. It is thought that the mutation probably eliminates or greatly reduces the ability of the protein to bind to DNA. A heterozygous child between a person with polyalanine expansion in the *HOXA13* gene and another one with missense mutation in *HOXD13* (causing synpolydactyly) manifested a more severe clinical phenotype, suggesting a synergistic effect of these two homeobox genes. [94]

Blepharophimosis, ptosis and epicanthus inversus syndrome (BPES) [95] is a form of dysplasia of the eyelids. BPES type I accompanies premature ovarian failure and BPES type II does not. Both are caused by mutations in the *FOXL2* gene, a novel putative winged helix/forkhead transcription factor gene at 3q23, which has been positionally cloned. [96] The *FOXL2* gene has been highly conserved through the course of evolution, the number of alanine residues in particular being strictly conserved among the mammals, suggesting significant functional and/or structural constraints. FOXL2 is a nuclear protein specifically expressed in the eyelids and ovarian follicular cells. FOXL2 is one of the earliest markers of ovarian differentiation in mammals. Molecular analyses have revealed that polyalanine expansion within or downstream of the forkhead domain as well as a frameshift mutation downstream result in BPES type II, whereas mutations that result in a truncated protein either lacking or containing the forkhead domain led to BPES type I. Furthermore, a contraction of the polyalanine tract has been reported to cause the same disease phenotype. In addition, the polyalanine expansion has been suggested to cause premature ovarian failure (POF).

X-linked mental retardation associated with isolated growth hormone deficiency [97] was first reported in 1996 and mapped to Xq24-q27.3 by linkage analysis. The responsible mutation was determined to be an in-frame duplication of 33 base pairs encoding 11 alanines in a polyalanine tract of the *SRY-BOX 3* (*SOX3*) [98] gene, which was mapped to Xq26-q27. *SRY* [99] is the testis-determining gene at the sex determining region on Y chromosome, and *SOX3* is also one of the SRY-related transcription factor genes. The homology is restricted to the region of *SRY* that encodes a DNA-binding motif of the HMG-box class. The DNA-binding domain is called HMG for 'high mobility group'. The expression pattern during neural and pituitary development suggests that dysfunction of the *SOX3* gene caused by the polyalanine expansion might disturb transcription pathways and the regulation of genes involved in cellular processes and certain functions required for both cognitive and pituitary development. [100]

The X-linked infantile spasm syndrome (ISSX), [101] also known as West syndrome, is characterized by early-onset generalized seizures, hypsarrhythmia, and mental retardation. The responsible gene is the aristaless-related homeobox gene, *ARX*, [102] mapped to Xp22.3-p21.1. The mutation was identified as an expansion of polyalanine tracts in the ARX protein. The expansion was an additional stretch of 7 tandem GCG repeats within the normal stretch of 10 GCG triplet repeats in exon 2, possibly resulting in protein aggregation, similar to other polyalanine and polyglutamine disorders. In addition, a missense mutation within the *ARX* homeodomain and a truncation mutation has been identified as well.

Partington X-linked mental retardation syndrome [103] can also be caused by a polyalanine repeat expansion in the aristaless-related homeobox gene, *ARX*. The polyA expansion is due to a 24-bp duplication. This syndrome was first described by Partington et al., [104] in which case a family with 10 males widely distributed in the pedigree and connected through females showed a syndrome of mild to moderate mental retardation and episodic dystonic movements of the hands. Inheritance is X-linked recessive. All the affected individuals are male.

X-linked mental retardation 36 (MRX36) [105] is also caused by an in-frame duplication of 24-bp in the *ARX* gene. The affected males manifest a nonspecific form of mental retardation with normal physical and neurologic findings. Also females show an intelligence level in the low to normal range, possibly as a result of heterozygous carrier status.

Idiopathic congenital central hypoventilation syndrome (CCHS), [106] also known as 'Ondine's curse,' is a rare disorder characterized by abnormal control of respiration in the absence of neuromuscular or lung disease, or an identifiable brainstem lesion. A deficiency in autonomic control of respiration results in inadequate or negligible ventilatory and arousal responses to hypercapnia and hypoxemia. The patients breathe normally while awake but hypoventilate with normal respiratory rates but shallow breathing during sleep. More severely affected patients hypoventilate while both awake and asleep. CCHS can be caused by heterozygous mutations in the *PHOX2B* gene [107] as well as in several other genes, including *RET*, [108] *GDN,F* [109] *EDN3*, [110] and *BDNF*. [111] The mutations in the *PHOX2B* gene include polyalanine expansions ranging from 5 to 13 residues as well as frame-shift mutations. Inheritance of CCHS is likely to be autosomal dominant with reduced penetrance, since asymptomatic individuals carrying polyalanine expansions have been identified. The molecular mechanism for this reduced penetrance is yet to be elucidated.

CTG Triplet Diseases

CTG triplet repeat diseases are caused by dynamic mutations which occur either in the untranslated regions or in the alternatively spliced exon of the responsible genes. The molecular mechanism for the clinical manifestations are yet to be elucidated.

Myotonic dystrophy (DM) is a multisystem disorder and the most common form of muscular dystrophy in adults. Myotonic dystrophy type1 (Dystrophia miotonica 1, DM1) [112] is an autosomal dominant disorder characterized by myotonia, muscular dystrophy, cataracts, hypogonadism, frontal balding, and changes in electrocardiogram (ECG) pattern. The responsible CTG triplet repeat expansion has been identified in the 3'-untranslated region (3'-UTR) of *dystrophia miotonica protein kinase* (*DMPK*) [113] gene located at 19q13.2-q13.3. The dynamic mutation of the CTG repeat explains the anticipation observed in the affected pedigrees. Immunohistochemical staining analyses have revealed that DMPK protein is localized predominantly at sites of neuromuscular and myotendinous junctions in human and rodent skeletal muscles. The protein has also been demonstrated in the neuromuscular junctions of muscular tissues of adults in congenital cases of DM with no gross changes in structural organization. The physiological function of the DMPK protein is unknown. The discovery of type 2 myotonic dystrophy (DM2), caused by an untranslated CCTG expansion, indicates that the clinical features common to both diseases are caused by a gain-of-function RNA mechanism in which the CUG and CCUG repeats alter cellular function, including the alternative splicing of various genes. A description of DM2 will be given separately.

The SCA8 [114] phenotype corresponds to ADCA III. SCA8 patients show trunk and limb incoordination, ataxic dysarthria, impaired smooth pursuit and horizontal nystagmus, as well as significant atrophy of the cerebellar vermis and hemispheres on MRI. The responsible repeat expansion has been identified in the 3'-untranslated region (3'-UTR) of the *SCA8* gene [115]at 13q21. The CTG repeat is exclusively expressed in the anti-sense transcript, not in the sense transcript as the CAG repeat. The CTG repeat in the normal *SCA8* transcript is thought to play a role in regulating the sense transcript.

Huntington disease-like-2 (HDL2) [116] is clinically similar to Huntington disease, such as having onset in the fourth decade, involuntary movements and abnormalities of voluntary movements, psychiatric symptoms, weight loss, dementia, and a relentless course with death at approximately 20 years after disease onset. HDL2 is caused by an expanded CAG/CTG repeat in an alternatively spliced exon of the *Junctophilin 3* (*JPH3*) [117] gene localized at 16q24.3. JPH3 is a member of the junctophilin family involved in junctional complexes and cross talk between cell surface and intracellular ion channels. The exon containing the genetically unstable CAG/CTG repeat has multiple splice acceptor sites. Recently, a magnetic reasonance imaging (MRI) scanning technique has been introduced for the diagnosis of neurological disorders, in conjunction with autopsy, and HDL2 has been reported to exhibit marked striatal atrophy and moderate cortical atrophy, with striatal neurodegeneration in a dorsal-to-ventral gradient and occasional intranuclear inclusions. [118]

OTHER REPEAT DISEASES

Pseudoachondroplasia (PSACH) [119] is one of the most frequent skeletal dysplasias, and characterized by the most marked shortening of the limbs. The responsible dynamic mutation has been identified at the conserved GAC repeat region for polyaspartate in cartilage oligomeric matrix protein (COMP). [120] Dynamic mutation in *COMP* gene is also responsible for multiple epiphyseal dysplasia (MED), [121] which characteristic feature includes short stature and brachydactyly. Both disorders can be manifested by the conventional static mutations. Cartilage oligomeric matrix protein is a 524-kD protein that is expressed at high levels in the territorial matrix of chondrocytes. Interestingly, GAC repeat expansion as well as contraction of the repeat cause the same diseases.

Friedreich ataxia 1 (FRDA1) [122] is one of the most common forms of autosomal recessive ataxia, and the spinocerebellar tracts, dorsal columns, pyramidal tracts and, to a lesser extent, the cerebellum and medulla are all involved. The disorder is usually manifested before adolescence and is generally characterized by uncoordinated limb movements, dysarthria, nystagmus, diminished or absent tendon reflexes, Babinski sign, impairment of position and vibratory senses, scoliosis, *pes cavus*, and hammer toe. The triad of hypoactive knee and ankle jerks, signs of progressive cerebellar dysfunction, and preadolescent onset is commonly regarded as sufficient for diagnosis. FRDA 1 was the first triplet repeat disease observed to have a recessive inheritance. The majority (97-98%) of FRDA1 cases are due to the expansion of a GAA trinucleotide repeat in the first intron of the *FRDA* [123] gene, although a distinct minority of cases (2-3%) is due to point mutations. Larger GAA expansions are correlated with earlier onset. The GAA repeats are unstable during transmission, and contractions outnumber expansions. The specific tendency to contract or expand does not seem to be associated with particular haplotypes. Thus, the *FRDA1* gene variability appears to be different from that found in the other triplet diseases. For instance, anticipation has not been observed in FRDA1. The possibility that the GAA triplet repeat expansion may result in an unusual yet stable DNA structure that interferes with transcription, and ultimately leading to a cellular deficiency of frataxin protein, has been proposed. This hypothesis proposes a novel DNA structure, "sticky DNA", [124] formed by (GAA-TTC)n found in intron 1 of the frataxin gene of patients. Sticky DNA is formed by the association of 2 purine-purine-pyrimidine (R-R-Y) triplexes in negatively supercoiled plasmids at neutral pH. GAA-TTC repeats of more than 59 copies have been shown to form sticky DNA and inhibit transcription *in vivo* and *in vitro*. An *in vitro* study showed that the amount of RNA synthesized decreased as the copy number of (GAA-TTC)n repeat increased.

As previously described, myotonic dystrophy (DM) is an autosomal dominant multisystem disorder and the most common form of muscular dystrophy in adults. The classic form of the disorder, myotonic dystrophy-1 (DM1), is caused by an expanded CTG repeat in the *dystrophia myotonica protein kinase* gene (*DMPK*) at 19q13. Myotonic dystrophy type2 (Dystrophia Myotonica 2, DM2) [125] was identified in patients with clinical characteristics consistent with classic myotonic dystrophy, but without the CTG repeat expansion in the *DMPK* gene. The syndrome was originally reported as 2 distinct disorders, myotonic dystrophy-2 and proximal myotonic myopathy, but these two disorders are now referred to collectively as DM2. DM2 is caused by the expansion of a CCTG repeat in intron 1 of the *zinc finger protein-9* gene (*ZNF9*). [126] The ZNF9 protein contains 7 zinc finger domains and is believed to function as an RNA-binding protein, since the zinc finger domains show remarkable sequence similarity to the finger domains of the retroviral nucleic acid-binding proteins (NBPs). The viral NBPs have a strong preference for single-stranded DNA. It is interesting that children usually have shorter repeats than their parents. [127] It is possible that the time-dependent somatic variation of repeat size may complicate interpretation of the difference. A significant correlation between the age of onset and expansion size has not been observed in DM2.

Also as previously described, the autosomal dominant cerebellar ataxias (ADCAs) are a clinically and genetically heterogeneous group of disorders characterized by ataxia, dysarthria, dysmetria, and intention tremor, and are divided into 3 different groups based on the presence or absence of associated symptoms. The clinical phenotypes of spinocerebellar ataxia 10 (SCA10) [128] are characterized by predominantly cerebellar symptoms and signs and classified as ADCA III, which is a pure form of late-onset cerebellar ataxia. Since the anticipation phenomenon and autosomal dominant inheritance have been observed in the affected pedigree, an involvement of repeat expansion has been suggested as needed for disease progression at the molecular level. Eventually, ATTCT pentanucleotide repeat expansion in intron 9 of the *ataxin-10* (*ATXN10*) gene was identified as the cause of the clinical manifestation of SCA10. ATXN10 product has been shown to be essential for the survival of cerebellar neurons, and knock down of ATXN10 in primary neuronal cells by small interfering RNAs has been shown to result in an increased apoptosis of cerebellar neurons. [129] SCA10 pentanucleotide repeat expansion ranges from 920 to 4,140 repeats, the largest microsatellite expansion found in the human genome.

Discussion

In this chapter, brief descriptions have been provided on human repetitive sequences focusing on tandem repeats. These descriptions are focused on the disease manifestations caused by changes in the copy number of tandem repeats. This type of mutation is called a " dynamic mutation", since the copy number of the responsible tandem repeat varies among patients affected by the same syndrome. However, the responsible mutations are regarded as being essentially the same ones. In the case of the fragile X syndrome, the copy number of responsible CCG repeat varies even in an affected individual. Usually the somatic cells derived from an mentally affected fragile X patient are mosaic in the copy number of the fragile X CCG repeat. Thus, a dynamic mutation is obviously quite distinct from the conventional static mutations such as point mutation, insertion, and deletion. Dynamic mutation could be regarded as a specific type of insertion or deletion of tandem repeats ; however, the molecular mechanisms for mutagenesis and clinical manifestation are unlikely to be the same as in the classic insertion or deletion examples. Furthremore, dynamic mutations occur repeatedly in affected pedigrees, whereas a static mutation does not. Once a static mutation has occurred, it is inherited in that form in successive generations without an occurrence of additional mutations.

As described here, specific types of tandem repeats have been identified as causing certain human genetic diseases. A genetically unstable CCG trinucleotide repeat has been identified in the 5'-UTR of the FMR-1 gene, and the resulting abnormal expansion and the subsequent hypermethylation can cause the fragile X syndrome, one of the most common forms of familial mental impairment. A major clinical manifestation of the fragile X syndrome is mental retardation. Fragile X patients are fertile, and there are examples of the affected patients transmitting the mutation to their obligate daughters. When the majority of people lived in an agricultural setting typically requiring group work, as was long the case in Japan, the local society would have been better positioned to accommodate a mentally handicapped individual with an appropriate job and sometimes a healthy spouse. However, in modern society, people are expected to meet high expectations of achievement at both school and work. Therefore, in recent, much more competitive societies, it is getting harder for mentally handicapped individuals to live "normal" lives. The mentally handicapped fragile X individuals who had previously transmitted their mutations to their obligate daughters are now grandfathers, and it may become impossible to find any in the near future, at least in a hyperindustrial socitiety such as Japan.

CAG trinucleotide repeat expansions in coding regions result in the expansion of a polyglutamine tract in the resultant proteins, and then in the neurodegenerative disorders, such as Huntington disease and spinocerebellar ataxias, as well as myotonic dystrophy and SBMA, as shown in Table 1. Huntington disease and the spinocerebellar ataxias are late-onset disorders, and the clinical symptoms usually begin to manifest after the transmission of the mutation to the successive generations. Also, the mutations can be transmitted through non-symptomatic carriers.

Abnormal expansion of a polyalanine tract in transcription factor genes causes certain malformation syndromes, and these disorders are often referred to as the polyalanine diseases (Table 1). The majority of the polyalanine diseases involve various types of malformation, however, most of the clinical manifestations are not life-threatening. Most of the cases afflicted with polydactyly and/or syndactyly usually receive appropriate surgeries when the affected individuals are infants. Thus, there seem to be many asymptomatic carriers unaware that they carry the responsible mutations. Furthermore, as observed in the case of the cleidocranial dysplasia (CCD) with the characteristic head malformation, such as that referred to as 'Arnold head', the symptom may be inherited as a mere physical characteristic without any discernible impact on survival or fertility. Therefore, it is quite likely that most of the polyalanine diseases have been accepted, as it were, and survived in the local community without discrimination and stigmatization in spite of their autosomal dominant inheritance. In the polyalanine diseases, the tandem repeats that code for the polyalanine tract in the protein do not exhibit genetic instability. Repeat expansion is likely to be the molecular mechanism by which the responsible polyalanine tract was expanded, however, base-substitutions seem to have interrupted the perfect repeats, possibly by the introduction of point mutations. Such an introduction of new mutation after the repeat expansion of polyalanine tract is theoretically possible ; however, it may be thought unlikely to have actually occurred, because of the very low frequencies of trans-generational mutations even among the atomic bomb survivors as have been observed at Hiroshima and Nagasaki. [130] In this study, 64 children from 50 exposed families and 60 from 50 control families have been analyzed for the occurrence of new mutations at six minisatellite loci (*Pc-1*, *lambda TM-18*, *ChdTC-15*, *p lambda 3*, *lambda MS-1*, and *CEB-1*) by Southern blot analysis. The mean mutation rate per locus per gamete in these six minisatellite loci was calculated to be 1.5% in the exposed parents and 2.0% in the unexposed parents, i.e., there was no significant difference in mutation rates in the children of the exposed and unexposed parents. On the other hand, controversial results have been obtained on the effect of ionizing radiations on germline mutation, which would eventually be inherited by successive generations. For instance, an elevated mutation rate in germ cells has been reported in the offspring of irradiated male mice. [131] Similar analyses on the relationship between the occurrence of germline mutations and environmental stress are in the process of being carried out.

The existence of tandem repeat sequences in the human genome has been regarded as 'junk DNA' for a long time, mainly because the biological function of the non-coding DNA has not been recognized. Recently, the biological significance of tandem repeats has become widely acknowledged due to the discovery of "dynamic mutations" as a cause for human genetic diseases. In plants, a similar genetic instability of repetitive DNA sequences has been observed. The plant genome undergoes rapid genomic changes when the plant has been grown under a variety of stress conditions, including drought, heat, cold and freezing, high salinity, nutrient deficiency, toxic concentration of heavy metals, oxidative stress as well as oxygen shortage, and mechanical stress. Stress adaptation in crop plants has proved to be one of the most important issues for the agricultural sciences, especially the understanding of the interaction between plants and the environment. It would be of great interest to determine whether the molecular mechanisms underlying the genetic instability in tandem repeat DNA sequences share any common metabolic pathways in plants and animals. Given the progress in the field, such investigations may be forthcoming in the near future.

REFERENCES

[1] Cullis CA. Phenotypic consequences of environmentally induced changes in plant DNA. *Trends Genet*, 1986, 2, 307-309.

[2] Cullis CA. Mechanisms and Control of Rapid Genomic Changes in Flax. *Annals of Botany*, 2005, 95(1), 201-206.

[3] Birgit Arnholdt-Schmitt. Stress-Induced Cell Reprogramming. A Role for Global Genome Regulation? *Plant Physiology*, 2004, 136, 2579-2586.

[4] Ferrigno O, Virolle T, Djabari Z, Ortonne J-P, White RJ, Aberdam D. Transposable B2 SINE elements can provide mobile RNA polymerase II promoters. *Nature Genet*, 2001, 28, 77-81.

[5] Yulug IG, Yulug A, Fisher EM. The frequency and position of Alu repeats in cDNAs, as determined by database searching. *Genomics*, 1995, 27(3), 544-548.

[6] Sadamoto S, Suzuki S, Kamiya K, Kominami R, Dohi K, Niwa O. Radiation induction of germline mutation at a hypervariable mouse minisatellite locus. *Int J Radiat Biol,* 1994, 65(5), 549-557.

[7] Ruggiero BL, Topal MD. Triplet repeat expansion generated by DNA slippage is suppressed by human flap endonuclease 1. *J Biol Chem*, 2004, 279(22), 23088-23097.

[8] Yamauchi M, Nishimura M, Tsuji S, Terada M, Sasanuma M, Shimada Y. Effect of SCID mutation on the occurrence of mouse Pc-1 (Ms6-hm) germline mutations. *Mutat Res*, 2002, 503(1-2), 43-49.

[9] McKusick, VA. +309550 fragile site mental retardation 1 gene; FMR1; Fragile X mental retardation protein; FMRP, Fragile X syndrome, included; Fragile X mental retardation syndrome, included; Mental retardation, X-Linked, associated with marXq28, included; X-Linked mental retardation and macroorchidism, included; Marker X syndrome, included; Martin-Bell syndrome, included; fragile site, Folic acid type, rare, FRA(X)(q27.3), included; FRAXA, included; Fragile X tremor/ataxia syndrome, included; FXTAS, included [Online Mendelian Inheritance in Man (OMIM)] 1986 [last updated on 2005/01/27]. URL: *http://www.genome.jp/dbget-bin/www_bget?mim:309550.*

[10] Kremer EJ, Pritchard M, Lynch M, Yu S, Holman K, Baker E, Warren ST, Schlessinger D, Sutherland GR, Richards RI. Mapping of DNA instability at the fragile X to a trinucleotide repeat sequence p(CCG)n. *Science*, 1991, 252(5013), 1711-1714.

[11] Hori T, Yamauchi M, Seki N, Tsuji S, Kondo I. Heritable unstable DNA sequences and hypermethylation associated with fragile X syndrome in Japanese families. *Clin Genet*, 1993, 43(1), 34-38.

[12] Yamauchi M, Nagata S, Seki N, Toyama Y, Harada N, Niikawa N, Masuno I, Kajii T, Hori T. Prenatal diagnosis of fragile X syndrome by direct detection of the dynamic mutation due to an unstable DNA sequence. *Clin Genet*, 1993, 44(4), 169-172.

[13] Suzumori K, Yamauchi M, Seki N, Kondo I, Hori T. Prenatal diagnosis of a hypermethylated full fragile X mutation in chorionic villi of a male fetus. *J Med Genet,* 1993, 30(9), 785-787.

[14] Oudet C, von Koskull H, Nordstrom AM, Peippo M, Mandel JL. Striking founder effect for the fragile X syndrome in Finland. *Eur J Hum Genet*, 1993, 1(3), 181-189.

[15] Malmgren H, Gustavson KH, Oudet C, Holmgren G, Pettersson U, Dahl N. Strong founder effect for the fragile X syndrome in Sweden. *Eur J Hum Genet,* 1994, 2(2), 103-109.

[16] Richards RI, Kondo I, Holman K, Yamauchi M, Seki N, Kishi K, Staples A, Sutherland GR, Hori T. Haplotype analysis at the FRAXA locus in the Japanese population. *Am J Med Genet*, 1994, 51(4), 412-416.

[17] Chen SH, Schoof JM, Buroker NE, Scott CR. The identification of a (CGG)6AGG insertion within the CGG repeat of the FMR1 gene in Asians. *Hum Genet*, 1997, 99(6), 793-5.

[18] Hirst MC, Arinami T, Laird CD. Sequence analysis of long FMR1 arrays in the Japanese population: insights into the generation of long (CGG)n tracts. *Hum Genet*, 1997, 101(2), 214-218.

[19] McKusick, VA. +309548 fragile site, Folic acid type, rare, FRA(X)(q28); FRAXE ;Mental retardation, X-linked, Associated with fragile site fraxe fragile site mental retardation 2 gene, included; FMR2, included [Online Mendelian Inheritance in Man (OMIM)] 1993 [last updated on 2003/04/16] URL: *http://www.genome.jp/dbget-bin/www_bget?mim:309548.*

[20] Knight SJ, Flannery AV, Hirst MC, Campbell L, Christodoulou Z, Phelps SR, Pointon J, Middleton-Price HR, Barnicoat A, Pembrey ME, Holland J, Oostra BA, Bobrow M, Davies KE. Trinucleotide repeat amplification and hypermethylation of a CpG island in FRAXE mental retardation. *Cell*, 1993, 74(1), 127-134.

[21] Gecz J, Gedeon AK, Sutherland GR, Mulley JC. Identification of the gene FMR2, associated with FRAXE mental retardation. *Nat Genet*, 1996, 13(1), 105-108.

[22] Gu Y, Shen Y, Gibbs RA, Nelson DL. Identification of FMR2, a novel gene associated with the FRAXE CCG repeat and CpG island. *Nat Genet*, 1996, 13(1), 109-113.

[23] Hirst MC, Barnicoat A, Flynn G, Wang Q, Daker M, Buckle VJ, Davies KE, Bobrow M. The identification of a third fragile site, FRAXF, in Xq27--q28 distal to both FRAXA and FRAXE. *Hum Mol Genet*, 1993, 2(2), 197-200.

[24] Ritchie RJ, Knight SJ, Hirst MC, Grewal PK, Bobrow M, Cross GS, Davies KE. The cloning of FRAXF: trinucleotide repeat expansion and methylation at a third fragile site in distal Xqter. *Hum Mol Genet*, 1994, 3(12), 2115-2121.

[25] Parrish JE, Oostra BA, Verkerk AJ, Richards CS, Reynolds J, Spikes AS, Shaffer LG, Nelson DL. Isolation of a GCC repeat showing expansion in FRAXF, a fragile site distal to FRAXA and FRAXE. *Nat Genet*, 1994, 8(3), 229-235.

[26] Hartz, PA. *300483 family with sequence similarity 11, Member A; FAM11A [Online Mendelian Inheritance in Man (OMIM)] 2004 [last updated on 2004/03/08] URL: http://www.genome.jp/dbget-bin/www_bget?mim:300483.

[27] Shaw MA, Chiurazzi P, Romain DR, Neri G, Gecz J. A novel gene, FAM11A, associated with the FRAXF CpG island is transcriptionally silent in FRAXF full mutation. *Eur J Hum Genet*, 2002, 10(11), 767-772.

[28] McKusick, VA. #147791 jacobsen syndrome; JBS; 11q Deletion syndrome; Partial 11q monosomy syndrome [Online Mendelian Inheritance in Man (OMIM)] 1990 [last updated on 2005/01/14] URL: *http://www.genome.jp/dbgetbin/www_bget?-mim:147791.*

[29] Le Beau MM, Rassool FV, Neilly ME, Espinosa R 3rd, Glover TW, Smith DI, McKeithan TW. Replication of a common fragile site, FRA3B, occurs late in S phase

and is delayed further upon induction: implications for the mechanism of fragile site induction. *Hum Mol Genet*, 1998, 7(4), 755-761.

[30] Sutherland GR, Baker E, Richards RI. Fragile sites still breaking. *Trends Genet*, 1998, 14(12), 501-506.

[31] Ishii H, Furukawa Y. Alterations of common chromosome fragile sites in hematopoietic malignancies. *Int J Hematol.* 2004, 79(3), 238-242.

[32] Thorland EC, Myers SL, Gostout BS, Smith DI. Common fragile sites are preferential targets for HPV16 integrations in cervical tumors. *Oncogene*. 2003, 22(8), 1225-1237.

[33] Mishmar D, Rahat A, Scherer SW, Nyakatura G, Hinzmann B, Kohwi Y, Mandel-Gutfroind Y, Lee JR, Drescher B, Sas DE, Margalit H, Platzer M, Weiss A, Tsui LC, Rosenthal A, Kerem B. Molecular characterization of a common fragile site (FRA7H) on human chromosome 7 by the cloning of a simian virus 40 integration site. *Proc Natl Acad Sci USA*. 1998, 95(14), 8141-8146.

[34] Cesari R, Martin ES, Calin GA, Pentimalli F, Bichi R, McAdams H, Trapasso F, Drusco A, Shimizu M, Masciullo V, D'Andrilli G, Scambia G, Picchio MC, Alder H, Godwin AK, Croce CM. Parkin, a gene implicated in autosomal recessive juvenile parkinsonism, is a candidate tumor suppressor gene on chromosome 6q25-q27. *Proc Natl Acad Sci USA*. 2003, 100(10), 5956-5961.

[35] Denison SR, Callahan G, Becker NA, Phillips LA, Smith DI. Characterization of FRA6E and its potential role in autosomal recessive juvenile parkinsonism and ovarian cancer. *Genes Chromosomes Cancer*. 2003, 38(1), 40-52.

[36] Picchio MC, Martin ES, Cesari R, Calin GA, Yendamuri S, Kuroki T, Pentimalli F, Sarti M, Yoder K, Kaiser LR, Fishel R, Croce CM. Alterations of the tumor suppressor gene Parkin in non-small cell lung cancer. *Clin Cancer Res*. 2004, 10(8), 2720-2724.

[37] Arlt MF, Miller DE, Beer DG, Glover TW. Molecular characterization of FRAXB and comparative common fragile site instability in cancer cells. *Genes Chromosomes Cancer*. 2002, 33(1), 82-92.

[38] Becker NA, Thorland EC, Denison SR, Phillips LA, Smith DI. Evidence that instability within the FRA3B region extends four megabases. *Oncogene*. 2002, 21(57), 8713-8722.

[39] Krummel KA, Denison SR, Calhoun E, Phillips LA, Smith DI., The common fragile site FRA16D and its associated gene WWOX are highly conserved in the mouse at Fra8E1. *Genes Chromosomes Cancer*. 2002, 34(2), 154-167.

[40] Rozier L, El-Achkar E, Apiou F, Debatisse M. Characterization of a conserved aphidicolin-sensitive common fragile site at human 4q22 and mouse 6C1: possible association with an inherited disease and cancer. *Oncogene*. 2004, 23(41), 6872-6880.

[41] Nancarrow JK, Kremer E, Holman K, Eyre H, Doggett NA, Le Paslier D, Callen DF, Sutherland GR, Richards RI. Implications of FRA16A structure for the mechanism of chromosomal fragile site genesis. *Science*. 1994, 264(5167), 1938-1941.

[42] Nancarrow JK, Holman K, Mangelsdorf M, Hori T, Denton M, Sutherland GR, Richards RI. Molecular basis of p(CCG)n repeat instability at the FRA16A fragile site locus. *Hum Mol Genet*. 1995, 4(3), 367-372.

[43] Hartz, PA, +608866 fragile site, Folic acid-type, rare, FRA(10)(q23.3), Candidate gene 1; FRA10A, Candidate gene 1; FRA10AC1; Chromosome 10 open reading frame 4; C10ORF4 fragile site, Folic acid-type, rare, FRA(10)(q23.3), included; FRA10A, included; Fragile site 10q23.3, included [Online Mendelian Inheritance in Man

(OMIM)] 2004 [last updated on 2004/08/23] URL: *http://www.genome.jp/dbget-bin/www_bget?mim:608866.*

[44] Sarafidou T, Kahl C, Martinez-Garay I, Mangelsdorf M, Gesk S, Baker E, Kokkinaki M, Talley P, Maltby EL, French L, Harder L, Hinzmann B, Nobile C, Richkind K, Finnis M, Deloukas P, Sutherland GR, Kutsche K, Moschonas NK, Siebert R, Gecz J; European Collaborative Consortium for the Study of ADLTE. Folate-sensitive fragile site FRA10A is due to an expansion of a CGG repeat in a novel gene, FRA10AC1, encoding a nuclear protein. *Genomics*. 2004, 84(1), 69-81.

[45] Jones C, Penny L, Mattina T, Yu S, Baker E, Voullaire L, Langdon WY, Sutherland GR, Richards RI, Tunnacliffe A. Association of a chromosome deletion syndrome with a fragile site within the proto-oncogene CBL2. *Nature*. 1995, 376(6536), 145-149.

[46] McKusick, VA. #147791 JACOBSEN SYNDROME; JBS; 11q DELETION SYNDROME; PARTIAL 11q MONOSOMY SYNDROME [Online Mendelian Inheritance in Man (OMIM)] 1994 [last updated on 2005/01/14] URL: *http://www.genome.jp/dbget-bin/www_bget?mim:147791.*

[47] Michaelis RC, Velagaleti GV, Jones C, Pivnick EK, Phelan MC, Boyd E, Tarleton J, Wilroy RS, Tunnacliffe A, Tharapel AT. Most Jacobsen syndrome deletion breakpoints occur distal to FRA11B. *Am J Med Genet*. 1998, 76(3), 222-228.

[48] Jones C, Mullenbach R, Grossfeld P, Auer R, Favier R, Chien K, James M, Tunnacliffe A, Cotter F. Co-localisation of CCG repeats and chromosome deletion breakpoints in Jacobsen syndrome: evidence for a common mechanism of chromosome breakage. *Hum Mol Genet*. 2000, 9(8), 1201-1208.

[49] Wang YH, Griffith J. Methylation of expanded CCG triplet repeat DNA from fragile X syndrome patients enhances nucleosome exclusion. *J Biol Chem*. 1996, 271(38), 22937-22940.

[50] Wang YH, Gellibolian R, Shimizu M, Wells RD, Griffith J. Long CCG triplet repeat blocks exclude nucleosomes: a possible mechanism for the nature of fragile sites in chromosomes. *J Mol Biol*. 1996, 263(4), 511-516.

[51] Mulvihill DJ, Wang YH. Two breakpoint clusters at fragile site FRA3B form phased nucleosomes. *Genome Res*. 2004, 14(7), 1350-1357.

[52] Yu S, Mangelsdorf M, Hewett D, Hobson L, Baker E, Eyre HJ, Lapsys N, Le Paslier D, Doggett NA, Sutherland GR, Richards RI. Human chromosomal fragile site FRA16B is an amplified AT-rich minisatellite repeat. *Cell*. 1997, 88(3), 367-374.

[53] Yamauchi M, Tsuji S, Mita K, Saito T, Morimyo M. A novel minisatellite repeat expansion identified at FRA16B in a Japanese carrier. *Genes Genet Syst*. 2000, 75(3), 149-154.

[54] Hsu YY, Wang YH. Human fragile site FRA16B DNA excludes nucleosomes in the presence of distamycin. *J Biol Chem*. 2002, 277(19), 17315-17319.

[55] van Dyke M.W., Hertzberg R.P., Dervan P.B. Map of distamycin, netropsin, and actinomycin binding sites on heterogeneous DNA. DNA cleavage inhibition patterns with methidiumpropyl-EDTA·Fe(II). *Proc. Natl Acad. Sci. USA*, 1982, 79, 5470–5474.

[56] Dolenc J, Oostenbrink C, Koller J, van Gunsteren WF. Molecular dynamics simulations and free energy calculations of netropsin and distamycin binding to an AAAAA DNA binding site. *Nucleic Acids Res*. 2005, 33(2), 725-733.

[57] Fagan P, Wemmer DE. Cooperative binding of distamycin-A to DNA in the 2-1 mode. *J Am Chem Soc*. 1992, 114, 1080-1081.

[58] Chen FM, Sha F. Circular dichroic and kinetic differentiation of DNA binding modes of distamycin. *Biochemistry*, 1998, 37, 11143–11151.

[59] Hocking T, Feichtinger W, Schmid M, Haan EA, Baker E, Sutherland GR. Homozygotes for FRA16B are normal. *Chromosome Res*. 1999, 7(7), 553-556.

[60] Hewett DR, Handt O, Hobson L, Mangelsdorf M, Eyre HJ, Baker E, Sutherland GR, Schuffenhauer S, Mao JI, Richards RI. FRA10B structure reveals common elements in repeat expansion and chromosomal fragile site genesis. *Mol Cell*. 1998, 1(6), 773-781.

[61] Sulek A, Hoffman-Zacharska D, Bednarska-Makaruk M, Szirkowiec W, Zaremba J. Polymorphism of trinucleotide repeats in non-translated regions of SCA8 and SCA12 genes: allele distribution in a Polish control group. *J Appl Genet*. 2004, 45(1), 101-105.

[62] Matsuura T, Yamagata T, Burgess DL, Rasmussen A, Grewal RP, Watase K, Khajavi M, McCall AE, Davis CF, Zu L, Achari M, Pulst SM, Alonso E, Noebels JL, Nelson DL, Zoghbi HY, Ashizawa T. Large expansion of the ATTCT pentanucleotide repeat in spinocerebellar ataxia type 10. *Nat Genet*. 2000, 26(2), 191-194.

[63] McKusick, VA. #313200 spinal and bulbar muscular atrophy, X-Linked 1; SMAX1; Spinal and bulbar muscular atrophy; SBMA; Kennedy disease; KD; Kennedy spinal and bulbar muscular atrophy; Bulbospinal muscular atrophy, X-Linked; Bulbospinal neuronopathy, X-Linked recessive; XBSN spinal muscular atrophy, Benign, with hypertrophy of calves, Included NOD [Online Mendelian Inheritance in Man (OMIM)] 1986 [last updated on 2004/12/15] URL: *http://www.genome.jp/dbgetbin/-www_bget?mim:313200.*

[64] McKusick, VA. *313700 androgen receptor; AR; Dihydrotestosterone receptor; DHTR [Online Mendelian Inheritance in Man (OMIM)] 1986 [last updated on 2004/08/02] URL: *http://www.genome.jp/dbget-bin/www_bget?omim+313700.*

[65] McKusick, VA. #125370 dentatorubral-pallidoluysian atrophy; DRPLA; Myoclonic epilepsy with choreoathetosis; Naito-Oyanagi disease; NOD [Online Mendelian Inheritance in Man (OMIM)] 1986 [last updated on 2004/12/15] URL: *http://www.genome.jp/dbget-bin/www_bget?mim:125370.*

[66] Kniffin, CL. *607462 DRPLA gene; DRPLA NOD; Atrophin 1 [Online Mendelian Inheritance in Man (OMIM)] 2003 [last updated on 2003/11/13] URL: *http://www.genome.jp/dbget-bin/www_bget?omim+607462.*

[67] McKusick, VA. +143100 huntington disease; HD; Huntington chorea huntingtin, included; HD, included; HTT, included; IT15, included [Online Mendelian Inheritance in Man (OMIM)] 1986 [last updated on 2005/02/08] URL: *http://www.genome.jp/-dbget-bin/www_bget?mim:143100.*

[68] McKusick, VA. #164400 spinocerebellar ataxia 1; SCA1; Spinocerebellar atrophy I; Olivopontocerebellar atrophy I; OPCA1; OPCA I; Menzel type OPCA [Online Mendelian Inheritance in Man (OMIM)] 1986 [last updated on 2004/07/26] URL: *http://www.genome.jp/dbget-bin/www_bget?mim:164400.*

[69] Jackson, JF. #183090 spinocerebellar ataxia 2; SCA2; Spinocerebellar atrophy II; Olivopontocerebellar atrophy, Holguin type; Olivopontocerebellar atrophy 2; Spinocerebellar ataxia, Cuban type;; Cerebellar degeneration with slow eye movements; Wadia-swami syndrome; Spinocerebellar degeneration with slow eye movements; SDSEM [Online Mendelian Inheritance in Man (OMIM)] 1995 [last updated on 2004/12/29] URL: *http://www.genome.jp/dbgetbin/www_bget?-mim:183090.*

[70] McKusick, VA. #109150 machado-joseph disease; MJD; Spinocerebellar ataxia 3; SCA3; Spinocerebellar atrophy III; Azorean neurologic disease; Spinopontine atrophy; Nigrospinodentatal degeneration [Online Mendelian Inheritance in Man (OMIM)] 1986 [last updated on 2004/12/15] URL: *http://www.genome.jp/dbgetbin/www_bget?-mim:109150.*

[71] Tiller, GE, *605440 Ataxin-1 ubiquitin-like interacting protein; A1U; C1ORF6 [Online Mendelian Inheritance in Man (OMIM)] 2000 [last updated on 2002/04/25] URL: *http://www.genome.jp/dbget-bin/www_bget?mim:605440.*

[72] Smith, M. *601517 Ataxin 2; ATXN2 [Online Mendelian Inheritance in Man (OMIM)] 1996 [last updated on 2003/03/21] URL: *http://www.genome.jp/dbget-bin/www_bget?omim+601517.*

[73] Kniffin, CL. *607047 Ataxin 3; ATXN3 [Online Mendelian Inheritance in Man (OMIM)] 2002 [last updated on 2004/07/27] URL: *http://www.genome.jp/dbget-bin/www_bget?omim+607047.*

[74] McKusick, VA. *601011 calcium channel, Voltage-dependent, P/Q type, Alpha-1A subunit; CACNA1A; Calcium channel, L type, Alpha-1 polypeptide, Isoform 4; CACNL1A4 [Online Mendelian Inheritance in Man (OMIM)] 1996 [last updated on 2005/01/19] URL: *http://www.genome.jp/dbget-bin/www_bget?omim+601011.*

[75] McKusick, VA. #183086 spinocerebellar Ataxia 6; SCA6 [Online Mendelian Inheritance in Man (OMIM)] 1994 [last updated on 2004/12/15] URL: *http://www.genome.jp/dbget-bin/www_bget?mim:183086.*

[76] McKusick, VA. #164500 spinocerebellar Ataxia 7; SCA7; olivopontocerebellar atrophy III; OPCA3; OPCA III; OPCA with retinal degeneration; OPCA with macular degeneration and external phthalmoplegia; Autosomal dominant cerebellar ataxia, Type II; ADCA, Type II [Online Mendelian Inheritance in Man (OMIM)] 1986 [last updated on 2005/01/04] URL: *http://www.genome.jp/dbgetbin/www_bget?-mim:164500.*

[77] Kniffin, CL. *607640 Ataxin 7; ATXN7 [Online Mendelian Inheritance in Man (OMIM)] 2003 [last updated on 2004/10/26] URL: *http://www.genome.jp/dbget-bin/www_bget?omim+607640.*

[78] Kniffin, CL. #607136 spinocerebellar Ataxia 17; SCA17; Huntington disease-like 4; HDL4 [Online Mendelian Inheritance in Man (OMIM)] 2002 [last updated on 2004/12/15] URL: *http://www.genome.jp/dbget-bin/www_bget?mim:607136.*

[79] McKusick, VA. *600075 Tata box-binding protein; TBP; SCA17 GENE [Online Mendelian Inheritance in Man (OMIM)] 1994 [last updated on 2004/05/19] URL: *http://www.genome.jp/dbget-bin/www_bget?omim+600075.*

[80] McKusick, VA. *313650 TAF1 RNA Polymerase II, Tata box-binding protein-associated factor, 250-KD; TAF1; Tata box-binding protein-associated factor 2A; TAF2A; TBP-Associated factor, RNA Polymerase II, 250-KD; TAFII250; Cell cycle, G1 phase defect; CCG1; Cell cycle gene 1; Temperature-sensitive mutation, Mouse and hamster, complementation of; Cell cycle block, G1-TO-S; CCGS; BA2R transcription factor IID, included; TFIID, included; TF2D, included; Tata-binding protein-containing cmplex, included gene [Online Mendelian Inheritance in Man (OMIM)] 1986 [last updated on 2004/06/09] URL: *http://www.genome.jp/dbget-bin/www_bget?omim+313650.*

[81] McKusick, VA. #604326 spinocerebellar Ataxia 12; SCA12 GENE [Online Mendelian Inheritance in Man (OMIM)] 1999 [last updated on 2002/10/28] URL: *http://www.genome.jp/dbget-bin/www_bget?mim:604326.*

[82] McKusick, VA. *604325 protein phosphatase 2, regulatory subunit B, BETA; PPP2R2B; PP2APR55-BETA; PP2AB55-BETA; PP2AB-BETA; PR55-BETA Gene [Online Mendelian Inheritance in Man (OMIM)] 1999 [last updated on 2002/02/22] URL: *http://www.genome.jp/dbget-bin/www_bget?omim+604325.*

[83] McKusick, VA. #164300 oculopharyngeal muscular dystrophy; OPMD; Muscular dystrophy, Oculopharyngeal included [Online Mendelian Inheritance in Man (OMIM)] 1986 [last updated on 2004/06/21] URL: *http://www.genome.jp/dbgetbin/www_bget?-mim:164300.*

[84] McKusick, VA. *602279 polyadenylate-binding protein, Nuclear, 1; PABPN1; Polyadenylate-binding protein 2; PABP2; PAB2; Poly(A)-Binding protein 2 included [Online Mendelian Inheritance in Man (OMIM)] 1998 [last updated on 2003/01/29] URL: *http://www.genome.jp/dbget-bin/www_bget?omim+602279.*

[85] McKusick, VA. #257950 muscular dystrophy, Oculopharyngeal, Autosomal recessive; Oculopharyngeal muscular dystrophy, Autosomal recessive; OPMD, Autosomal recessive [Online Mendelian Inheritance in Man (OMIM)] 1986 [last updated on 1998/01/23] URL: *http://www.genome.jp/dbget-bin/www_bget?omim+257950.*

[86] McKusick, VA. #186000 syndactyly, Type II; Synpolydactyly; SPD Synpolydactyly with foot anomalies, included [Online Mendelian Inheritance in Man (OMIM)] 1986 [last updated on 2003/09/17] URL: *http://www.genome.jp/dbgetbin/www_bget?-mim:186000.*

[87] McKusick, VA. *142989 Homeobox D13; HOXD13; Homeobox 4I; HOX4I [Online Mendelian Inheritance in Man (OMIM)] 1991 [last updated on 2004/06/22] URL: *http://www.genome.jp/dbget-bin/www_bget?omim+142989.*

[88] McKusick, VA. #119600 Cleidocranial dysplasia; CCD; Ceidocranial dysostosis; CLCD cleidocranial dysplasia, Forme fruste, with brachydactyly, included; Cleidocranial dysplasia, Forme fruste, Dental anomalies only, included gene [Online Mendelian Inheritance in Man (OMIM)] 1986 [last updated on 2004/08/23] URL: *http://www.genome.jp/dbget-bin/www_bget?mim:119600.*

[89] Jackson WP. Osteo-dental dysplasia (cleido-cranial dysostosis); the "Arnold head". *Acta Med Scand.* 1951, 139(4), 292-307.

[90] McKusick, VA. *600211 Runt-Related transcription factor 2; RUNX2; Core-Binding factor, Runt domain, Alpha subunit 1; CBFA1; AML3 GENE; AML3; PEBP2-Alpha-A; OSF2 Gene [Online Mendelian Inheritance in Man (OMIM)] 1994 [last updated on 2005/01/18] URL: *http://www.genome.jp/dbget-bin/www_bget?omim+600211.*

[91] McKusick, VA. +603073 Zinc finger protein of cerebellum, 2; ZIC2; ZIC family member 2 Holoprosencephaly 5, included; HPE5, included gene [Online Mendelian Inheritance in Man (OMIM)] 1998 [last updated on 2001/08/24] URL: *http://www.genome.jp/dbget-bin/www_bget?mim:603073.*

[92] McKusick, VA. #140000 Hand-Foot-uterus syndrome; HFU syndrome; Hand-Foot-genital syndrome; HFG syndrome [Online Mendelian Inheritance in Man (OMIM)] 1986 [last updated on 1997/02/02] URL: *http://www.genome.jp/dbgetbin/www_bget?-omim+140000.*

[93] McKusick, VA. *142959 Homeobox A13; HOXA13; Homeobox 1J; HOX1J [Online Mendelian Inheritance in Man (OMIM)] 1990 [last updated on 2004/05/04] URL: *http://www.genome.jp/dbget-bin/www_bget?omim+142959.*
[94] Debeer P, Bacchelli C, Scambler PJ, De Smet L, Fryns JP, Goodman FR. Severe digital abnormalities in a patient heterozygous for both a novel missense mutation in HOXD13 and a polyalanine tract expansion in HOXA13. *J Med Genet,* 2002, 39(11), 852-856.
[95] McKusick, VA. #110100 blepharophimosis, Ptosis, and epicanthus inversus; BPES, BPES Type I, included; BPES Type II, included; BPES with ovarian failure, included; BPES without ovarian failure, included, [Online Mendelian Inheritance in Man (OMIM)] 1986 [last updated on 2004/10/01] URL: *http://www.genome.jp/dbget-bin/www_bget?mim:110100.*
[96] McKusick, VA. *605597 forkhead transcription factor FOXL2; FOXL2; Pituitary forkhead factor, Mouse, homolog of; PFRK, Mouse, homolog of; PFRK [Online Mendelian Inheritance in Man (OMIM)] 2001 [last updated on 2004/10/20] URL: *http://www.genome.jp/dbget-bin/www_bget?omim+605597.*
[97] McKusick, VA. #300123 mental retardation, X-Linked, with isolated growth hormone deficiency; MRGH [Online Mendelian Inheritance in Man (OMIM)] 1998 [last updated on 2003/01/08] URL: *http://www.genome.jp/dbgetbin/www_bget?-mim:300123.*
[98] McKusick, VA. *313430 SRY-BOX 3; SOX3; SRY-Related HMG-BOX Gene 3 [Online Mendelian Inheritance in Man (OMIM)] 1994 [last updated on 2004/10/12] URL: *http://www.genome.jp/dbget-bin/www_bget?omim+313430.*
[99] McKusick, VA. *480000 SEX-Determining region Y; SRY; Testis- determining factor; TDF; Testis-determining factor on Y; TDY [Online Mendelian Inheritance in Man (OMIM)] 1986 [last updated on 2004/11/15] URL: *http://www.genome.jp/dbget-bin/www_bget?omim+480000.*
[100] Laumonnier F, Ronce N, Hamel BC, Thomas P, Lespinasse J, Raynaud M, Paringaux C, Van Bokhoven H, Kalscheuer V, Fryns JP, Chelly J, Moraine C, Briault S. Transcription factor SOX3 is involved in X-linked mental retardation with growth hormone deficiency. *Am J Hum Genet.* 2002, 71(6), 1450-1455.
[101] McKusick, VA. #308350 infantile spasm syndrome, X-Linked; ISSX; Infantile spasms, X-Linked; West syndrome, X-Linked [Online Mendelian Inheritance in Man (OMIM)] 1986 [last updated on 2003/06/19] URL: *http://www.genome.jp/dbgetbin/www_bget?-mim:308350.*
[102] McKusick, VA. *300382 aristaless-related homeobox, X-Linked; ARX [Online Mendelian Inheritance in Man (OMIM)] 2002 [last updated on 2004/10/08] URL: *http://www.genome.jp/dbget-bin/www_bget?omim+300382.*
[103] McKusick, VA. #309510 partington X-Linked mental retardation SYNDROME; PRTS; Partington syndrome; Mental retardation, X-Linked, Syndromic 1; MRXS1; Mental Retardation, X-Linked, with dystonic movements, Ataxia, and seizures [Online Mendelian Inheritance in Man (OMIM)] 1986 [last updated on 2003/05/29] URL: *http://www.genome.jp/dbget-bin/www_bget?mim:309510.*
[104] Partington MW, Mulley JC, Sutherland GR, Hockey A, Thode A, Turner G. X-linked mental retardation with dystonic movements of the hands. *Am J Med Genet.* 1988, 30(1-2), 251-262.

[105] Kniffin, CL. #300430 mental retardation, X-Linked 36; MRX36 [Online Mendelian Inheritance in Man (OMIM)] 2003 [last updated on 2003/03/27] URL: *http://www.genome.jp/dbget-bin/www_bget?mim:300430.*

[106] McKusick, VA. #209880 Autonomic control, Congenital failure of; Ondine curse, Ongenital; Central hypoventilation syndrome, Congenital; CCHS ondine-hirschsprung disease, included; OHD, included; CCHS with hirschsprung disease, included; HADDAD syndrome, included [Online Mendelian Inheritance in Man (OMIM)] 1986 [last updated on 2005/02/09] URL: *http://www.genome.jp/dbgetbin/www_bget?-mim:209880.*

[107] Hamosh, A. *603851 paired-like homeobox 2B; PHOX2B; Paired mesoderm homeobox 2B; PMX2B; Neuroblastoma paired-type homeobox gene; NBPHOX [Online Mendelian Inheritance in Man (OMIM)] 1999 [last updated on 2005/02/09] URL: *http://www.genome.jp/dbget-bin/www_bget?omim+603851.*

[108] McKusick, VA. *164761 rearranged during transfection protooncogene; RET; RET protooncogene RET/ELKS fusion gene, included [Online Mendelian Inheritance in Man (OMIM)] 1990 [last updated on 2004/10/12] URL: *http://www.genome.jp/dbget-bin/www_bget?mim:164761.*

[109] Scott, AF. *600837 glial cell line-derived neurotrophic factor; GDNF [Online Mendelian Inheritance in Man (OMIM)] 1995 [last updated on 2004/06/14] URL: *http://www.genome.jp/dbget-bin/www_bget?mim:600837.*

[110] McKusick, VA. *131242 Endothelin 3; EDN3; ET3 [Online Mendelian Inheritance in Man (OMIM)] 1990 [last updated on 2002/06/28] URL: *http://www.genome.jp/dbget-bin/www_bget?mim:131242.*

[111] McKusick, VA. *113505 Brain-derived neurotrophic factor; BDNF [Online Mendelian Inheritance in Man (OMIM)] 1990 [last updated on 2005/02/01] URL: *http://www.genome.jp/dbget-bin/www_bget?mim:113505.*

[112] McKusick, VA. #160900 dystrophia myotonica 1; DM1; Dystrophia myotonica; DM; Myotonic dystrophy 1; Steinert disease gene [Online Mendelian Inheritance in Man (OMIM)],1986 [last updated on 2004/08/27] URL: *http://www.genome.jp/dbget-bin/www_bget?mim:160900.*

[113] Hamosh, A. *605377 dystrophia myotonica protein kinase; DMPK; DM Kinase; DMK; DM protein kinase; Myotonin-protein kinase gene [Online Mendelian Inheritance in Man (OMIM)] 2000 [last updated on 2004/04/27] URL: *http://www.genome.jp/dbget-bin/www_bget?omim+605377.*

[114] Kniffin, CL. #608768 spinocerebellar Ataxia 8; SCA8 GENE [Online Mendelian Inheritance in Man (OMIM)] 2002 [last updated on 2004/06/30] URL: *http://www.genome.jp/dbget-bin/www_bget?mim:608768.*

[115] Hamosh, A. *603680 SCA8 GENE; SCA8 [Online Mendelian Inheritance in Man (OMIM)] 1999 [last updated on 2004/07/02] URL: *http://www.genome.jp/dbget-bin/www_bget?omim+603680.*

[116] McKusick, VA. #606438 huntington disease-like 2; HDL2 [Online Mendelian Inheritance in Man (OMIM)] 2001 [last updated on 2004/01/28] URL: *http://www.genome.jp/dbget-bin/www_bget?mim:606438.*

[117] Antonarakis, SE. *605268 Junctophilin 3; JPH3; JP3 [Online Mendelian Inheritance in Man (OMIM)] 2000 [last updated on 2001/11/05] URL: *http://www.genome.jp/dbget-bin/www_bget?omim+605268.*

[118] Margolis RL, O'Hearn E, Rosenblatt A, Willour V, Holmes SE, Franz ML, Callahan C, Hwang HS, Troncoso JC, Ross CA. A disorder similar to Huntington's disease is associated with a novel CAG repeat expansion. *Ann Neurol.* 2001, 50(6), 373-380.

[119] McKusick, VA. #177170 pseudoachondroplastic dysplasia; Pseudoachondroplasia; PSACH; Spondyloepiphyseal dysplasia, Pseudoachondroplastic; Pseudoachondroplastic dysplasia III, formerly; Pseudoachondroplastic dysplasia IV, formerly included [Online Mendelian Inheritance in Man (OMIM)] 1986 [last updated on 2005/01/28] URL: *http://www.genome.jp/dbget-bin/www_bget?mim:177170.*

[120] McKusick, VA. *600310 cartilage oligomeric matrix protein; COMP [Online Mendelian Inheritance in Man (OMIM)] 1995 [last updated on 2005/01/11]. URL: *http://www.genome.jp/dbget-bin/www_bget?omim+600310.*

[121] McKusick, VA. #132400 epiphyseal dysplasia, multiple, 1; EDM1; Multiple epiphyseal dysplasia, Comp-related, Epiphyseal dysplasia, fairbank type, included; Epiphyseal dysplasia, ribbing type, included [Online Mendelian Inheritance in Man (OMIM)] 1986 [last updated on 2005/01/28] URL: *http://www.genome.jp/dbget-bin/www_bget?omim+132400.*

[122] McKusick, VA. #229300 friedreich Ataxia 1; FRDA; FRDA1; FA friedreich ataxia with retained reflexes, included; FARR, included [Online Mendelian Inheritance in Man (OMIM)] 1986 [last updated on 2004/08/26]. URL: *http://www.genome.jp/dbget-bin/www_bget?mim:229300.*

[123] Kniffin, CL. *606829 FRDA Gene; FRDA; Frataxin; X25 included [Online Mendelian Inheritance in Man (OMIM)] 2002 [last updated on 2004/08/25]. URL: *http://www.genome.jp/dbget-bin/www_bget?omim+606829.*

[124] Sakamoto, N, Chastain, PD, Parniewski, P, Ohshima, K, Pandolfo, M, Griffith, JD, Wells, RD. Sticky DNA: self-association properties of long GAA-TTC repeats in R-R-Y triplex structures from Friedrich's ataxia. *Molec. Cell*, 1999, 3, 465-475.

[125] McKusick, VA. #602668 dystrophia myotonica 2; DM2; Myotonic dystrophy 2; Proximal myotonic myopathy; PROMM; Myotonic myopathy, Proximal; Ricker syndrome [Online Mendelian Inheritance in Man (OMIM)] 1998 [last updated on 2003/09/11]. URL: *http://www.genome.jp/dbget-bin/www_bget?mim:602668.*

[126] McKusick, VA. *116955 zinc finger protein 9; ZNF9; Cellular retroviral nucleic acid-binding protein 1; CNBP1 [Online Mendelian Inheritance in Man (OMIM)] 1990[last updated on 2003/10/07]. URL: *http://www.genome.jp/dbgetbin/www_bget?-omim+116955.*

[127] Liquori, CL, Ricker, K, Moseley, ML, Jacobsen, JF, Kress, W, Naylor, SL, Day, JW, Ranum,LPW. Myotonic dystrophy type 2 caused by a CCTG expansion in intron 1 of ZNF9. *Science*, 2001, 293, 864-867.

[128] McKusick, VA. +603516 spinocerebellar Ataxia 10; SCA10, Ataxin 10, included; ATXN10, included [Online Mendelian Inheritance in Man (OMIM)] 1999[last updated on 2005/05/21] URL: *http://www.genome.jp/dbget-bin/www_bget?mim:603516.*

[129] Marz P, Probst A, Lang S, Schwager M, Rose-John S, Otten U, Ozbek S. Ataxin-10, the spinocerebellar ataxia type 10 neurodegenerative disorder protein, is essential for survival of cerebellar neurons. *J Biol Chem.* 2004, 279(34), 35542-35550.

[130] Kodaira M, Satoh C, Hiyama K, Toyama K. Lack of effects of atomic bomb radiation on genetic instability of tandem-repetitive elements in human germ cells. *Am J Hum Genet.* 1995, 57(6), 1275-1283.

[131] Barber R, Plumb MA, Boulton E, Roux I, Dubrova YE. Elevated mutation rates in the germ line of first- and second-generation offspring of irradiated male mice. *Proc Natl Acad Sci USA*. 2002, 99(10), 6877-6882.

In: New Developments in Mutation Research
Editor: Charles L. Valon, pp. 71-96

ISBN 1-59454-664-9

Chapter 4

PROGRESS IN TRANSGENIC RODENT MUTATION ASSAYS*

Tao Chen,** Nan Mei and Robert H. Heflich

Division of Genetic and Reproductive Toxicology, National Center for Toxicological Research, U.S. Food and Drug Administration, Jefferson, AR 72079, USA

ABSTRACT

Transgenic mouse and rat mutation assays have provided a tool for analyzing *in vivo* mutation in any tissue from which sufficient amounts of DNA can be recovered. Thus these assays provide a method for directly comparing cancer incidence with mutant frequency. In these systems, chromosomally-integrated transgenes are used as the targets for mutation. Following recovery of the transgenes, the target DNA can be assayed *in vitro* to identify and quantify mutations produced *in vivo*. In this article, we review progress in the development, validation, and use of transgenic *in vivo* mutation models.

INTRODUCTION

Mutations are heritable changes in the nucleotide sequence of DNA. Mutations that confer some advantage to the cell in which they occur are rare, although the frequency is sufficient to provide the variation necessary to drive evolution. The majority of mutations are either silent (produce no phenotype) or are deleterious to cells. Among the phenotypic consequences of mutations in somatic cells are neoplasms and possibly aging; mutations in germ cells are responsible for a large number of genetic diseases, including birth defects, and predispositions to disease, including cancer. Mutations are thought to be involved in the

---

* The views presented in this article do not necessarily reflect those of the Food and Drug Administration.

** *Correspondence:* HFT-130, NCTR, 3900 NCTR Rd., Jefferson, AR 72079; Tel: 870-543-7954 (Tao Chen); 870-543-7386 (Nan Mei); 870-543-7493 (Robert H. Heflich); Fax: 870-543-7682; Email: tchen@nctr.fda.gov; nmei@nctr.fda.gov; rheflich@nctr.fda.gov

etiology of cancer because multiple genetic events are involved in different stages of carcinogenesis [1]. Recent developments in molecular cancer genetics have identified inherited and somatic cell mutations associated with cancer in oncogenes, tumor suppressor genes, DNA repair genes and other related genes [2-4].

The possible impact of environmental chemicals on human health and well-being has been clearly recognized. There are more than 60,000 synthetic chemicals in use, and the carcinogenic properties of many of these are uncertain. In addition, new chemicals are introduced at a rate of about 1,000 per year. If some of these chemicals are mutagenic, exposure to them could cause genetic changes leading to either inherited or somatic genetic diseases, such as cancer. The observation that most carcinogens are mutagens and most mutagens are carcinogens [5, 6] provides the rationale for assessing the risk of exposure to carcinogenic substances in our environment through analysis of their mutagenicity. Since evaluating the mutagenic potential of environmental agents is considerably less expensive and time consuming than directly testing their carcinogenic potential, analyzing for mutations can be of practical value in identifying agents that are potentially carcinogenic. Mutational analysis also has value in identifying individuals that are at risk for tumorigenesis because of their genetic background or their exposure to genotoxic agents.

As a result, there is widespread use of in *vitro* and *in vivo* somatic cell mutation assays as predictive tools for identifying potentially carcinogenic chemicals in the environment, and as biomarkers for carcinogenesis. *In vitro* systems such as the *Salmonella* reverse mutation assay [7, 8], bacterial forward mutation assays [9-11] and measurement of gene mutations in cultured mammalian cells [12-15] have provided invaluable information from their use as screens for potential carcinogens and their application to understanding the mechanisms of carcinogenesis. However, *in vitro* mutagenicity assays have significant shortcomings. One major problem that is common to all *in vitro* tests is that they cannot provide information about the effects of the integration of pharmacokinetics, biotransformation, DNA repair and tissue susceptibility on mutagens that occurs *in vivo*. It has been found that, either as individual tests or when used in batteries of tests, *in vitro* assays are imperfect predictors of whether or not a chemical is a carcinogen [16]. Accordingly, attention has turned to the development of *in vivo* somatic cell mutation assays to identify putative carcinogens.

Rodent models provide the opportunity for testing chemical tumorigenic potential and for evaluating hypotheses that cannot be tested easily or ethically in man. Accordingly, the standard National Toxicological Program (NTP) tumorigenicity bioassays and other cancer assays have been established and a large number of chemicals have been tested for tumorigenicity. In addition, several *in vivo* gene mutation tests have been developed using rodent endogenous genes as target genes. All the assays use reporter genes with phenotypes that make measuring mutations reasonably convenient. These assays, however, are limited to certain tissues. For example, the hypoxanthine-guanine phosphoribosyl transferase (*Hprt*) assays [17, 18], adenine phosphoribosyl transferase (*Aprt*) mouse assay [19, 20], and thymidine kinase (*Tk*) mouse assay [21, 22] have been conducted mainly on lymphocytes, while the *Dlb-1* specific locus mouse assay scores mutations in the small intestine [23] and the mouse spot test targets embryos at gestation day 9-11 for mutation detection [24].

In many carcinogenic exposures, mutations are known to be generated as the result of organ- or tissue-specific metabolic activity (*e.g.*, in the liver) or as the result of damage by direct-acting mutagens at the site of entry (*e.g.*, the lung). This may result in the accumulation of DNA damage in particular tissues or organs and may contribute to the targeting of tumor induction to specific sites. *In vivo* test systems utilizing endogenous genes are generally limited to analyzing mutations in a single tissue that most likely is not the target tissue for carcinogenesis. Important target organs, such as the liver, bladder, lung, and mammary gland are not used for assaying mutations because of the difficulty in culturing cells from these organs. This limits the ability of these assays to study mutations in the most important organs for carcinogenesis. In terms of the use of *in vivo* mutational assays as screening systems for carcinogens, this also may reduce the sensitivity of these assays and prevent their application for predicting the target organs for carcinogenesis.

In contrast, transgenic rodent mutation assays are not constrained to particular tissues and organs. These *in vivo* mutagenicity assays provide unique opportunities for studying the induction of tissue-specific mutations that reflect the integration of endogenous factors such as pharmacokinetics and tissue specificity. Transgenic rodents are capable of providing data on tissue-specific mutant frequency (MF) and mutational spectra, both of which may be useful for regulatory purposes or for evaluating hypotheses on the mechanisms responsible for mutagenesis and carcinogenesis. This information may ultimately lead to improvements in human risk assessments based on extrapolation from animal data and on mechanism of action.

Here, we review progress in the development and validation of different transgenic rodent mutation models, and the development of testing protocols for these assays. We also consider the application of transgenic rodent models for detecting carcinogens via the evaluation of tissue-specific mutations, and the utilization of the transgenic rodent mutation assays for mechanic studies.

TRANSGENIC RODENT MUTATION SYSTEMS

Development of Transgenic Rodents for Mutation Assays

Transgenic animals are developed using molecular genetic techniques to produce animals whose cells contain foreign target genes (reporter genes) for detecting *in vivo* mutations. In these *in vivo* models, reporter genes are normally located on so-called shuttle vectors that are derivatives of bacteriophage or plasmids. The vectors recover the mutational target from one species for mutant detection in the second. Typically, one host is a rat or mouse where the mutations are induced and the second host is an appropriate bacterial strain where the mutations are assayed.

A shuttle vector containing reporter gene(s) is constructed using recombinant DNA technologies. For example, the shuttle vector (Figure 1) used in Big Blue mice and rats is approximately 45.6 kb, and contains the *lacI* and *cII* reporter genes. The vector is flanked at each end by *cos* sites, which allow excision and packaging into phage heads [25]. Multiple genomic copies of the vector are tandemly concatenated in a head-to-tail array. These recombinant DNA constructions then are inserted into a germ cell chromosome and replicated as part of the genome, and thus are present in all the cells of the transgenic animal. Each animal that develops after receiving the transgene DNA is referred to as the founder of a new transgenic lineage. If the germ cells of the founder transmit the transgene stably, then all descendants of this animal are members of a unique transgenic line. Since the new transgenic locus is present in only one of a pair chromosomes, the genotype of the founder is hemizygous for the transgene. A homozygous genotype, in which transgene alleles are present in both copies of the recipient chromosome, can be produced by mating a pair of hemizygous F1 siblings.

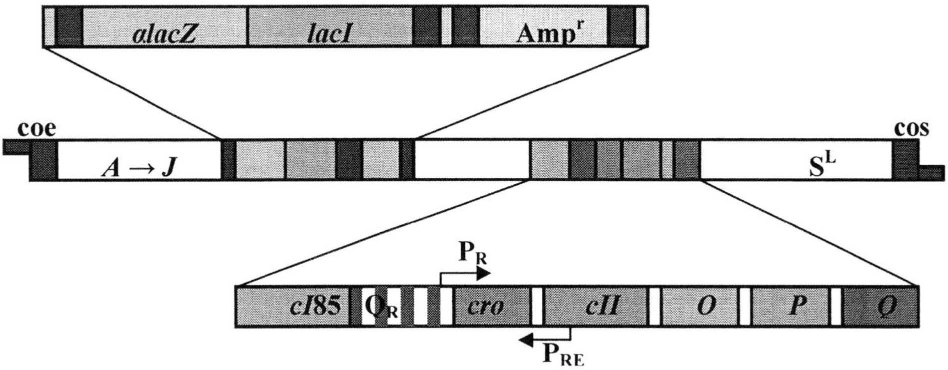

Figure 1. The λLIZ shuttle vector is approximately 45.6 kb in length. The expanded regions show the areas important for both the plaque color-screening (*lacI*) and λSelect-*cII* (*cII*) assays.

The transgenic animals can be used for mutation studies. After exposure of transgenic animals to a test substance, DNA can be isolated from individual organs or tissues, and single copies of the vector DNA can be excised from the high molecular weight rodent DNA and transferred into infectious particles (virus or plasmid). If appropriate *E. coli* host cells are infected, plated and incubated, plaques or colonies containing individual transgene vectors become visible on the plates. MFs can be determined by comparing the number of mutant plaques/colonies to the total number of plaques/colonies. The types of mutations also can be evaluated by sequencing the mutants.

A number of transgenic mice and rats have been developed for mutagenesis studies (Table 1). Among these transgenic animals, the Big Blue mouse and rat, and Muta mouse are the most widely used and are commercially available. When first developed, the Big Blue mouse and Big Blue rat models used the bacterial *lacI* gene as the primary mutational target [26, 27]. Each cell of the animal contains multiple copies of the λLIZ shuttle vector that carries the *lacI* (1080 bp) and *cII* (294 bp) genes as mutational targets (Figure 1) in a head-to-tail fashion. The C57BL/6 homozygous mouse has 80 copies of the λLIZ shuttle vector, while the C57BL/6 hemizygous mouse and B6C3F1 hybrid mouse have 40 copies of the vector; the Fischer 344 homozygous rat has 30-40 copies of the vector. These animals are commercially available from Stratagene (La Jolla, CA). Muta mouse was originally constructed by micro-injection of the bacteriophage λ gt10-*lacZ* shuttle vector into fertilized CD2 mouse oocytes. The vector is approximately 47 kb in length and contains *lacZ* (3100 bp) and *cII* as reporter genes. This transgenic mouse carries 40 copies of the shuttle vector in a head-to-tail array at a single site in chromosome 3 [28]. Homozygous mice with 80 copies of the shuttle vector are available as the Muta Mouse from Covance Research Products (Denver, PA).

Table 1. Transgenic rodent gene mutation models

Mutation assay (Mutation identification method)	Target gene	Shuttle vector	*E. coli* Host	Type of animal (strain background)	Major mutations detected	Reference
lacI assay (Color selection)	*lacI* gene	λ bacteriophage	SCS-8	Big Blue mouse (C57BL/6 and B6C3F1) Big Blue rat (Fisher 344)	Point mutations and small deletions	[26, 27]
lacZ assay (Color selection and P-gal positive selection)	*lacZ* gene	λ bacteriophage	3021 (*gal*E$^-$)	Muta mouse (CD2F1)	Point mutations and small deletions	[28]
cII assay (Temperature positive selection)	λ *cII* gene	λ bacteriophage	G1225 (*hfl*$^-$)	Big Blue mouse (C57BL/6 and B6C3F1); Big Blue rat (Fisher 344); Muta mouse (CD2F1)	Point mutations and small deletions	[77, 100]
gpt assay (6-thioguanine positive selection)	*gpt* gene	λEG10 phage	YG6020	*gpt* delta mouse (C571/6J); *gpt* delta rat (Sprague Dawley)	Point mutations and small deletions	[29, 30]
Spi$^-$ assay (Spi$^-$ positive selection)	*gam* and *red* genes	λEG10 phage	XL1 (P_2 lysogen)	*gpt* delta mouse (C57BL/6J); *gpt* delta rat (Sprague Dawley)	Deletions up to 10,000 base pair	[29, 30]
λ*supF* assay (Color selection)	λ*supF* gene	λ bacteriophage		λ*supF* transgenic mouse (C57BL/6)	Point mutations and small deletions	[101]
ΦX174 assays (reversion and forward positive selection)	*am3, gene A*	Bacteriophage ΦX174*am3cs70*		ΦX174 transgenic mouse (C57BL/6J)	Base pair substitutions	[31, 33]
pKZ1 assay (PCR detection)	*lacZ* inversion			pKZ1 mouse (C57BL/6JXDBA2)	Intrachromosomal recombination	[35, 36]
lacZ plasmid assay (Color selection and P-gal positive selection)	*lacZ*	pUR288 plasmid		Plasmid *lacZ* transgenic mouse (C57BL/6J)	Point mutations, and small & large deletions	[34]
rpsL assay (streptomycin positive selection)	*rpsL*	pML4 plasmid		*rpsL* transgenic mouse (C57BL/6J)	Point mutations and deletions	[102]

Other transgenic rodent mutation models have not been as widely used and evaluated. The *gpt* delta mouse and rat, which carry λEG10 phage DNA (48 kb) as transgene, have been used primarily in Japan. After recovery of λEG10 phage, point mutations in the *gpt* gene and deletions in the *red/gam* genes can be identified by 6-thioguanine and Spi$^-$ selection, respectively [29, 30]. The ΦX174 transgenic mouse reversion [31, 32] and forward [31, 33] mutation assays identify mouse *in vivo* mutation among a background of bacterial mutation, and may prove to be more sensitive to induced mutation than other models; the plasmid *lacZ* transgenic mouse model can detect large deletions and insertions with a very high rescue efficiency [34]; and the pKZ1 transgenic mouse detects a specific type of somatic intrachromosomal recombination [35, 36].

Protocol for Transgenic Rodent Mutation Assays

Although there are a number of different transgenic models, the overall procedure for conducting the assays with these models is similar; the major differences are in the methods of identifying or selecting mutants. The various steps in the transgenic rodent mutation assays involve treatment of animals, a period for mutation manifestation, the collection of tissues, the isolation of genomic DNA, excision of the vectors and packaging the vector DNA into phages or constructing plasmids, plating the infected bacteria, enumerating mutant and wild type plaques or colonies, and calculating MF (Figure 2). Although there are no specific rules for conducting transgenic animal mutation assays, the International Workshop on Genotoxicity Test Procedures (IWGT) has discussed most of the important issues relating to the assays and reached an agreement upon a general approach for the design of transgenic rodent mutation assays [37, 38]. These consensus reports are the basis for the following description of experimental procedures for conducting transgenic rodent mutation assays.

Treatment

The species, gender, and the treatment route, dose and schedule used in transgenic rodent studies are mainly dictated by the purpose of the study. A single dose treatment may be sufficient for detecting mutations induced by potent mutagens. For regulatory safety assessment, the IWGT recommends repeat treatment of animals for 28 days with the maximum tolerated dose (MTD) [38]. Two further dose groups receiving 1/3 and 2/3 the MTD are desirable. The two lower dose groups should be analyzed if the number of animals in the high dose group is reduced by agent toxicity so that the statistical power is below an acceptable level [37]. If there are no significant differences between the sexes in terms of toxicity or metabolism, the IWGT recommends that male animals should normally be used, consistent with guideline recommendations for other *in vivo* genotoxicity tests [38].

Mutant Manifestation Time

Mutant manifestation time is the period between the last treatment and the time of sampling tissues for mutation. It is a very important factor for transgenic mutation assays because the significance of a negative result is uncertain unless the mutant manifestation time is known. The manifestation time is dependent on the tissue that is assayed; it can be a few days for tissues with fast turnover rates such as bone marrow or it can be several months for tissues with low turnover rates such as liver [38-40]. In rapidly dividing tissues like bone marrow, longer sampling times might decrease the MF due to the loss of differentiated cells from the tissue. For this reason, the IWGT recommends a general schedule of sampling 3 days after 28 days of repeated treatment [38].

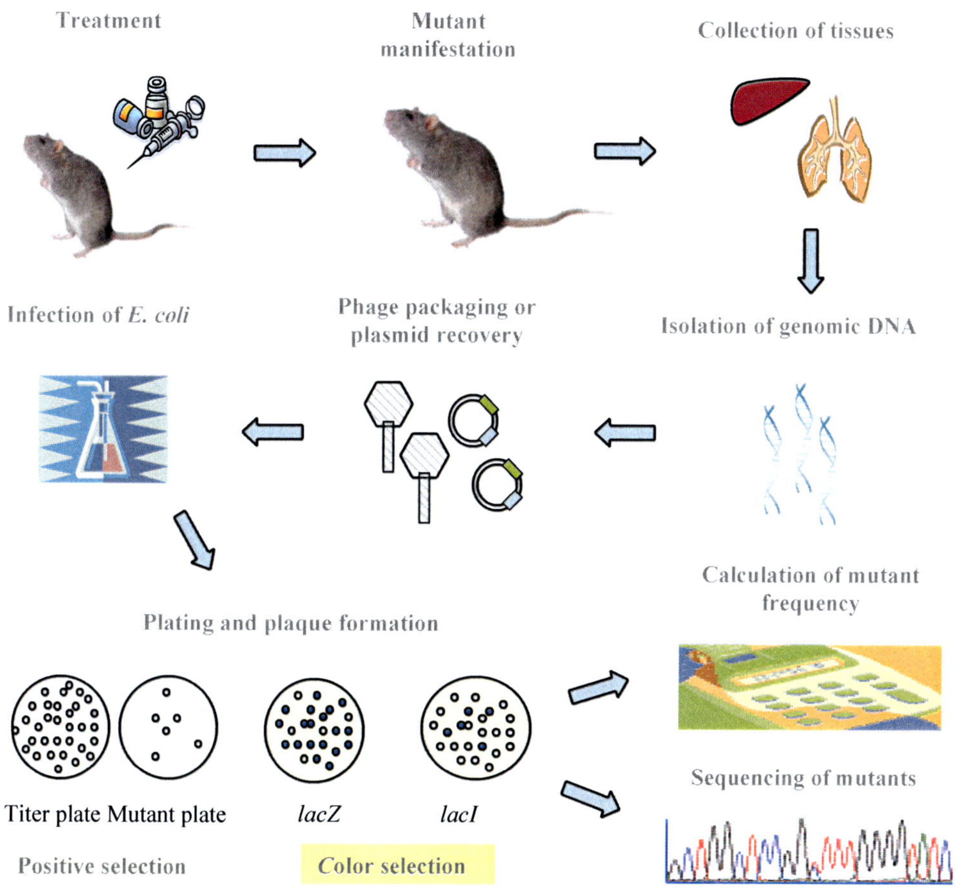

Figure 2. Scheme of procedures involved in transgenic rodent mutation assays.

Collection of Tissues

To the extent feasible, all potential target tissues from which reasonable amounts of DNA can be recovered should be collected. Alternatively, tissues can be chosen according to the purpose of the particular study. To minimize the in situ degradation of the DNA, each tissue should be dissected, wrapped, and flash-frozen in liquid nitrogen as rapidly as possible. Store the frozen tissues at –80°C. Frozen tissues are stable for at least one year and probably much longer.

Isolation of Genomic DNA

The isolation of high molecular weight DNA is extremely important for optimum phage packaging efficiency. Genomic DNA should be 100–300 kb or larger for efficient phage rescue from transgenic rodents containing λ shuttle vectors. The RecoverEase DNA isolation kit (Stratagene) is recommended for the isolation of high molecular genomic DNA. The extraction procedure involves isolation and digestion of nuclei in a Proteinase K solution followed by overnight dialysis to purify the DNA. The DNA isolated by this method is completely solubilized and ready for immediate use. For those tissues whose nuclei are not easily isolated, such as bladder and skin, a phenol–chloroform extraction method can be used. The DNA pellet precipitated by ethanol must be rehydrated for 1–4 days before it can be packaged efficiently. Genomic DNA can be stored in a TE buffer at 4°C, protected from light, for up to one month.

Phage Packaging or Plasmid Rescue

For λ shuttle vectors, the genomic DNA is mixed with a λ packaging extract and incubated at 30°C for 90 minutes. The vectors (full length λ DNA) are selectively excised from genomic DNA and a single copy is inserted into a λ phage head to produce an infectious phage particle. The packaging extract can be prepared following the instructions described previously [41]. It is also commercially available as Transpack (Stratagene). For plasmid vectors, the genomic DNA is enzymatically cut to release the monomeric plasmid sequences. The plasmid DNA is then purified from the genomic DNA, and circularized by ligation.

Infection of E. Coli, Plating, and Vector Selection

Specific bacterial *E. coli* strains are required for each mutation assay (see Table 1). For virus infection, the packaged phages are mixed with freshly cultured bacteria and the mixture is incubated at room temperature for 30 minutes to allow the phage to adsorb to the host cells. For plasmid assays, plasmids made from the genomic DNA are electroporated in *E. coli* deficient for the transgene. The infected bacteria are plated on specific medium according to the mutant selection or identification method (see next section, 2.3).

MF Calculation

The MF is the ratio of the number of mutant plaques (colonies) to the total number of plaques (colonies) screened.

Statistics and Criteria

It has been estimated that at least 200,000 transgenes per animal from at least 5 animals per treatment must be screened to detect a 2-fold increase in the MF [42]. If an appropriate statistical analysis is performed and there is a significant difference between the MFs from the treated and control groups, the agent is judged positive. If a suitable protocol was used and the appropriate tissues were sampled, a result that does not demonstrate a significant increase of MF indicates that the agent is negative.

Sequencing of Mutants

Generally, the sequencing of mutants is not necessary for testing drugs or chemicals for regulatory applications. In certain situations, however, the types of mutations obtained from sequencing of mutants may provide useful information. First, sequencing data can be used to correct MFs for clonal expansion of mutants and for the calculation of mutation frequencies. It has been suggested that sequencing up to 10 mutants per tissue from each animal is sufficient for simply identifying clonality, but that sequencing as many as 25 mutants may be necessary for correcting MFs for clonality [38]. Second, sequencing data can be used to clarify equivocal results. However, if a specific mutation does not result in a significant induction in MF, sequencing data are insufficient to conclude a positive response. Finally, sequencing data can be useful for understanding biological mechanisms underlying mutation induction by specific mutagens. In this case, a certain number of mutations are required for establishing a mutation spectrum. Although sequencing at least 50 mutants per dose group is recommended [38], the number of mutations necessary to produce an adequate spectrum is dependent upon the fold-increase of a MF over its negative control, the nature of the induced mutations, and the total number of target sites. The characterization of the induced mutational changes may lead to an understanding of the mechanism of mutation induction.

Mutant Identification and Selection Methods for Transgenic Mutation Assays

Although the methods for assaying mutation in the various transgenic mutation models are different, the available data suggest that there is significant agreement among these assays [38]. This section will be limited to methods for the *lacI* and *lacZ* color assays, and the *lacZ* and *cII* positive-selection assays because these systems are the most frequently used.

lacI Color Mutation Assay

The λLIZ shuttle vector (Figure 1) used in Big Blue mice and rats contains the *lacI* target gene and the α*lacZ* reporter gene. The α*lacZ* gene codes for the amino portion of β-galactosidase, which complements the ω*lacZ* gene provided by the *E. coli* host to produce full β-galactosidase activity. β-galactosidase activity can be measured in *E. coli* by plating on medium containing the chromogenic substrate X-gal (5-bromo-4-chloro-3-indolyl-β-D-galactopyranoside). The presence of β-galactosidase activity results in a blue colony or plaque, whereas the absence of activity results in a white colony or clear plaque. Mutations in the *lacI* gene are detected through the use of the *lacZ* reporter gene. Wild-type lac repressor protein binds to the *lacO* operator sequence, which prevents expression of the α*lacZ* gene. Infection of *E. coli* containing ω*lacZ* with λLIZ containing wild-type *lacI* results in the production of clear plaques. If a mutation inactivates *lacI*, the repressor will fail to bind to *lacO* and α*lacZ* will be expressed and cleave the X-gal which is incorporated in the agar medium resulting in a blue plaque. The ratio of the number of blue plaques to the number of total plaques is the MF [25, 27].

lacZ Color Mutation Assay

The λgt10*lacZ* shuttle vector is excised from genomic DNA and packaged into phage heads using a packaging extract. The resultant phages are absorbed onto *E. coli C* (*lacZ*) cells, and the infected bacteria are plated onto medium containing X-Gal. Bacteria infected by phage with a wild type *lacZ* gene will have β-galactosidase activity. The galactosidase protein cleaves X-gal and the plaques that eventually form become blue. Bacteria infected by phage with a mutated *lacZ* gene have a non-functional β-galactosidase, and the plaques that develop will remain clear. Thus, the *lacZ* MF is quantified as the ratio of clear or light blue plaques (mutants) to blue plaques (non-mutants) [28].

LacZ Positive-Selection Assay

Due to difficulties in the identification of clear plaques, a positive selection assay in which only mutant particles form plaques was developed using phenyl-beta-D-galactoside (P-gal) [43-45]. In this positive-selection system, *galE⁻lacZ E. coli* cells receiving a wild-type *lacZ* gene are unable to grow because the normal gene converts the P-gal into toxic UDP-galactose and results in cell death with no phage propagation. Only cells containing a *lacZ*-mutated phage are able to form plaques on selective plates with P-gal. The total number of phage-transfected bacteria is determined separately under non-selective conditions.

cII Positive Selection Assay

The λ phage *cII* gene encodes a repressor protein that controls the λ lysogenic/lytic cycle. Mutations in the *cII* gene can be positively selected on a bacterial strain that is deficient for hfl protease. In *hfl⁻ E. coli,* phages with active *cII* genes cannot enter a lytic cycle and only phages with a mutated inactive *cII* gene can make plaques. Therefore, the plating of the phage is performed with an *hfl⁻ E. coli* host strain. When incubated at 37°C, λ phage with both wild-type and mutated *cII* genes undergo the lytic cycle, resulting in plaque formation. For mutant selection, the packaged phage are mixed with the *hfl⁻ E. coli* host strain, plated, and incubated at 24°C for 42 hours. Under these conditions, λ phage with a wild-type *cII* gene undergo lysogenization and become part of the developing bacterial lawn, whereas phage carrying a mutant *cII* gene undergo a lytic cycle and give rise to plaques.

VALIDATION OF TRANSGENIC MUTATION ASSAYS

A basic assumption in using transgenic animals is that the transgenes reflect most of the important parameters affecting the mutation burden in native genes. There is, however, concern that transgenes may not respond to mutagens in a similar manner to endogenous genes. The prokaryotic nature of the transgenes may have affected DNA repair and thus altered the MF or even the mutation spectrum in transgenes compared to host genes. Some transgenes like the *lacI* gene contain a relatively high number of sites for endogenous methylation (CpG sites) compared to mammalian genes, and this may affect both spontaneous and induced mutagenesis [46, 47]. Transgenes are presumably non-expressed and thus may be repaired differently than transcribed endogenous genes [26, 46]. Finally, the different sizes of the target genes and phenotypic selection methods used for isolating transgene and endogenous gene mutants may contribute to the differences in sensitivity of the two systems [48, 49]. Differences between endogenous genes and transgenes do not preclude the utility of transgenic assays as general screening tools, but the differences should be identified and considered when interpreting results from transgenic rodent mutation assays.

In this section, we will present results from studies comparing spontaneous and induced mutation in transgenes and endogenous genes from transgenic rodents. We will also discuss the relationship between responses in transgenic mutation assays and carcinogenicity assays and the major advantages and disadvantages of transgenic rodent mutation assays.

Comparison of Induced and Spontaneous Mutations in Transgenes and Endogenous Genes in Transgenic Rodents

Several studies have evaluated transgenic rodent mutation assays by comparing the sensitivity and specificity of transgenic assays to assays based on endogenous genes [46, 47, 49-58]. Direct comparisons in transgenic animals indicate that despite differences in their mutational properties, acute doses of various model mutagens produce responses in transgenes (*lacI*, *lacZ*) that are qualitatively similar to responses in endogenous genes (*Dlb-1*, *Hprt*).

Table 2 summarizes results from studies that compare mutational responses in the *lacI* transgene and the endogenous *Hprt* gene of Big Blue mice and rats. Possible factors that may have affected the MFs observed in these studies are listed in Table 3. This summary does not include studies where no mutation sequence data is available (e.g., studies employing the *Dlb*-1 gene). Mutagens like *N*-ethyl-*N*-nitrosourea (ENU), 7,12-dimethylbenz[*a*]anthracene (DMBA), and thiotepa produce similar mutation spectra in both the lymphocyte *lacI* gene and *Hprt* gene. These data indicate that the types of mutations induced by mutagens in the *lacI* transgene generally reflect those observed in the native gene, at least for point mutations, frameshifts and small deletions and insertions. The *lacI* mutation spectra produced from mutagen-exposed Big Blue mice and rats are consistent with the proposed mutagenic mechanisms of these agents and correlate with the *in vivo* spectra observed in the lymphocyte *Hprt* gene.

The data summary in Table 2 indicates that the magnitude of the MFs induced by many carcinogens, as well as the spontaneous MF, differs for the *lacI* gene and the *Hprt* gene. Most of these differences can be understood by considering the nature of the DNA damage responsible for the mutations and by considering the nature of mutation induction in the transgenic and endogenous reporter genes (Table 3). The target for substitution mutation at G:C base pairs in the *lacI* gene is approximately twice that of the *Hprt* gene [46]. In addition, approximately 33% of mutable G:C sites in the *lacI* gene are part of CpG sequences, as compared with only 4% in the *Hprt* gene. Therefore, agents that specifically target CpG sites for mutation have an advantage for inducing mutations in the *lacI* gene relative to the *Hprt* gene. Agents that cause mutations mainly at A:T base pairs have similarly-sized targets for a mutation in the *lacI* and *Hprt* genes. However, the *Hprt* gene is expressed in most tissues, while the *lacI* gene is not. If promutagenic lesions induced by chemicals are actively repaired by transcription-coupled DNA repair (TCR), the target size of the *Hprt* gene relative to the *lacI* gene is reduced by a factor of approximately two. Finally, large genomic alterations such as deletions are recovered much more efficiently by endogenous genes like *Hprt* than by *lacI*.

Mutation spectra reveal the main targets of a mutagen and give an indication of the effects of TCR on promutagenic lesions. If a mutagen primarily produces point mutations at G:C base pairs in CpG sequences and is subject to TCR, the preceding analysis predicts that it would produce a higher MF in the *lacI* gene than in the *Hprt* gene. If a mutagen primarily causes large DNA alterations, it follows that it should produce a higher MF in the endogenous *Hprt* gene than in the *lacI* gene. With these considerations in mind, it is possible to account for the relative MFs produced in the *lacI* and *Hprt* genes in the various studies with Big Blue animals shown in Table 2.

Table 2. Comparisons between the frequencies and types of mutations induced in the *Hprt* and *lacI* genes of Big Blue rats and mice.

Mutagen	ENU [a]		DMBA [b]		B[a]P [c]		Thiotepa [d]		EO [e]		None [f]	
Model	Mouse		Rat		Mouse		Rat		Mouse		Rat	
Gene	*Hprt*	*lacI*	*Hprt*	*lacI*	*Hprt*	*lacI*	*Hprt*	*lacI*	*Hprt*	*lacI*	*Hprt*	*lacI*
Induced mutant frequency (x 10^{-6})	56	80	160	316	9	138	39	106	12	0	0	0
% base pair substitution at G:C sites	9	41	38	41	ND	92	75	88	36	72	75	80
% at CpG sites	0	13	3	6	ND	79	17	65	ND	44	22	70
% frameshifts/small deletions	18	4	3	13	ND	7	28 [g]	34	39	14	17	10
Strand bias of mutations	yes	no	yes	no	yes	no	yes	no	ND	ND	no	no

ND, not determined.

[a] Data from [54]. [b] Data from [57, 58]. The induced mutant frequencies shown are the highest determined. [c] Mutant frequency data from [55]. The induced mutant frequencies shown are the highest determined. The data for mutation types come from [103]. [d] Data from [46]. [e] Data from [62, 104]. [f] Data from [46, 58]. [g] Additionally there were 34% genomic deletions [46].

Table 3. Factors influencing the relative mutant frequencies in the *lacI* and *Hprt* genes of untreated and carcinogen-treated Big Blue mice and rats

Mutagen	Factors affecting mutant frequency					
	G:C target size	TCR	Global repair	Targeting to CpGs	Recovery of large deletions	*lacI/Hprt* [1] MF ratio
EO	-	ND	-	-	++++ *Hprt*	~0/1
ENU	+ *lacI*	++ *lacI*	+++ *Hprt*	-	-	~ 1/1
DMBA	+ *lacI*	++ *lacI*	-	-	-	~ 2/1
Thiotepa	+++ *lacI*	++ *lacI*	-	++ *lacI*	++ *Hprt*	~ 3/1
None	+++ *lacI*	++ *lacI*	-	+++ *lacI*	+ *Hprt*	~ 10/1
B[a]P	++++ *lacI*	++ *lacI*	-	++ *lacI*	++ *Hprt*	~15/1

[1] When mutagen employed, ratio calculated from induced mutant fractions.

ND, not determined; TCR, transcription-coupled DNA repair; MF, mutant frequency; -, no advantage; +, small advantage; ++, moderate advantage; +++, large advantage; ++++, very large advantage.

Benzo[*a*]pyrene (B[*a*]P) mutagenesis has all the characteristics necessary for high MFs in the *lacI* gene (Table 3). Ninety two percent of B[a]P mutations were substitutions at G:C base pairs, and 79% of these were at CpG sites. Although not demonstrated for the *Hprt* gene *in vivo*, other studies have repeatedly shown that the DNA damage produced by B[a]P is subject to TCR [59]. Given all these factors in favor of mutation induction in the *lacI* gene, it is not surprising that, of the mutagens thus far examined, B[a]P produced the highest frequency of *lacI* mutants relative to *Hprt* mutants, about 15/1.

Thiotepa also targets G:C base pairs for base pair substitution, although not to the extent that B[a]P does (Table 3). Its substitutions at G:C base pairs are targeted to CpG sites, but, again, not to the extent of B[a]P. The promutagenic DNA damage produced by thiotepa in the *Hprt* gene (but not the *lacI* gene) appears to be subject to TCR. All these factors favor mutation induction in the *lacI* gene. One factor favoring the induction of *Hprt* mutants over *lacI* mutants is the induction of a fraction of mutations in the *Hprt* gene due to large deletions that are not recovered by the *lacI* gene, and these events may account for about one-third of all *Hprt* mutations. On sum it would appear that mutation induction in the *lacI* gene is favored over the *Hprt* gene and finding a three-fold higher induced MF in the *lacI* gene of thiotepa-treated Big Blue rats is consistent with this conclusion [46].

Metabolites of DMBA bind to both dGs and dAs in rat lymphocyte DNA and cause mutations at both G:C and A:T base pairs (Table 3) [60, 61]. This mutagen does not target mutations to CpG sites (only 6%). Since only some of the mutations are at G:C base pairs, with no CpG targeting, this would produce only a small advantage for mutation induction in the *lacI* gene. The occurrence of mutated purines is biased to the non-transcribed DNA strand for *Hprt* mutations, but not *lacI* mutations, presumably because of TCR in the *Hprt* gene. This favors *lacI* mutagenesis [58]. The advantage for *lacI* mutations because of TCR in the *Hprt* gene and the partial advantage because of the fraction of mutations at G:C base pairs is consistent with DMBA producing about twice as many mutations in the *lacI* gene as the *Hprt* gene.

The lack of a detectable mutational response to ethylene oxide (EO) in the *lacI* gene of Big Blue mouse lymphocytes is likely due to the mechanism of EO-induced mutation [62]. Molecular characterization of EO-induced *HPRT* mutations in diploid human fibroblasts *in vitro* [63] indicates that as many as 50% of the mutations are due to large deletions, often involving the loss of the entire *HPRT* gene. Although a low level of point mutations is also found, the point mutations are not highly targeted to G:C base pairs or CpG sequences, as is the case with thiotepa. In addition, EO is an extremely weak mutagen for the *Hprt* gene *in vivo* (Table 2). Even if twice as many *lacI* mutations than *Hprt* mutations were induced by EO in Big Blue mice, the high background *lacI* MF would make it difficult to resolve such a low induced MF from the spontaneous MF. The same situation also has been found for another strong clastogen/weak point mutagen, cyclophosphamide [53]. These observations are consistent with the experimental finding indicating that EO is more mutagenic for the *Hprt* gene than the *lacI* gene.

The relatively equal mutagenicity of ENU in the *lacI* and *Hprt* genes (Table 2) is perhaps the most difficult response to explain. ENU is a strong base pair substitution mutagen, and it produces nearly all its substitutions at A:T base pairs in the mouse *Hprt* gene and a little more than half its base substitution at A:T in the *lacI* gene (Table 3). The relatively small proportion of G:C base pair substitutions among *lacI* mutations gives *lacI* only a small advantage for ENU mutagenesis. It is known that the initial number of O^{et}-dG and O^{et}-dT adducts formed by ENU is about equal, but that O^{et}-dT adducts are relatively resistant to 'global' repair in the mouse (and rat) *Hprt* gene, resulting in about 90% of ENU mutations being induced by O^{et}-dT adducts [64, 65]. It is clear, however, from the orientation of mutated dTs in the non-transcribed DNA strand that dT adducts leading to mutations in the *Hprt* gene are subject to TCR. In the Big Blue mouse *lacI* gene, the mutant spectra data suggest that O^{et}-dG and O^{et}-dT adducts produce almost equal numbers of mutations.

It can be hypothesized that this is due to dG and dT adducts being repaired equally well in the *lacI* gene. If this hypothesis is true, it follows that the inefficient 'global' repair of dT adducts in the *Hprt* gene may counterbalance the effects of TCR of those adducts and the small advantage to *lacI* given by the fraction of ENU mutations due to base pair substitution at G:C.

Finally, the spontaneous MF in the *lacI* gene is approximately one order of magnitude higher than in the *Hprt* gene (Table 3). Two factors may contribute to this difference, the much higher number of CpG sites in the *lacI* gene as compared to the *Hprt* gene and the absence of TCR in the non-expressed *lacI* gene. Most spontaneous mutations in the *lacI* gene of Big Blue rats and mice are located at CpG sequences (~70%, Table 2). It is known that non-expressed transgenes can be heavily methylated at CpG sequences [25]. The deamination of 5-methylcytosine residues located at CpG sequences result in G:C→A:T mutations, the major type of spontaneous *lacI* mutation found for Big Blue mice and rats. Thus, the MF in highly methylated CpG sequences is higher than in normal DNA sequences, and the 10-fold higher spontaneous MF in the *lacI* gene compared with the *Hprt* gene may be a consequence of this.

From above detailed analysis of mutations in the *lacI* gene and *Hprt* gene, it is clear that there are differences between the foreign gene and the native gene. For example, The *Hprt* gene is more sensitive to mutagens that produce large DNA alterations than the *lacI* gene. It is not realistic to expect that two different genes will react to mutagens in exactly the same manner; each target gene always will be different than other genes with respect to location in the genome, size, expression, and DNA sequence, whether or not they are endogenous genes or transgenes. The important conclusions from this comparison are that mutations induced by chemicals in the transgenes reflect the biological activity of the mutagens that induce them and that these mutations can be related to mutations produced in native genes. These observations support the usefulness of transgenic rodent mutation models for evaluating the tissue-specific mutagenicity of potential carcinogens.

Correlation of Transgenic Rodent Mutation Assays with Rodent Carcinogenicity Assays

One of the crucial questions for the use and acceptance of transgenic rodent systems in toxicological testing is whether or not these test systems are capable of predicting carcinogenic potential. A good concordance has been found between the results of carcinogenicity bioassays and transgenic rodent mutation assays [38, 66]. Most of the agents that were positive for carcinogenicity gave positive results in the transgenic rodent mutation assays, and most noncarcinogens gave negative results. Seventy-one out of 90 carcinogens gave at least one positive result in at least one transgenic mutation system. Fourteen out of 18 noncarcinogens evaluated in these assays produced negative results; the remaining four gave weak positive mutagenicity responses [38].

There are substances for which transgenic rodent mutation assays indicate evidence for a genotoxic mechanism of carcinogenicity, although conflicting or negative results were found in other genotoxicity assays. Good examples are found in assays conducted with tamoxifen and asbestos. Tamoxifen is a nonsteroidal antiestrogen that is widely used as an adjuvant therapy for breast cancer and is undergoing clinical trials as a chemopreventative agent in healthy women at risk of breast cancer [67]. This drug induces liver tumors in rats but is inactive in the Ames' *Salmonella typhimurium* gene mutation assay and other short-term genotoxicity tests [68, 69]. Treatment of Big Blue rats with tamoxifen, however, caused a significant, 3-fold increase in *lacI* MF in liver DNA and induced a distinct spectrum of mutations characterized by an increased proportion of G:C → T:A transversions [70, 71]. Thus, tamoxifen was confirmed both as a mutagen and a carcinogen in rat liver. Epidemiological evidence indicates that asbestos induces lung cancer in humans. However, the underlying mechanism of asbestos carcinogenesis is unknown; the carcinogenic activity of asbestos might be mediated via genotoxicity or by an epigenetic mechanism. A study was conducted to examine the mutagenic potential of asbestos in Big Blue transgenic rats [72]. Sixteen weeks after treatment with asbestos, the MF was significantly increased in the lung *lacI* gene of Big Blue rats. Therefore, asbestos was identified as a genotoxic carcinogen. Other experimental results indicate that persistent inflammation and cell proliferation produced by asbestos in lung possibly played a major role in the mutagenic effect.

Carcinogens with a presumed nongenotoxic mode of action should not be (strongly) mutagenic in target tissues. Transgenic rodent mutation models provide an approach by which the mechanism of action of these nongenotoxic carcinogens can be examined. Chloroform is a rodent carcinogen and induces kidney and liver tumors in mice and rats. A Big Blue mouse mutation assay has been conducted using a similar dose and route of administration as was used for assaying carcinogenicity. No mutation induction was detected in the liver of Big Blue mice treated with chloroform for up to 6 months. It was found that the same treatment induced liver necrosis and karyomegaly as well as regenerative cell proliferation [73]. This study confirms that chloroform is not a genotoxic carcinogen. 2,3,7,8,-Tetrachlorodibenzo-p-dioxin (TCDD) is a significant environmental contaminant resulting from industrial processes such as pulp and paper production. It is a multisite and multispecies carcinogen that induces liver tumors in mice and rats; it also is a suspected human carcinogen. Its mechanism of tumor induction, however, is not well understood and *in vitro* mutagenicity studies have yielded inconsistent results. Big Blue transgenic rats were used to assess the mutagenicity of TCDD in both male and female animals [74]. No increase in mutation frequency or any change in mutation spectrum was observed in the animals exposed to 2 μg/kg TCDD for 6 weeks. This study is consistent with the previous evaluation of TCDD as a nongenotoxic carcinogen [75].

Major Advantages and Disadvantages of Transgenic Rodent Mutation Assays

A major advantage of transgenic mouse/rat mutation assays compared to other *in vivo* mutagenicity tests, is that mutations can be detected in any organ, and that target organs of carcinogens may be evaluated. In most cases, mutations have been found in the target organs for tumor induction. Carcinogens with a presumed nongenotoxic mode of action generally produce negative results in transgenic rodent mutation assays.

However, the *lacI*, *lacZ* and *cII* systems detect primarily base pair substitutions, frame shifts, and small deletions and insertions. Although the Spi$^-$ and plasmid *lacZ* systems can detect relatively large deletions, mutations involving multiple genes, recombination, gene conversion, non-disjunction or translocation that can be detected by mutation assays based on endogenous autosomal genes like the *Tk* and *Aprt* genes cannot be recovered by transgenic rodent mutation assays. Agents like EO that primarily produce large-scale gene alterations will be detected with low efficiency. Transgenic mutation models should, therefore, be used with caution when evaluating agents suspected of causing large-scale genomic alterations. The sensitivity of transgenic mutation systems is another concern due to the generally high spontaneous MFs in transgenic rodent mutation systems. The higher the background in a mutation system, the more difficult it is to detect a significant induced frequency. Consequently, certain weak mutagens that produce a significant response in an endogenous gene like the *Hprt* gene may go undetected with a transgenic marker.

Use of Transgenic Rodent Mutation Assays

Regulatory Use of Transgenic Rodent Mutation Models for *in vivo* Genotoxicity Testing

Transgenic rodent mutation systems provide a unique opportunity to assay substances for their capacity to induce gene mutation. Unlike endogenous genes like oncogenes or tumor suppressor genes, the transgenes confer no selective growth advantage or disadvantage to mammalian cells by their presence or absence [46, 51, 76, 77]. The neutrality of the transgenes allows the accumulation and persistence of mutations *in vivo* and makes the selection of expression times for the assays somewhat easier. Furthermore, experience with these assays indicates that the results are reproducible and the assays are transferable between different laboratories [78, 79]. Therefore, the assays are suitable for testing chemical compounds for regulatory purposes and transgenic rodent mutation assays are beginning to be incorporated into current testing strategies.

Along with the IWGT efforts, the World Health Organization (WHO)/International Program on Chemical Safety (IPCS) is preparing an Environmental Health Criteria document on transgenic rodent mutation assays and their use in toxicity assessment (http://www.who.int/ipcs/publications/ehc/methodology_alphabetical/en/). Also, the Organization for Economic Co-operation and Development (OECD) is developing guidelines for use of transgenic rodent mutation assays. It is anticipated that transgenic rodent mutation assays may fill a gap in regulatory testing strategies and that these assays can result in fewer unresolved conflicts, and lead to lower costs and the use of fewer animals.

Current *in vivo* genotoxicity assays that are recommended by regulatory agencies are the chromosomal aberration, micronucleus and UDS assays. The chromosomal aberration and micronucleus assays typically use bone marrow for testing, while the UDS assay uses liver. These two tissues frequently are not the target tissues or are remote from the site of exposure.

It has been suggested that the UDS assay should be replaced by the transgenic rodent mutation assay [66]. The chromosomal aberration and micronucleus assays mainly detect genotoxic agents that have clastogenic characteristics while transgenic rodent mutation assays primarily detect gene mutations. Thus, these assays may complement each other in safety assessments.

Use of Transgenic Rodent Mutation Assays for Mechanistic Studies

Transgenic rodent mutation assays allow not only *in vivo* gene mutation testing but also mechanistic studies of *in vivo* mutagenesis and carcinogenesis. These assays have been widely used to determine mechanisms of action and to verify hypotheses. Primarily using data from our own laboratory, the following are several examples of the use of transgenic rodent mutation assays in mechanistic studies.

Determination of Relationships among DNA Adducts, Cell Proliferation and Mutagenesis

Muta mice were used to study the relationship between the conversion of DNA adducts into mutations and cell proliferation in the liver [80]. Animals were exposed to a single topical application of 10 or 90 mg/kg 5,9-dimethyldibenzo[c,g]carbazole (DMDBC). Both doses induced a similar level of persistent DNA adducts; however, MF was increased 2-fold by the low dose and 44-fold by the high dose. The low dose treatment did not change the proliferation rate of liver cells; however, the high dose treatment resulted in regenerative cell proliferation. The authors concluded that the cell proliferation induced by the high dose of DMDBC facilitated the conversion of DNA adducts into mutations. In a subsequent study to confirm this hypothesis [81], carbon tetrachloride and phenobarbital were used to induce regenerative cell proliferation after the 10 mg/kg dose (low dose) of DMDBC. This treatment protocol resulted in a greater than 15-fold increase in MF.

In a recent study [82], we treated neonatal and adult Big Blue mice with an equivalent amount of 4-aminobiphenoly (4-ABP) on a mg/kg body weight basis. The treatments produced similar levels of DNA adducts in neonatal and adult mice. In contrast, neonatal treatment with 4-ABP resulted in a significant increase of liver *cII* MF whereas the 4-ABP treatment did not increase the MF in adults. These observations suggest that the fixation of DNA damage into mutations is the major factor accounting for the sensitivity of neonatal mice to genotoxic carcinogens like 4-ABP. DNA replication is a prerequisite for DNA adduct-induced mutagenesis [83]. It has been reported that each neonatal liver cell divides an average of 3 times, whereas adult liver cells divide less than once every three months [84-86]. Therefore, the higher rate of cell proliferation in the neonatal liver compared to the adult liver is likely to potentate the mutagenic effects of 4-ABP. Rapid cell division makes a cell population more vulnerable to the genotoxic effects of chemicals and leaves less time for DNA repair prior to fixation of the damage as a mutation [87, 88].

Age-Related Sensitivity to Genotoxic Carcinogens

The incidence of childhood cancer is increasing and recent evidence suggests an association between childhood cancer and environmental exposure to genotoxins. Previously published analysis of carcinogenicity studies using genotoxic carcinogens indicate that juvenile animals are 3- to 10-fold more sensitive than adult animals for tumor induction in a wide range of tissues, including liver, lung, kidney, mammary tissue, the hematopoeitic system and nervous tissues [89]. To determine whether specific periods in early life represent windows of vulnerability to mutation induction by genotoxins, studies [90, 91] were conducted by treating Big Blue mice with ENU using a similar treatment protocol as has been used in the carcinogenesis studies [92, 93]. The cII mutation assay was performed to determine the MFs in the livers and brains of the mice.

In liver, cII MFs were dependent on the ages at which the animals were treated. Perinatal treatment with ENU induced relatively high MFs. However, ENU treatment of neonates resulted in the highest mutation induction. The lowest mutation induction occurred in animals treated as adults. In brain, a significant increase in MFs over the control was found in the prenatal and neonatal groups whereas there was no significant difference between the adult group and its control. Molecular analysis of the mutants also indicated that the mutational spectra from the ENU-treated mice were age-dependent. The percentage of A:T→T:A transversion, a typical mutation induced by ENU, was inversely related to the treatment age, whereas G:C→A:T transition, a typical spontaneous mutation was the predominant mutation in mice treated as adults.

The mutagenicity results indicate that infancy is a period when mouse liver is the most sensitive to mutation induction by genotoxins while the perinatal period is particularly susceptible to mutation induction in mouse brain. These periods correlate well with the age-dependent sensitivity to carcinogenicity in mouse liver and brain [92, 93], suggesting that mutation is an important rate-limiting factor for age-related carcinogenesis. Also the results suggest that rapid cell division in the earlier stages of development is a determining factor for the greater mutagenic sensitivity of early life. Therefore, exposures in early life should not simply be prorated over the entire life span when assessing the cancer risks of genotoxic chemicals.

Determination of Active Components for the Mutagenicity and Carcinogenicity of Mixtures

Genotoxic carcinogens appear to induce mutations with very specific patterns or spectra. An increase in mutations caused by a carcinogen is not just an increase in random genetic changes, but occurs in a pattern specific to the inducing agent. These spectra have been termed as the fingerprints of carcinogens [94]. Assuming an assay provides an accurate representation of these fingerprints, the molecular analysis of mutations in a reporter gene can provide important biological insights into carcinogenesis such as hotspots for mutagenesis and chemical mutational specificity. The transgenic animal models appear to be an good tool for evaluating mutational spectra because the transgenes are relatively easy to isolate and sequence.

Comfrey is a plant that has been used as a vegetable and a herbal remedy. It is, however, hepatotoxic and carcinogenic. There is little known about the mechanism of tumor induction by comfrey because comfrey is a mixture of many different substances and it is difficult to determine which substance(s) is responsible for the carcinogenesis. To examine the mutagenicity of comfrey and determine its underlying mechanisms of action, transgenic Big Blue rats were fed with comfrey diet using a protocol that results in liver tumors [95]. Comfrey treatment significantly increased the liver *cII* MF over the control, indicating that comfrey acts as genotoxic carcinogen in rat liver. Sequence analysis of mutants induced by comfrey indicated that comfrey produced a distinctive pattern of mutations that was similar to the pattern produced by riddelliine, a typical pyrrolizidine alkaloid (PA) [96] in rat liver. G:C → T:A transversion was the major type of mutation in comfrey-fed rats, whereas G:C → A:T transition was the predominant mutation in the controls. In addition, an unusually high frequency of tandem base substitutions (17%) was observed among the mutations from comfrey-fed rats. Tandem base substitution and G:C → T:A transversion have been suggested as a mutational signature for the genetic damage of PAs [96]. Therefore, these mutational data support the hypothesis that the mutations induced by comfrey in rat liver are due to PAs in the comfrey.

Determination of Active Metabolites

Many genotoxic chemicals are metabolized *in vivo* to DNA-reactive derivatives. Determination of active metabolites can provide evidence as to the likelihood of a particular metabolic pathway being associated with the biological activity of the agent. The transgenic rodent mutation assays can be used to identify the active metabolites. A good example of this is the studies on α-hydroxytamoxifen [97, 98]. The antiestrogen tamoxifen is widely used for the treatment of breast cancer and more recently for the prevention of breast cancer. A concern over the use of tamoxifen as a chemopreventive agent is its carcinogenicity in rat liver. To determine if α-hydroxylation is the major metabolic pathway for tamoxifen activation and carcinogenicity, mutation induction by tamoxifen and α-hydroxytamoxifen in the liver *cII* and *lacI* genes of Big Blue rats were determined and the types of mutations induced by α-hydroxytamoxifen in the liver *lacI* and *cII* genes were characterized and compared with those induced by tamoxifen [70, 71]. The results showed that α-hydroxytamoxifen is a much stronger mutagen than tamoxifen in rat liver. The induced MF produced by α-hydroxytamoxifen was 35-fold greater than that of tamoxifen in the *lacI* gene and 27-fold greater in the *cII* gene. Given the low rate of conversion of tamoxifen to α-hydroxytamoxifen [about 1%, [99]], it is not unexpected that administration of tamoxifen and α-hydroxytamoxifen at equimolar doses will result in the induction of far more mutations by α-hydroxytamoxifen. Molecular analysis of the *lacI* and *cII* mutants indicated that the α-hydroxytamoxifen-induced mutational spectra differed significantly from the control spectra, but were very similar to the spectra induced by tamoxifen. G:C → T:A transversion was the major mutation induced by α-hydroxytamoxifen and tamoxifen, while G:C → A:T transition was the main type of mutation in the control. These results support the hypothesis that α-hydroxytamoxifen is a major proximate tamoxifen metabolite causing the initiation of tumors in the liver of rats treated with tamoxifen.

CONCLUSION

Over 15 years of experience with transgenic *in vivo* mutation models indicate that they are relatively quick and statistically reliable assays for evaluating tissue specific gene mutation *in vivo* compared to other *in vivo* mutation assays. Validation studies have found that the mutations induced by chemicals in the transgenes reflect the biological activity of the mutagens and that these mutations can be related to mutations produced in native genes.

These models have the potential to become a part of genotoxicity evaluations to predict the carcinogenicity of environmental compounds. They are also powerful approaches to explore the various steps involved in spontaneous and induced mutagenesis and carcinogenesis.

REFERENCES

[1] Hennings H, Glick AB, Greenhalgh DA, Morgan DL, Strickland JE, Tennenbaum T, et al. Critical aspects of initiation, promotion, and progression in multistage epidermal carcinogenesis. *Proc Soc Exp Biol Med* 1993; 202(1): 1-8.

[2] Knudson AG. Antioncogenes and human cancer. Proc Natl Acad Sci U S A 1993; 90(23): 10914-21.

[3] Moolgavkar SH, Knudson AG, Jr. Mutation and cancer: a model for human carcinogenesis. *J Natl Cancer Inst* 1981; 66(6): 1037-52.

[4] Loeb LA. A mutator phenotype in cancer. Cancer Res 2001; 61(8): 3230-9.

[5] Ames BN, McCann J, Yamasaki E. Proceedings: carcinogens are mutagens: a simple test system. *Mutat Res* 1975; 33(1 Spec No): 27-8.

[6] Casciano DA. Introduction: historical perspectives of genetic toxicology. In: Li AP, Heflich RH, editors. *Genetic Toxicology*. Boca Raton: CRC; 1991. p. 1-12.

[7] Ames BN, McCann J, Yamasaki E. Methods for detecting carcinogens and mutagens with the *Salmonella*/mammalian-microsome mutagenicity test. *Mutat Res* 1975; 31(6): 347-64.

[8] McCann J, Choi E, Yamasaki E, Ames BN. Detection of carcinogens as mutagens in the Salmonella/microsome test: assay of 300 chemicals. *Proc Natl Acad Sci U S A* 1975; 72(12): 5135-9.

[9] Ruiz-Vazquez R, Pueyo C, Cerda-Olmedo E. A mutagen assay detecting forward mutations in an arabinose-sensitive strain of *Salmonella typhimurium*. *Mutat Res* 1978; 54(2): 121-9.

[10] Pueyo C. Forward mutations to arabinose resistance in *Salmonella typhimurium* strains: a sensitive assay for mutagenicity testing. *Mutat Res* 1978; 54(3): 311-21.

[11] Skopek TR, Liber HL, Krolewski JJ, Thilly WG. Quantitative forward mutation assay in *Salmonella typhimurium* using 8-azaguanine resistance as a genetic marker. *Proc Natl Acad Sci U S A* 1978; 75(1): 410-4.

[12] Hsie AW, Brimer PA, Mitchell TJ, Gosslee DG. The dose-response relationship for ultraviolet-light-induced mutations at the *hypoxanthine-guanine phosphoribosyl-transferase* locus in Chinese hamster ovary cells. *Somatic Cell Genet* 1975; 1(4): 383-9.

[13] Clive D, Johnson KO, Spector JF, Batson AG, Brown MM. Validation and characterization of the L5178Y/$TK^{+/-}$ mouse lymphoma mutagen assay system. *Mutat Res* 1979; 59(1): 61-108.

[14] Liber HL, Thilly WG. Mutation assay at the *thymidine kinase* locus in diploid human lymphoblasts. *Mutat Res* 1982; 94(2): 467-85.

[15] Bradley MO, Bhuyan B, Francis MC, Langenbach R, Peterson A, Huberman E. Mutagenesis by chemical agents in V79 Chinese hamster cells: a review and analysis of the literature. A report of the Gene-Tox Program. *Mutat Res* 1981; 87(2): 81-142.

[16] Tennant RW, Margolin BH, Shelby MD, Zeiger E, Haseman JK, Spalding J, et al. Prediction of chemical carcinogenicity in rodents from in vitro genetic toxicity assays. *Science* 1987; 236(4804): 933-41.

[17] Jones IM, Burkhart-Schultz K, Carrano AV. A method to quantify spontaneous and in vivo induced thioguanine-resistant mouse lymphocytes. *Mutat Res* 1985; 147(3): 97-105.

[18] Aidoo A, Lyn-Cook LE, Mittelstaedt RA, Heflich RH, Casciano DA. Induction of 6-thioguanine-resistant lymphocytes in Fischer 344 rats following *in vivo* exposure to *N*-ethyl-*N*-nitrosourea and cyclophosphamide. *Environ Mol Mutagen* 1991; 17(3): 141-51.

[19] Stambrook PJ, Shao C, Stockelman M, Boivin G, Engle SJ, Tischfield JA. *APRT*: a versatile *in vivo* resident reporter of local mutation and loss of heterozygosity. *Environ Mol Mutagen* 1996; 28(4): 471-82.

[20] Wijnhoven SW, Van Sloun PP, Kool HJ, Weeda G, Slater R, Lohman PH, et al. Carcinogen-induced loss of heterozygosity at the *Aprt* locus in somatic cells of the mouse. *Proc Natl Acad Sci U S A* 1998; 95(23): 13759-64.

[21] Dobrovolsky VN, Chen T, Heflich RH. Molecular analysis of *in vivo* mutations induced by *N*-ethyl-*N*- nitrosourea in the autosomal *Tk* and the X-linked *Hprt* genes of mouse lymphocytes. *Environ Mol Mutagen* 1999; 34(1): 30-8.

[22] Dobrovolsky VN, Casciano DA, Heflich RH. $Tk^{+/-}$ mouse model for detecting *in vivo* mutation in an endogenous, autosomal gene. *Mutat Res* 1999; 423(1-2): 125-36.

[23] Winton DJ, Blount MA, Ponder BA. A clonal marker induced by mutation in mouse intestinal epithelium. *Nature* 1988;333(6172): 463-6.

[24] Fahrig R. The mammalian spot test (Fellfleckentest) with mice. *Arch Toxicol* 1977; 38(1-2): 87-98.

[25] Kohler SW, Provost GS, Kretz PL, Fieck A, Sorge JA, Short JM. The use of transgenic mice for short-term, *in vivo* mutagenicity testing. *Genet Anal Tech Appl* 1990; 7(8): 212-8.

[26] Kohler SW, Provost GS, Fieck A, Kretz PL, Bullock WO, Putman DL, et al. Analysis of spontaneous and induced mutations in transgenic mice using a lambda ZAP/*lacI* shuttle vector. *Environ Mol Mutagen* 1991; 18(4): 316-21.

[27] Dycaico MJ, Provost GS, Kretz PL, Ransom SL, Moores JC, Short JM. The use of shuttle vectors for mutation analysis in transgenic mice and rats. *Mutat Res* 1994; 307(2): 461-78.

[28] Gossen JA, de Leeuw WJ, Tan CH, Zwarthoff EC, Berends F, Lohman PH, et al. Efficient rescue of integrated shuttle vectors from transgenic mice: a model for studying mutations *in vivo*. *Proc Natl Acad Sci U S A* 1989; 86(20): 7971-5.

[29] Hayashi H, Kondo H, Masumura K, Shindo Y, Nohmi T. Novel transgenic rat for *in vivo* genotoxicity assays using 6-thioguanine and Spi- selection. *Environ Mol Mutagen* 2003; 41(4): 253-9.

[30] Nohmi T, Katoh M, Suzuki H, Matsui M, Yamada M, Watanabe M, et al. A new transgenic mouse mutagenesis test system using Spi- and 6-thioguanine selections. *Environ Mol Mutagen* 1996; 28(4): 465-70.

[31] Burkhart JG, Burkhart BA, Sampson KS, Malling HV. ENU-induced mutagenesis at a single A:T base pair in transgenic mice containing phi X174. *Mutat Res* 1993; 292(1): 69-81.

[32] Malling HV, Delongchamp RR. Direct separation of *in vivo* and *in vitro* am3 revertants in transgenic mice carrying the phiX174 am3, cs70 vector. *Environ Mol Mutagen* 2001; 37(4): 345-55.

[33] Valentine CR, Montgomery BA, Miller SG, Delongchamp RR, Fane BA, Malling HV. Characterization of mutant spectra generated by a forward mutational assay for gene *A* of Phi X174 from ENU-treated transgenic mouse embryonic cell line PX-2. *Environ Mol Mutagen* 2002; 39(1): 55-68.

[34] Gossen JA, Martus HJ, Wei JY, Vijg J. Spontaneous and X-ray-induced deletion mutations in a *LacZ* plasmid-based transgenic mouse model. *Mutat Res* 1995; 331(1): 89-97.

[35] Matsuoka M, Nagawa F, Okazaki K, Kingsbury L, Yoshida K, Muller U, et al. Detection of somatic DNA recombination in the transgenic mouse brain. *Science* 1991; 254(5028): 81-6.

[36] Sykes PJ, Hooker AM, Harrington CS, Jacobs AK, Kingsbury L, Morley AA. Induction of somatic intrachromosomal recombination inversion events by cyclophosphamide in a transgenic mouse model. *Mutat Res* 1998; 397(2): 209-19.

[37] Heddle JA, Dean S, Nohmi T, Boerrigter M, Casciano D, Douglas GR, et al. *In vivo* transgenic mutation assays. *Environ Mol Mutagen* 2000; 35(3): 253-9.

[38] Thybaud V, Dean S, Nohmi T, de Boer J, Douglas GR, Glickman BW, et al. *In vivo* transgenic mutation assays. *Mutat Res* 2003; 540(2): 141-51.

[39] Wang J, Liu X, Heflich RH, Chen T. Time course of *cII* gene mutant manifestation in the liver, spleen, and bone marrow of *N*-ethyl-*N*-nitrosourea-treated Big Blue transgenic mice. *Toxicol Sci* 2004; 82(1): 124-8.

[40] Douglas GR, Jiao J, Gingerich JD, Gossen JA, Soper LM. Temporal and molecular characteristics of mutations induced by ethylnitrosourea in germ cells isolated from seminiferous tubules and in spermatozoa of *lacZ* transgenic mice. *Proc Natl Acad Sci U S A* 1995; 92(16): 7485-9.

[41] Gunther EJ, Murray NE, Glazer PM. High efficiency, restriction-deficient *in vitro* packaging extracts for bacteriophage lambda DNA using a new E.coli lysogen. *Nucleic Acids Res* 1993; 21(16): 3903-4.

[42] Gorelick NJ, Thompson ED. Overview of the workshops on statistical analysis of mutation data from transgenic mice. *Environ Mol Mutagen* 1994; 23(1): 12-6.

[43] Gossen JA, Molijn AC, Douglas GR, Vijg J. Application of galactose-sensitive *E. coli* strains as selective hosts for *LacZ-* plasmids. *Nucleic Acids Res* 1992; 20(12): 3254.

[44] Gossen JA, Vijg J. A selective system for *lacZ-* phage using a galactose-sensitive *E. coli* host. *Biotechniques* 1993; 14(3): 326, 330.

[45] Dean SW, Myhr B. Measurement of gene mutation *in vivo* using Muta Mouse and positive selection for *lacZ-* phage. *Mutagenesis* 1994; 9(3): 183-5.

[46] Chen T, Aidoo A, Manjanatha MG, Mittelstaedt RA, Shelton SD, Lyn-Cook LE, et al. Comparison of mutant frequencies and types of mutations induced by thiotepa in the endogenous *Hprt* gene and transgenic *lacI* gene of Big Blue rats. *Mutat Res* 1998; 403(1-2): 199-214.

[47] Chen T, Mittelstaedt RA, Aidoo A, Hamilton LP, Beland FA, Casciano DA, et al. Comparison of *hprt* and *lacI* mutant frequency with DNA adduct formation in *N*-hydroxy-2-acetylaminofluorene-treated Big Blue rats. *Environ Mol Mutagen* 2001; 37(3): 195-202.

[48] Skopek TR. Of mice and mutants: target size and sensitivity. *Mutat Res* 1995; 331(2): 225-8.

[49] Skopek TR, Kort KL, Marino DR. Relative sensitivity of the endogenous *hprt* gene and *lacI* transgene in ENU-treated Big Blue B6C3F1 mice. *Environ Mol Mutagen* 1995; 26(1): 9-15.

[50] Chen T, Mittelstaedt RA, Shelton SD, Dass SB, Manjanatha MG, Casciano DA, et al. Gene- and tissue-specificity of mutation in Big Blue rats treated with the hepatocarcinogen *N*-hydroxy-2-acetylaminofluorene. *Environ Mol Mutagen* 2001; 37(3): 203-14.

[51] Tao KS, Urlando C, Heddle JA. Comparison of somatic mutation in a transgenic versus host locus. *Proc Natl Acad Sci U S A* 1993; 90(22): 10681-5.

[52] Tao KS, Urlando C, Heddle JA. Mutagenicity of methyl methanesulfonate (MMS) in vivo at the *Dlb*-1 native locus and a *lacI* transgene. *Environ Mol Mutagen* 1993; 22(4): 293-6.

[53] Walker VE, Andrews JL, Upton PB, Skopek TR, deBoer JG, Walker DM, et al. Detection of cyclophosphamide-induced mutations at the *Hprt* but not the *lacI* locus in splenic lymphocytes of exposed mice. *Environ Mol Mutagen* 1999; 34(2-3): 167-81.

[54] Walker VE, Gorelick NJ, Andrews JL, Craft TR, deBoer JG, Glickman BW, et al. Frequency and spectrum of ethylnitrosourea-induced mutation at the *hprt* and *lacI* loci in splenic lymphocytes of exposed *lacI* transgenic mice. *Cancer Res* 1996; 56(20): 4654-61.

[55] Skopek TR, Kort KL, Marino DR, Mittal LV, Umbenhauer DR, Laws GM, et al. Mutagenic response of the endogenous *hprt* gene and *lacI* transgene in benzo[a]pyrene-treated Big Blue B6C3F1 mice. *Environ Mol Mutagen* 1996; 28(4): 376-84.

[56] Monroe JJ, Kort KL, Miller JE, Marino DR, Skopek TR. A comparative study of *in vivo* mutation assays: analysis of *hprt, lacI, cII/cI* and as mutational targets for *N*-nitroso-*N*-methylurea and benzo[*a*]pyrene in Big Blue mice. *Mutat Res* 1998; 421(1): 121-36.

[57] Manjanatha MG, Shelton SD, Aidoo A, Lyn-Cook LE, Casciano DA. Comparison of *in vivo* mutagenesis in the endogenous *Hprt* gene and the *lacI* transgene of Big Blue(R) rats treated with 7, 12- dimethylbenz[*a*]anthracene. *Mutat Res* 1998; 401(1-2): 165-78.

[58] Mittelstaedt RA, Manjanatha MG, Shelton SD, Lyn-Cook LE, Chen JB, Aidoo A, et al. Comparison of the types of mutations induced by 7,12- dimethylbenz[*a*]anthracene in the *lacI* and *hprt* genes of Big Blue rats. *Environ Mol Mutagen* 1998; 31(2): 149-56.

[59] Denissenko MF, Venkatachalam S, Yamasaki EF, Wani AA. Assessment of DNA damage and repair in specific genomic regions by quantitative immuno-coupled PCR. *Nucleic Acids Res* 1994; 22(12): 2351-9.

[60] Manjanatha MG, Lyn-Cook LE, Culp SJ, Beland FA, Heflich RH, Aidoo A. Lymphocyte mutant frequency in relation to DNA adduct formation in rats treated with tumorigenic doses of the mammary gland carcinogen 7,12- dimethylbenz[a]anthracene. *Mutat Res* 1996; 357(1-2): 89-96.

[61] Heflich RH, Mittelstaedt RA, Manjanatha MG, Lyn-Cook LE, Aidoo A. DNA sequence analysis of *hprt* mutations in lymphocytes from Sprague- Dawley rats treated with 7,12-dimethylbenz[*a*]anthracene. *Environ Mol Mutagen* 1996; 28(1): 5-12.

[62] Sisk SC, Pluta LJ, Meyer KG, Wong BC, Recio L. Assessment of the *in vivo* mutagenicity of ethylene oxide in the tissues of B6C3F1 *lacI* transgenic mice following inhalation exposure. *Mutat Res* 1997; 391(3): 153-64.
[63] Bastlova T, Andersson B, Lambert B, Kolman A. Molecular analysis of ethylene oxide-induced mutations at the *HPRT* locus in human diploid fibroblasts. *Mutat Res* 1993; 287(2): 283-92.
[64] Skopek TR, Walker VE, Cochrane JE, Craft TR, Cariello NF. Mutational spectrum at the *Hprt* locus in splenic T cells of B6C3F1 mice exposed to *N*-ethyl-*N*-nitrosourea. *Proc Natl Acad Sci U S A* 1992; 89(17): 7866-70.
[65] Mittelstaedt RA, Smith BA, Heflich RH. Analysis of *in vivo* mutation induced by *N*-ethyl-*N*-nitrosourea in the *hprt* gene of rat lymphocytes. *Environ Mol Mutagen* 1995; 26(4): 261-9.
[66] Dean SW, Brooks TM, Burlinson B, Mirsalis J, Myhr B, Recio L, et al. Transgenic mouse mutation assay systems can play an important role in regulatory mutagenicity testing *in vivo* for the detection of site-of-contact mutagens. *Mutagenesis* 1999; 14(1): 141-51.
[67] Jordan VC. Overview from the International Conference on Long-Term Tamoxifen Therapy for Breast Cancer. *J Natl Cancer Inst* 1992; 84(4): 231-4.
[68] Tucker M, Adam H, Patterson J. Tamoxifen. In: Laurence D, McLean A, Weatherall M, editors. *Safety Testing of New Drugs*. New York: Academic Press; 1984. p. 125-161.
[69] White IN, de Matteis F, Davies A, Smith LL, Crofton-Sleigh C, Venitt S, et al. Genotoxic potential of tamoxifen and analogues in female Fischer F344/n rats, DBA/2 and C57BL/6 mice and in human MCL-5 cells. *Carcinogenesis* 1992; 13(12): 2197-203.
[70] Davies R, Oreffo VI, Bayliss S, Dinh PA, Lilley KS, White IN, et al. Mutational spectra of tamoxifen-induced mutations in the livers of *lacI* transgenic rats. *Environ Mol Mutagen* 1996; 28(4): 430-3.
[71] Davies R, Oreffo VI, Martin EA, Festing MF, White IN, Smith LL, et al. Tamoxifen causes gene mutations in the livers of lambda/*lacI* transgenic rats. *Cancer Res* 1997; 57(7): 1288-93.
[72] Topinka J, Loli P, Georgiadis P, Dusinska M, Hurbankova M, Kovacikova Z, et al. Mutagenesis by asbestos in the lung of lambda-*lacI* transgenic rats. *Mutat Res* 2004; 553(1-2): 67-78.
[73] Butterworth BE, Templin MV, Constan AA, Sprankle CS, Wong BA, Pluta LJ, et al. Long-term mutagenicity studies with chloroform and dimethylnitrosamine in female lacI transgenic B6C3F1 mice. *Environ Mol Mutagen* 1998; 31(3): 248-56.
[74] Thornton AS, Oda Y, Stuart GR, Glickman BW, de Boer JG. Mutagenicity of TCDD in Big Blue transgenic rats. *Mutat Res* 2001; 478(1-2): 45-50.
[75] IARC. Polychorinated dibenzo-para-dioxins. In: Polycholrinated Dibenzo-para-dioxins and Polychlorinated Dibenxofurans. In: IARC Monographs on the evaluation of carcinogenic risks to humans. *Lyon: IARC*; 1997. p. 194-219.
[76] Cosentino L, Heddle JA. Differential mutation of transgenic and endogenous loci *in vivo*. *Mutat Res* 2000; 454(1-2): 1-10.
[77] Swiger RR, Cosentino L, Shima N, Bielas JH, Cruz-Munoz W, Heddle JA. The *cII* locus in the MutaMouse system. *Environ Mol Mutagen* 1999; 34(2-3): 201-7.

[78] Hachiya N, Yajima N, Hatakeyama S, Yuno K, Okada N, Umeda Y, et al. Induction of *lacZ* mutation by 7,12-dimethylbenz[*a*]anthracene in various tissues of transgenic mice. *Mutat Res* 1999; 444(2): 283-95.
[79] Suzuki T, Itoh S, Nakajima M, Hachiya N, Hara T. Target organ and time-course in the mutagenicity of five carcinogens in MutaMouse: a summary report of the second collaborative study of the transgenic mouse mutation assay by JEMS/MMS. *Mutat Res* 1999; 444(2): 259-68.
[80] Tombolan F, Renault D, Brault D, Guffroy M, Perin F, Thybaud V. Effect of mitogenic or regenerative cell proliferation on lacz mutant frequency in the liver of Muta^{TM}Mice treated with 5, 9-dimethyldibenzo[c,g]carbazole. *Carcinogenesis* 1999; 20(7): 1357-62.
[81] Tombolan F, Renault D, Brault D, Guffroy M, Perin-Roussel O, Perin F, et al. Kinetics of induction of DNA adducts, cell proliferation and gene mutations in the liver of MutaMice treated with 5,9-dimethyldibenzo[c,g]carbazole. *Carcinogenesis* 1999; 20(1): 125-32.
[82] Chen T, Mittelstaedt RA, Beland FA, Heflich RH, Moore MM, Parsons BL. 4-Aminobiphenyl induces liver DNA adducts in both neonatal and adult mice but induces liver mutations only in neonatal mice. *Int. J. Cancer* 2005; 117(2): 182-7.
[83] Bielas JH, Heddle JA. From the cover: proliferation is necessary for both repair and mutation in transgenic mouse cells. *Proc Natl Acad Sci U S A* 2000; 97(21): 11391-6.
[84] Schultze B, Kellerer AM, Grossmann C, Maurer W. Growth fraction and cycle duration of hepatocytes in the three-week-old rat. *Cell Tissue Kinet* 1978; 11(3): 241-9.
[85] Cameron IL. *Cell proliferation and renewal in the mammalian body.* In: Cameron IL, Thrasher JD, editors. Cellular and Molecular Renewal in the Mammalian Body. New York: Academic Press; 1971. p. 45-85.
[86] Buetow DH. Cell numbers vs. age in mammalian tissues and organs. In: Cristofalo HV, editor. *CRC Handbook of Cell Biology of Aging*. Boca Raton, FL: CRC Press; 1985. p. 1-115.
[87] Bertram JS, Heidelberger C. Cell cycle dependency of oncogenic transformation induced by *N*-methyl-*N'*-nitro-*N*-nitrosoquanidine in culture. *Cancer Res* 1974; 34(3): 526-37.
[88] Rabes HM, Muller L, Hartmann A, Kerler R, Schuster C. Cell cycle-dependent initiation of adenosine triphosphatase-deficient populations in adult rat liver by a single dose of *N*-methyl-*N*-nitrosourea. *Cancer Res* 1986; 46(2): 645-50.
[89] Ginsberg GL. Assessing cancer risks from short-term exposures in children. *Risk Anal* 2003; 23(1): 19-34.
[90] Slikker W, 3rd, Mei N, Chen T. *N*-ethyl-*N*-nitrosourea (ENU) increased brain mutations in prenatal and neonatal mice but not in the adults. *Toxicol Sci* 2004; 81(1): 112-20.
[91] Mei N, Heflich RH, Moore MM, Chen T. Age-dependent sensitivity of Big Blue transgenic mice to the mutagenicity of *N*-ethyl-*N*-nitrosourea (ENU) in liver. *Mutat Res* 2005; 572(1-2): 14-26.
[92] Vesselinovitch SD, Rao KV, Mihailovich N. Neoplastic response of mouse tissues during perinatal age periods and its significance in chemical carcinogenesis. *Natl Cancer Inst Monogr* 1979(51):239-50.
[93] Vesselinovitch SD, Rao KV, Mihailovich N, Rice JM, Lombard LS. Development of broad spectrum of tumors by ethylnitrosourea in mice and the modifying role of age, sex, and strain. *Cancer Res* 1974; 34(10): 2530-8.

[94] Vogelstein B, Kinzler KW. Carcinogens leave fingerprints. *Nature* 1992; 355(6357): 209-10.

[95] Mei N, Guo L, Fu PP, Heflich RH, Chen T. Mutagenicity of comfrey (Symphytum Officinale) in rat liver. *Br J Cancer* 2005; 92(5): 873-5.

[96] Mei N, Heflich RH, Chou MW, Chen T. Mutations induced by the carcinogenic pyrrolizidine alkaloid riddelliine in the liver *cII* gene of transgenic big blue rats. *Chem Res Toxicol* 2004; 17(6): 814-8.

[97] Gamboa da Costa G, Manjanatha MG, Matilde Marques M, Beland FA. Induction of *lacI* mutations in Big Blue rats treated with tamoxifen and alpha-hydroxytamoxifen. *Cancer Lett* 2002; 176(1): 37-45.

[98] Chen T, Gamboa da Costa G, Marques MM, Shelton SD, Beland FA, Manjanatha MG. Mutations induced by alpha-hydroxytamoxifen in the *lacI* and *cII* genes of Big Blue transgenic rats. *Carcinogenesis* 2002; 23(10): 1751-7.

[99] White IN, Carthew P, Davies R, Styles J, Brown K, Brown JE, et al. Short-term dosing of alpha-hydroxytamoxifen results in DNA damage but does not lead to liver tumours in female Wistar/Han rats. *Carcinogenesis* 2001; 22(4): 553-7.

[100] Jakubczak JL, Merlino G, French JE, Muller WJ, Paul B, Adhya S, et al. Analysis of genetic instability during mammary tumor progression using a novel selection-based assay for *in vivo* mutations in a bacteriophage lambda transgene target. *Proc Natl Acad Sci U S A* 1996; 93(17): 9073-8.

[101] Leach EG, Narayanan L, Havre PA, Gunther EJ, Yeasky TM, Glazer PM. Tissue specificity of spontaneous point mutations in lambda supF transgenic mice. *Environ Mol Mutagen* 1996; 28(4): 459-64.

[102] Gondo Y, Shioyama Y, Nakao K, Katsuki M. A novel positive detection system of *in vivo* mutations in rpsL (strA) transgenic mice. *Mutat Res* 1996; 360(1): 1-14.

[103] Shane BS, de Boer J, Watson DE, Haseman JK, Glickman BW, Tindall KR. *LacI* mutation spectra following benzo[*a*]pyrene treatment of Big Blue mice. *Carcinogenesis* 2000; 21(4): 715-25.

[104] Walker VE, Sisk SC, Upton PB, Wong BA, Recio L. *In vivo* mutagenicity of ethylene oxide at the *hprt* locus in T-lymphocytes of B6C3F1 *lacI* transgenic mice following inhalation exposure. *Mutat Res* 1997; 392(3): 211-22.

In: New Developments in Mutation Research
Editor: Charles L. Valon, pp. 97-131
ISBN 1-59454-664-9

Chapter 5

THE ROLE OF OXYGEN FREE RADICALS IN DNA DAMAGE AND MUTATION INDUCTION

Metka Filipič and Bojana Žegura
National Institute of Biology, Ljubljana, Slovenia

ABSTRACT

Reactive oxygen species (ROS) are constantly being generated in all living organisms. Their origin may be endogenous, such as those arising from normal aerobic metabolism, or exogenous, produced as a consequence of exposure to environmental chemicals or irradiation. ROS play an essential function in performing inter- and intracellular signaling but, when they are in excess, their effects are deleterious as they modify cellular macromolecules, accumulation of which has been associated with numerous diseases. In DNA, ROS induce different types of base- and sugar-derived DNA lesions, of which the guanine modification, 7,8-dihydro-8-oxo-2′-deoxyguanosine (8-OH-dG) is best studied and is also most often used as a biomarker of oxidative DNA damage. If not repaired, these lesions can have a number of deleterious consequences, including induction of base pair mutations, rearrangements, deletions, insertions and sequence amplification. In addition to point mutations, ROS can produce gross chromosomal alterations, which may lead to the inactivation of tumor suppressor genes or expression of mutated proto-oncogenes during tumor promotion and progression.

DNA repair mechanisms are extremely important for maintaining DNA integrity. Numerous pathways are involved in the repair of oxidative DNA damage, however, we still do not know which pathways and proteins are responsible for removing specific types of oxidative DNA lesions.

There is evidence that many carcinogens and/or their metabolites, apart from inducing DNA damage by direct bonding to DNA, also induce ROS formation and oxidative DNA damage. However, when detecting mutagenic effects it is generally difficult to distinguish the contribution of direct from that of ROS mediated DNA damage. Compounds of arsenic, cadmium, chromium (VI), cobalt and nickel are all classified as human carcinogens. They have all been shown to induce intracellular ROS formation and oxidative DNA damage. They are also known to interfere with DNA repair mechanisms. Therefore the carcinogenicity of metals can, in part, be explained by inducing oxidative DNA damage and by decreasing the repair capacity towards DNA lesions, which may in turn increase the risk of tumor formation. Carcinogenesis is a multistep process requiring accumulation of changes through tumor initiation, promotion and progression, which include DNA mutation, activation of oncogenes and inactivation of tumor suppressor genes. Oxidative mechanisms have been shown to be involved in all these steps. The data from human studies obtained so far support the experimentally based indication of oxidative DNA damage as an important mutagenic and apparently carcinogenic factor. However, proof of a causal relationship in humans is still lacking. In the future the use of biomarkers may provide this evidence and allow further investigation of the qualitative and quantitative importance of oxidative DNA modification and carcinogenesis in humans, and also suggest possible preventive measures.

Introduction

Oxygen is an essential element of life, but can also participate in the destruction of tissue or impair its ability to function normally (Halliwell and Gutteridge, 1989; Kehrer, 1993). Oxidants are generated from normal cellular functions, including mitochondrial metabolism, neutrophil activation and activation of mixed function oxidation pathways. These processes generate free radicals such as superoxide anions and hydroxyl radicals, and non-radical species such as hydrogen peroxide, ozone and peroxynitrite. It has been estimated that the average person has around 10,000–20,000 free radicals attacking each body cell each day. Reactive oxygen species (ROS) are also generated by external factors, including visible and ultraviolet light, ionizing radiation and ozone. Heavy metals, man-made pollutants such as automobile emissions and chemical oxidants in air, water and the food chain, as well as life style (e.g., smoking, chronic excessive exercise) also contribute to the oxidant challenge.

Reactive oxygen species are transient and, due to their high chemical reactivity, prone to cause oxidative damage to cellular macromolecules such as lipids, proteins and DNA (Mates, 2000). In reactions with DNA, ROS induce DNA strand breaks, DNA-protein cross links, oxidation of purines and pyrimidines, and abasic (AP) sites (Cadet et al., 1999; Nordberg and Arner, 2001). Some of these DNA lesions are promutagenic and can induce point mutations, deletions, insertions, frameshifts, chromosomal aberration, sequence amplification and rearrangements. Depending on the type of lesion, the consequence of the persistence of the lesion in DNA is mutation or cell death. For this reason mammalian cells have multiple systems to prevent damage and, if the damage occurs, to remove the lesion. Prevention of damage comprises non-enzymatic (e.g. glutathione, thioredoxin) and enzymatic (e.g. superoxide dismutase, catalase, glutathione reductase, glutathione peroxidase) antioxidants (Sies, 1991; review in Wang et al., 1998; Evans et al., 2004), while removal of damage is achieved by the complex DNA repair systems.

DNA mutation is the crucial step in carcinogenesis. The fact that elevated levels of oxidative DNA damage have been detected in many tumors strongly implicates the involvement of ROS in the etiology of cancer.

ETIOLOGY OF REACTIVE OXYGEN RADICALS (ROS) (FIGURE 1)

Endogenous Sources of ROS

The major endogenous source of ROS is the electron leakage that occurs in the electron transport chain in mitochondria and in the endoplasmic reticulum. These leaking electrons interact with oxygen to produce superoxide ($O_2\bullet^-$) radicals. Under physiological conditions, about 1-3 % of the oxygen molecules in the mitochondria are converted into superoxide (Fridovich, 1986).

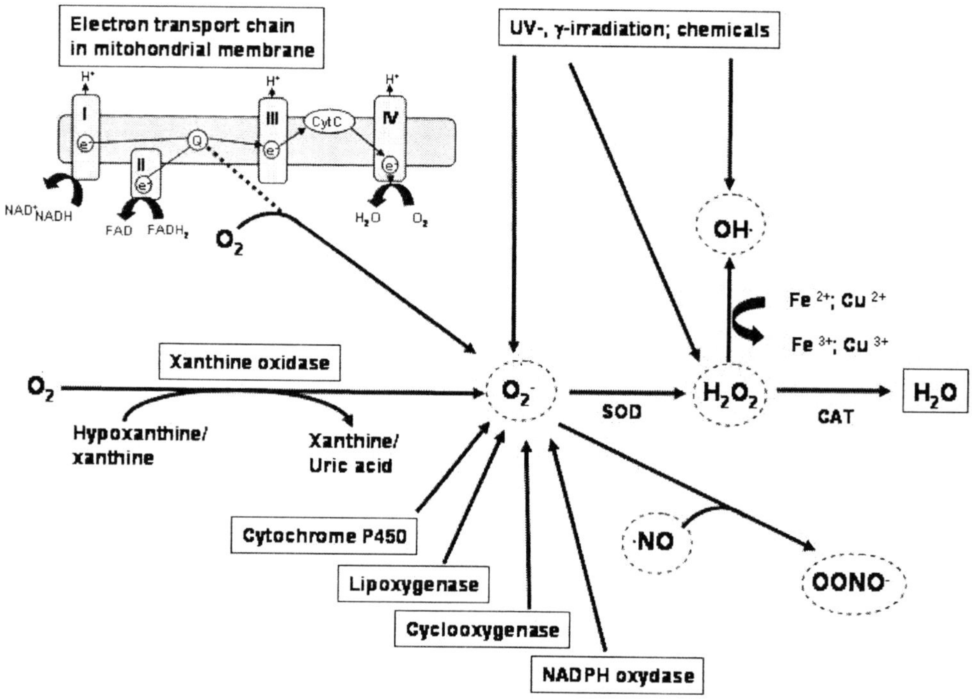

Figure 1. Pathways of ROS formation. O_2^- is generated by leakage of electrons from mitohondria, by NADPH cytochrome P450 reductase,, hypoxanthine/xanthine oksydase, NADPH oxidase, lipoxigenase and cyclooxygenase. Superoxide dismutase converts O_2^- into H_2O_2, which is mostly converted to H_2O by glutathione (GSH) peroxidase and catalase. Highly reactive •OH radicals are produced in the presence of Fe or Cu by Fenton or Haber-Weiss reaction. ROS are generated also by ionising and UV irradiation, exposure to chemicals and heavy metals.

Superoxide ($O_2\bullet^-$) is also generated by other enzymes, including cytochrome P-450 (Butler and Hoey, 1993), xanthine oxidase, NADPH oxidase, cyclooxygenase (Figure 1) and lipoxygenase. Cytochrome P-450 catalyzes oxidation and hydroxylation reactions during detoxification of xenobiotics and other toxic compounds. During these reactions electrons may be 'leaked' onto oxygen molecules, forming superoxide radicals (O2•-) (Butler and Hoey, 1993). Xanthine oxidase (XO) is a member of a group of enzymes known as molybdenum iron–sulfur flavin hydroxylases and catalyses the hydroxylation of purines (Harrison, 2002;

Borges et al., 2002), in particular hypoxanthine to xanthine and xanthine to uric acid. In both steps, molecular oxygen is reduced, forming the superoxide anion. The NADPH oxidases are a group of plasma membrane associated enzymes found in a variety of cells of mesodermal origin. The most thoroughly studied is the leukocyte NADPH oxidase, which is found in phagocytes and B-lymphocytes. When phagocytic cells, i.e. the neutrophils, are exposed to a stimulus, recognition of the foreign particle triggers a series of reactions called the *respiratory burst* (Borges et al., 2002). The respiratory burst enables the cell to provide oxidizing agents for the destruction of the target cells. When NADPH oxidase is activated, it takes NADPH from the cytoplasm and passes electrons to O_2, producing superoxide within the plasma membrane or on its outer surface (Vignais, 2002; Van Heerebeek et al., 2002). Myeloperoxidase converts peroxide produced in this system to microbiocidal products, probably hypochlorite (Proctor, 1989). Production of activated products by this system is thought to play a key role in cell-mediated immunity and microbiocidal activity. In non-inflammatory cells, NADPH oxidase may have a role in mediating cyclic nucleotide metabolism (Proctor, 1989).

In the reaction catalyzed by superoxide dismutase two molecules of superoxide rapidly dismutate to hydrogen peroxide and molecular oxygen. H_2O_2 is not a free radical, but is very important as it can penetrate biological membranes, in contrast to superoxide radical. It plays an important role as an intermediate in the production of the more reactive ROS-like hypochlorous acid (HOCl) and hydroxyl radicals ($^{\cdot}OH$). The latter are formed in a reaction mediated by transition metal ions (Fe^{+} or Cu^{+}) through Haber-Weiss and Fenton reactions (Nordberg and Arner, 2001).

Nitric oxide does not react with most biomolecules, however, it readily reacts with other free radicals. With the superoxide radical, a highly cytotoxic (Beckman and Koppenol, 1996) and extremely reactive oxidizing and nitrating agent (Richeson et al., 1998), peroxinitrite ($OONO^{-}$), is formed. In addition, NO can cross the membranes and act as a signaling molecule between cells (Nordberg and Arner, 2001).

Exogenous Sources of ROS

Numerous exogenous factors contribute to increased ROS production. Exposure to ionizing radiation can damage DNA directly, but the predominant pathway arises from radiolysis of H_2O, which results in the formation of ROS that can react with DNA. It is noteworthy that most of our knowledge regarding the interaction of ROS with DNA originates from studies of ionizing radiation.

There is evidence that many carcinogens and/or their metabolites, apart from inducing DNA damage by direct bonding to DNA, also induce ROS formation and oxidative DNA damage. Metabolites of many organic carcinogens cause ROS formation by NADH-dependent redox cycles. One of the most important intermediates of ROS formation are quinones, intermediates formed during the metabolism of many organic compounds, including carcinogens such as benzene, polyaromatic hydrocarbons, estrogens and catecholamines. Quinones can induce DNA damage through alkylation, however they are also highly redox reactive molecules, which can redox cycle with their semiquinone radicals, leading to formation of ROS including hydroperoxide, hydrogen peroxide, and ultimately hydroxyl radical (Bolton at al., 2000). Polynuclear compounds, such as 4-aminobiphenyl and heterocyclic amines, appear to induce cancer mainly through DNA adduct formation, although their *N*-hydroxy and nitroso metabolites can also cause oxidative DNA damage. On the other hand, mononuclear compounds, such as benzene metabolites, caffeic acid, and *o*-toluidine, appear to express their carcinogenicity through oxidative DNA damage (Kawanishi et al., 2002).

The list of metals whose compounds are definitely carcinogenic to humans includes As, Be, Cd, Cr and Ni and there are more candidates emerging from current research, including Co, Cu, Fe and Pt (Kasprzak and Buzard, 2000). The concept of ROS mediated metal carcinogenesis proposes that complexes formed by some of the above metals (Co, Cr, Cu, Fe, Ni) *in vivo* in the vicinity of DNA catalyze redox reactions whose products oxidize DNA. Metals that do not catalyze redox reactions, like arsenic and cadmium, have also been shown to induce ROS mediated gene mutations (Hei and Filipič, 2004, Filipič and Hei, 2004).

Oxidative DNA Damage

Oxidative damage to DNA includes a range of oxidized purines and pyrimidines, DNA cross links, alkali labile sites and oligonucleotide strand breaks. The latter two are formed directly by ROS attack or by the DNA repair processes (Dizdaroglu, 1991, 1994; Breen and Murphy, 1995; Cadet et al., 2000). Oxidative DNA damage can be formed through a direct insult on nucleotides and C-H bonds of 2'-deoxyribose in the DNA helix or through incorporation of damaged nucleoside triphosphates during replication (Moller and Wallin, 1998; Salles et al., 1999), such as incorporation of 8-oxodGTP (Tajiri et al., 1995).

The primary reactions of •OH with DNA are hydrogen abstraction from deoxyribose and addition to the π bonds of DNA bases (Breen and Murphy 1995). Analysis of the partitioning of damage in polymeric structures indicates that the reaction with the bases is favored by 3- to 4-fold over reaction with sugars (von Sonntag, 1987). Hydrogen atom abstraction from the deoxyribose moieties of DNA by •OH can lead to base loss and single-strand breaks due to fragmentation of the sugar (von Sonntag, 1987, Henner et al., 1983). Oxygen radical damage to the four nucleobases is highly complex, with many distinct lesions formed from each nucleobase (Table 1) (von Sonntag, 1987). *In vitro* studies of the reactions of oxidants (Douki et al., 1996; Luo et al., 1996; Henle et al., 1996) and ionizing radiation (von Sonntag, 1987) with individual nucleosides or nucleotides, as well as with polynucleotides, have enabled the identification of many oxidized base derivatives. At least 35 different oxidative base modifications have been identified. The chemical structures of the most important oxidized DNA bases, together with their biological effects, are shown in Table 1. A more comprehensive review of chemical structures of oxidized DNA can be found in Dizdaroglu (1991; 1992).

Oxidized Purines

Of the four nucleic acid bases, guanine has the lowest oxidation potential and is therefore the most easily oxidized (Kovacic and Wakelin, 2001). 7,8-dihydro-8-oxoguanine, more commonly named 8-oxoguanine (8oxoG), and its tautomeric form 8-hydroxyguanine (8-OH-

G), are the most extensively studied and appear to play a major role in mutagenesis and carcinogenesis. This lesion does not block DNA synthesis. It is, however, a highly mutagenic lesion (Floyd, 1990), because it can mispair with A leading to a GC→TA transversion at the site of modification (Grollman and Moriya, 1993), unless repaired prior to replication (Cheng et al., 1992). The mutation frequency of 8-oxoG in mammalian cells is 2,5-4,8 % (Moriya, 1993). If oxidatively modified, dGTP (8-oxo-dG) is misincorporated into DNA, it can lead to both AT→CG and GC→TA transversions (Tajiri et al., 1995; Wang et al., 1998).

Table 1. Oxidized nucleic bases in DNA, their mutagenic and lethal potential and proteins involved in their repair

Oxidized nucleic base in DNA	Biological effects		Repair proteins	
	Lethal	Mutagenic	E. coli	Human
Oxidized guanines				
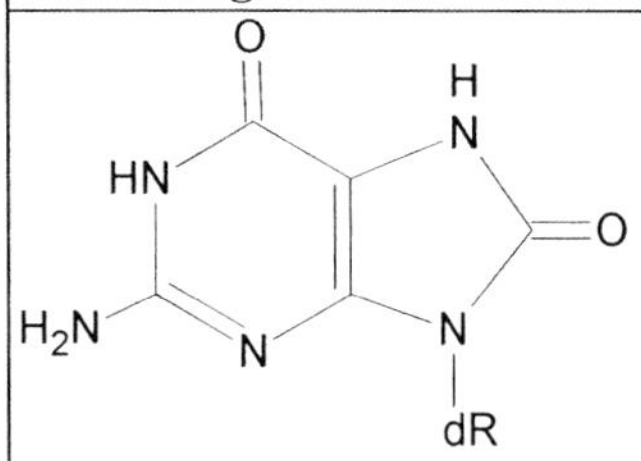 7,8-dihydro-8-oxoguanine (oxo8G)	No	Yes: G → T	Fpg, Nei, MutY, UvrABC	hOGG1, hNEIL1, hNEIL2, hMPG, hMYH, hMMR, hNER,
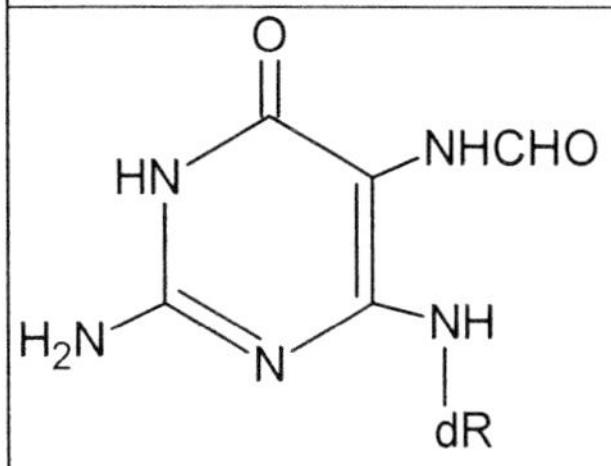 2,6-diamino-4-hydroxy-5-formamidopyrimidine (Fapy-G)	Yes: Blocking replication	No	Fpg, MutY	hNEIL1, hNEIL3
Oxidized adenines				
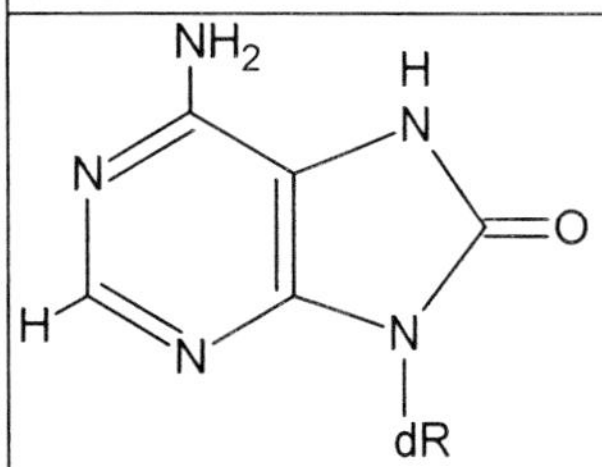 7,8-dihydro-8-oxoadenine (oxo8A)	No	Poor: A → G	Fpg, Nei, MutY, Xth	hOGG1
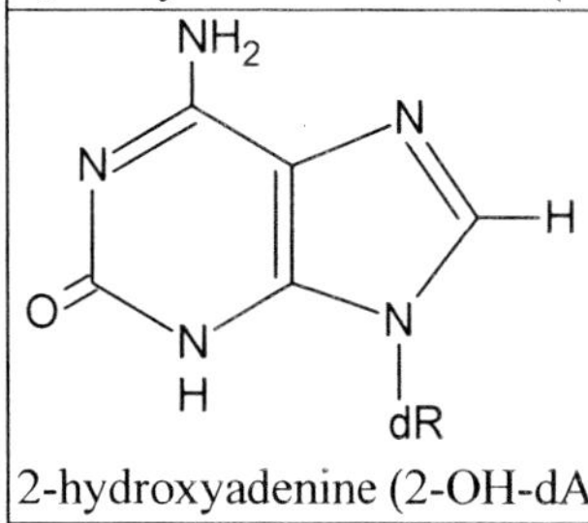 2-hydroxyadenine (2-OH-dA)	No	Yes: A → T A → G	MutY	hMYH

Table 1. Oxidized nucleic bases in DNA, their mutagenic and lethal potential and proteins involved in their repair (continued)

Oxidized nucleic base in DNA	Biological effects		Repair proteins	
	Lethal	Mutagenic	E. coli	Human
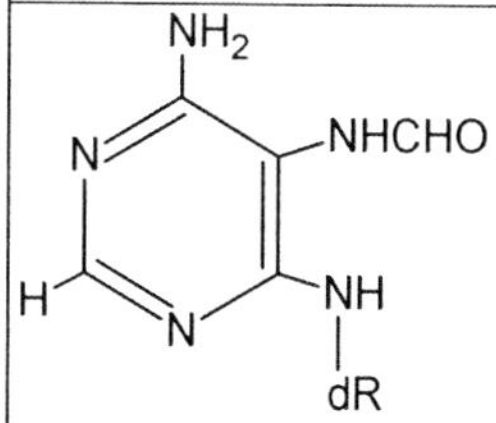 4,6-diamino-5-formamidopyrimidine (Fapy-A)	Yes: Blocking replication	Poor A → T	Fpg, Nth, Nei, DenV	hNEIL1
Oxidized thymines				
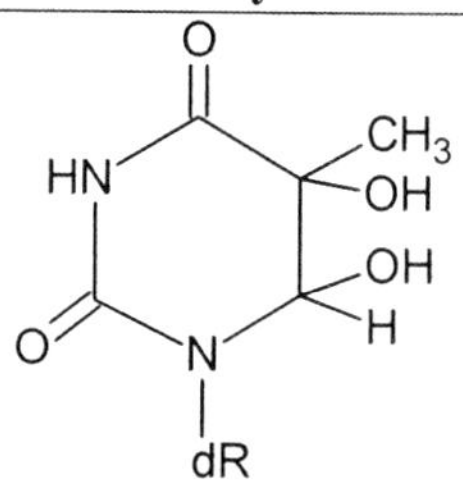 Thymine glycol (Tg)	Yes: Blocking replication	Poor: T → C	Nth, Nei, Fpg, UvrABC	hNTH1, ,hNEIL1, hNEIL2, hNER
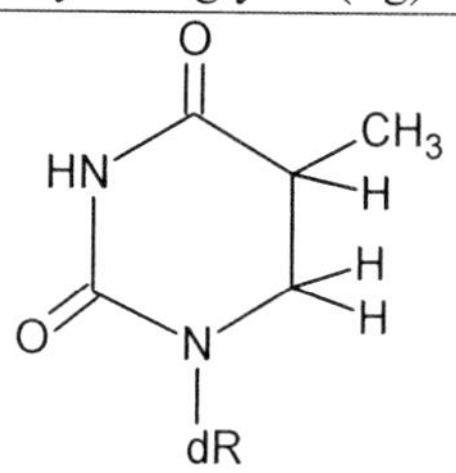 5,6-dihydrothymine (dHT)	No	No	Nth, Nei, Fpg, Nfo	hNTH1, hNEIL1, hNEIL2,
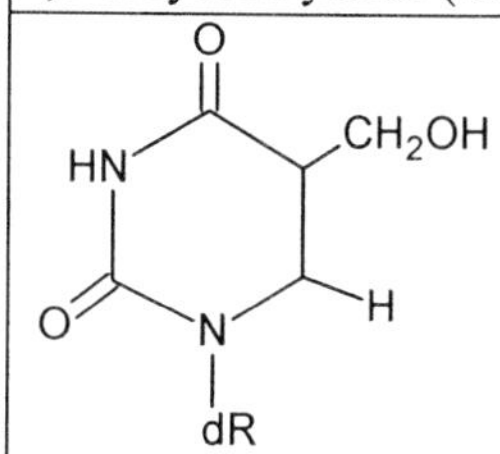 5-Hydroxymethyluracil (hm^5U)	No	No	Mug, AlkA, Fpg, Nth, Nei	hSMUG1,hMDB4,
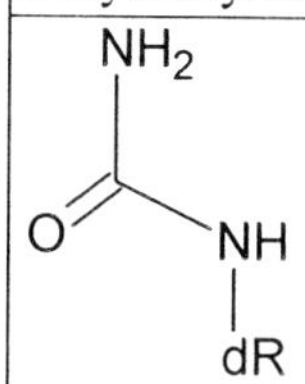 Urea	Yes: Blocking replication	Yes: T → C	Nth, Nei, Xth, Nfo, Nfi	hNTH1, hOGG1, hNER,

Table 1. Oxidized nucleic bases in DNA, their mutagenic and lethal potential and proteins involved in their repair (continued)

Oxidized nucleic base in DNA	Biological effects		Repair proteins	
	Lethal	Mutagenic	E. coli	Human
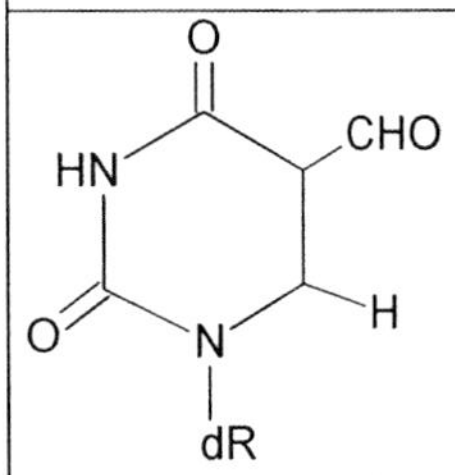 5-formyluracil (f^5U)	No	Yes: T → C, T → A	Mug, AlkA, Fpg, Nth, Nei, MutS	hSMUG1,hNTH1, hMDB4,
Oxidized cytosines				
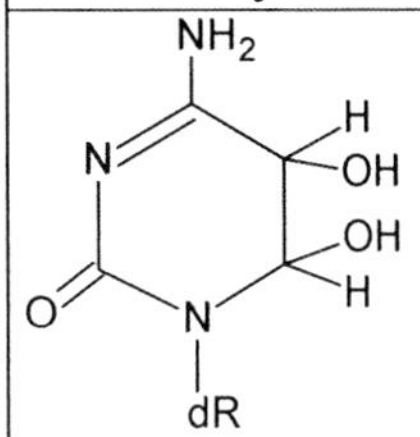 Cytosine glycol (Cg)	No	Yes: C → T	Nth	
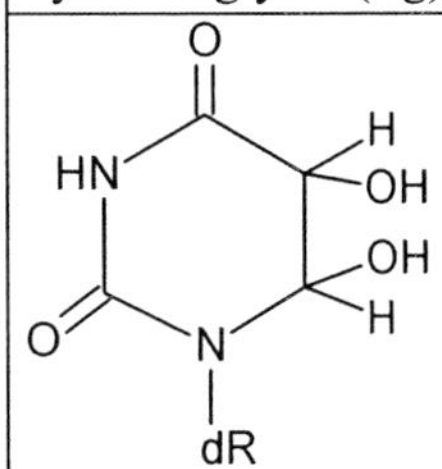 Uracil glycol (Ug)	No	Yes: C → T	Nth, Nei, Fpg	
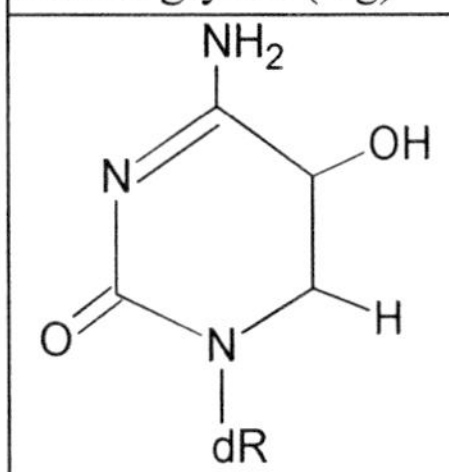 5-hydroxycytosine (5-hC)	No; Delaying replication	Yes: C → T C → G	Nth, Nei, Fpg, Mug	hNTH1, hNEIL1, hNEIL2
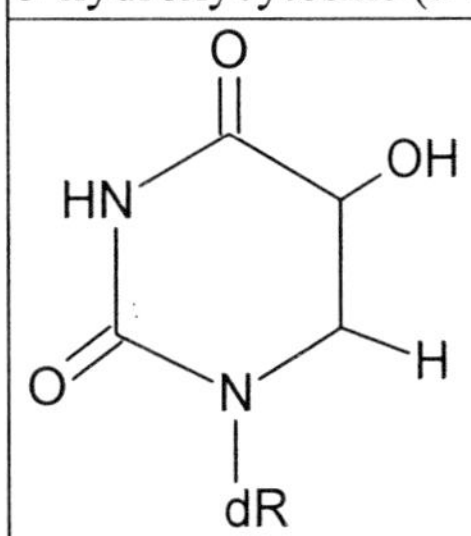 5-hydroxyuracil (5-hU)	No; Delaying replication	Yes: C → T	Nth, Nei, Fpg, Ung, Mug, Nfo	hNTH1, hNEIL1, hNEIL2, hUNG, hSMUG1,

Table 1. Oxidized nucleic bases in DNA, their mutagenic and lethal potential and proteins involved in their repair (continued)

Oxidized nucleic base in DNA	Biological effects		Repair proteins	
	Lethal	Mutagenic	E. coli	Human
Exocyclic adducts				
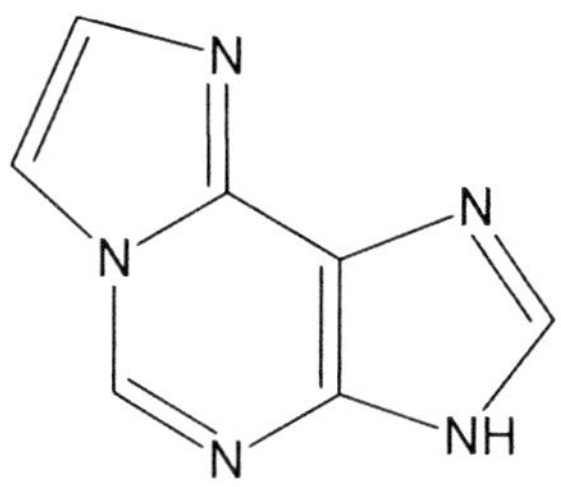 1,N^6-ethenoadenine (εdA)	No	Yes: A → G A → T A → C	AlkA	ANPG
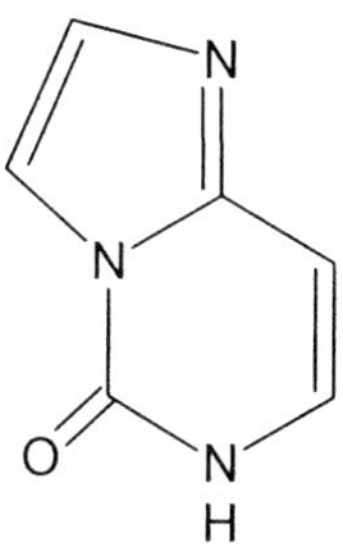 3,N^4-etehnocytosine (εdC)	No	Yes: C → T C → A	Mug	hTDG, MBD4/MED1, SMUG1
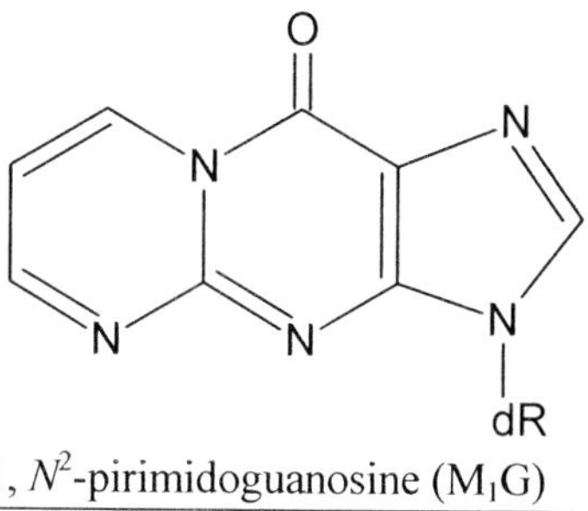 1, N^2-pirimidoguanosine (M_1G)	No	Yes: G → T G → A G → C	NER, Mug	

Oxygen-derived radicals are known to induce mutagenesis in hotspot codons of the human p53 (Hussain et al., 1994) and ras (Du et al., 1994) genes. Many GC→TA transversions have indeed been detected in p53 tumor suppressor genes (Hollstein et al., 1991, 1996; Hsu et al., 1991) and Ha-ras protooncogene that has been isolated from human lung carcinomas and primary liver cancer (Bos, 1988). However, GC→TA transversions are not only caused by 8-oxoG (Olinski et al., 2002).

The adenine analogue of 8-oxoG, 7,8,-dihydro-8-oxoadenine (8-oxoA), is 4-times less mutagenic than 8-oxoG. The lesion is not mutagenic in bacteria, although it may cause mispairing in mammalian cells (Wang et al., 1998). It can generate A→G transitions and predominantly A→C transversions (Tan et al., 1999; Kamiya et al., 1995); the mutation frequency is only 1% (Kamiya et al., 1995).

Another oxidation product of adenine, 2-hydroxyadenine (2-OH-dA), which is formed by •OH attack at the C2-position of adenine (Cooke et al., 2003) followed by oxidation, can lead to all the base substitution mutations involving A (A→G transition, and A→T and A→C transversions) (Wang et al., 1998; Kamiya and Kasai, 1997), but shows little or no genotoxicity in *E.coli* (Kamiya and Kasai, 1997).

The formamidopyrimidines (FapyA (4,6-diamino-5-formamidopyrimidine) and FapyG (2,6-diamino-4-hydroxy-5-formamidopyrimidine), ring-opened purine products, are major products of hydroxyl radical attack and are the second major oxidation products of adenine and guanine, respectively (Sluopphaug et al., 2003). They exert limited ability to terminate DNA synthesis (Graziewicz et al., 2000). There is a report that polymerase Klenow can mis-insert A opposite to FapyG, which suggests that FapyG is a mutagenic lesion (Wiederholt and Greenberg, 2002), while G can be misincorporated opposite to FapyA. FapyA is known to be more mutagenic than 8-oxoA (Delaney et al., 2002).

Oxidized Pyrimidines

Free radicals may attack the thymine moiety at two sites: the 5,6-double bond and the 5-methyl group (von Sonntag , 1987; Teebor et al., 1988). Free radical attack on the 5,6-double bond of thymidine, followed by addition of molecular oxygen, generates different diastereomers of 5-hydroxy-6-hydroperoxy-5,6-dihydrothymidine and 6-hydroxy-5-hydroperoxy-5,6-dihydrothymidine, which are rather unstable (Cadet et al., 1997; Wagner et al., 1994), and their presence and distribution in DNA are not known. The hydroperoxides are converted to more stable products in reactions where they may or may not ring-open, ring-contract or ring-fragment. The products of thymine that have been detected in DNA are: *cis*- and *trans*-5,6-dihydro-5,6-dihydroxythymine (thymine glycol (Tg)) (Téoule et al., 1977), 5,6-dihydrothymine (dHT), 5-hydroxy-5,6-dihydrothymine (5-Th) and *cis*- and *trans*-6-hydroxy-5,6-dihydrothymine (6-Th). Thymine glycol (Tg) is the most common oxidized pyrimidine and the consequence of its production is T→C transition (Basu et al., 1989). Tg does not have significant mutagenic potential, the mutation frequency being only 0,3-0,4% (Basu et al., 1989). However, thymine glycol and its products were reported to block DNA and RNA polymerases and thereby the replication one nucleotide before and one after the lesion (Rossman and Goncharova, 1998), and they are thought to be lethal (Ide et al., 1994). Urea, which is also a DNA lesion produced by oxidative damage, is a secondary product generated in substantial quantities by spontaneous hydrolytic cleavage of Tg (Reardon et al., 1997). Urea is toxic because it blocks DNA polymerase I (Evans et al., 1993). It also induces T→C mutations (Maccabee et al., 1994).

Free radical attack on the thymine 5-methyl group generates products with intact aromatic ring structure (von Sonntag, 1987; Teebor et al., 1988). Such a product is 5-(hydroperoxymethyl)uracil (Hpm^5U) which decomposes to more stable products 5-(hydroxymethyl)uracil (hm^5U) and 5-formyluracyl (5-FoU). The latter can potentially generate a 5-FoU:A pair and, possibly, also 5-FoU:G, resulting in a G→A mutation (Yoshida et al., 1997).

In cytosine the only primary target for oxidation is the 5,6-double bond. Contrary to most oxidized thymines, the cytosine derivatives are unstable due to deamination and/or dehydration. In aqueous solution 5,6-dihydroxy-5,6-dihydrocytosine (cytosine glycol (Cg)) will deaminate readily to form 5,6-dihydroxy-5,6-dihydrouracil (uracil glycol (Ug)) or dehydrate to form 5-hydroxycytosine (h^5C). However, within DNA the half-life of Cg is enhanced more than 30-fold and it is thus likely to have significant biological implications (Tremblay et al., 1999). 5-hydroxycytosine (5-hC) is known to induce CG→TA mutations with high frequency (Feig et al., 1994). The C→T transition is likely to result from mispairing with adenine instead of guanine during replication (McBride et al., 1991; Feig et al., 1994). Another important product, 5-hydroxyuracil (h^5U), can be formed from Ug by dehydration or from h^5C by deamination. It should be noted that dehydration restores the aromatic pyrimidine ring structure, thus making h^5C and h^5U rather more similar to natural bases in DNA.

Oxidative Sugar Damage

Hydroxyl radical can abstract a H atom from each of the carbon atoms of the sugar moiety in DNA (von Sonntag, 1987). The resulting C-centered radicals undergo further reactions to yield a variety of sugar modifications (Evans et al., 2004). Some sugar products are released from DNA as free modified sugars, whereas others remain within DNA or constitute end groups of broken DNA strands. A unique reaction of the C5′-centered sugar radical is the addition to the C8-position of the same purine nucleoside. This intramolecular cyclization, followed by oxidation, yields 8,5′-cyclo-2′-deoxyguanosine (cyclo-dG) and 8,5′-cyclo-2′-deoxyadenosine (cyclo-dA) (Keck, 1968; Raleigh et al., 1976; Dizdaroglu, 1986; Dirksen et al., 1988). 8,5′-cyclo-2′-deoxyguanosine has been identified in DNA of γ-irradiated human cells (Dizdaroglu , 1987). These tandem lesions constitute concomitant damage to both the base and sugar moieties of DNA. Oxygen inhibits their formation because of its diffusion-controlled reaction with the C5′-centered sugar radical prior to cyclization.

DNA-protein Cross-links

By exposing cells *in vivo* or chromatin *in vitro* to free radical-generating systems, e.g., ionizing radiation, covalent DNA–protein cross-links are formed (reviewed in Dizdaroglu, 1998; Dizdaroglu et al., 1987). Mechanisms involve the addition of a DNA base radical to an aromatic amino acid of proteins. The formation of a thymine–tyrosine cross-link has been observed in mammalian chromatin *in vitro* and *in vivo* on exposure to ionizing radiation, H_2O_2, metal ions or carcinogenic compounds (Dizdaroglu et al., 1989; Nackerdien et al., 1991; Olinski et al., 1992; Altman et al., 1995; Toyokuni et al., 1995). Chemical structures of other DNA base-amino acid cross-links have been identified in mammalian chromatin *in vitro*, but not in cells (reviewed in Dizdaroglu, 1998).

Etheno-DNA Adducts

ROS induce lipid peroxidation (LPO) of polyunsaturated fatty acids, which are the major components of phospholipid cell membranes, and oxidize low-density lipoproteins (LDL) present in the bloodstream (Marnett, 1999; Steinberg, 1997). The increased interest in exocyclic DNA adducts has been triggered by the observation that they can be formed by endogenous processes through the interaction of lipid peroxidation (LPO)-derived aldehydes and hydroxyalkenals with DNA (Chung et al., 1996; el-Ghissassi et al., 1995). Lipid peroxidation products form exocyclic adducts with either five-member (etheno adduct) or six-member rings (propano adduct). The etheno adducts are 1,N^6-etheno-2′-deoxyadenosine (εdA), 3,N^4-etheno-2′-deoxycytidine (εdC), 1,N^2-etheno-2′-deoxyguanosine (1,N^2εdG), N^2,3-etheno-2′-deoxyguanosine (N^2,3-εdG) and glyoxal-dG. The propano adducts are two isomers of 1,N^2-propano-2′-deoxyguanosine (AdG), 1,N^2-propano-2′-deoxyguanosine (CdG) and pyrimido[1,2-*α*]purine-10(3*H*)-one-2′-deoxyribose (M_1dG) (Moller and Wallin, 1998). Propano adducts can be formed directly by reactive lipid peroxidation products, whereas etheno adducts, other than glyoxal-dG, are formed after a further epoxidation of double bonds in lipid peroxidation products (Chung et al., 1996). Glyoxal-dG is formed directly by the glyoxal (Shapiro and Hachmann, 1966; Kasai et al., 1984) that is produced when ROS reacts with DNA (Murata-Kamiya et al., 1995) and also as a stable product of lipid peroxidation (Nishiyama et al., 1994). Propano adducts are produced by malondialdehyde, crotonaldehyde and acrolein, which are known to be mutagenic in bacteria and mammalian cells (Basu and Marnett, 1983; Yau, 1979; Kawanishi et al., 1998). Malondialdehyde produces M_1dG (Seto et al., 1983; Basu et al., 1988), crotonaldehyde produces CdG adducts, and acrolein produces two different PdG isomers (AdG) (Chung et al., 1984). *Trans*-4-hydroxy-2-nonenal can also form adducts *in vitro* that have a structures similar to those of the other propano adducts (Winter et al., 1986). AdG and CdG have been detected as common endogenous lesions in untreated rodents and humans (Nath and Chung, 1994; Nath et al., 1996). The mutagenic potential of exocyclic adducts is usually higher than that of small oxidative base lesions such as 8-oxodG and thymine glycol. The prevalent *in vivo* mutations induced by the propano adducts M_1dG, AdG and CdG are G → T and G → A, and by M_1dG G → C also (Moriya et al., 1994; Burcham and Marnett, 1994; Fink et al., 1997). The etheno adduct, εdA, induced A → G, A → T and A → C (Basu et al., 1993; Pandya and Moriya, 1996), while εdC induced C → T and C → A (Moriya et al., 1994; Basu et al., 1993; Palejwala et al., 1991). It was found that εdA and εdC were strongly mutagenic in mammalian cells but only weakly in *E. coli* (Moriya et al., 1994) The formation of etheno adducts by endogenous mechanisms and their role in carcinogenesis has been the subject of several reviews (Marnett and Burcham, 1993; Chung et al., 1996, Bartsch et al., 1994).

Repair of Oxidative DNA Lesions

As described in the previous section, large numbers of different types of DNA lesions, having mutagenic and cytotoxic effects, can be produced following oxidative attack on DNA. The removal of these lesions by cellular repair processes is crucial in limiting their adverse effects. Persistence of oxidative DNA lesions can cause arrest or induction of transcription, induction of signal transduction pathways, replication errors, and genomic instability, which consequently can all lead to increased mutation rate (Figure 2). These molecular changes are seen in both cancer and aging, supporting the hypothesis that persistent DNA damage in the genome could be a cause of both.

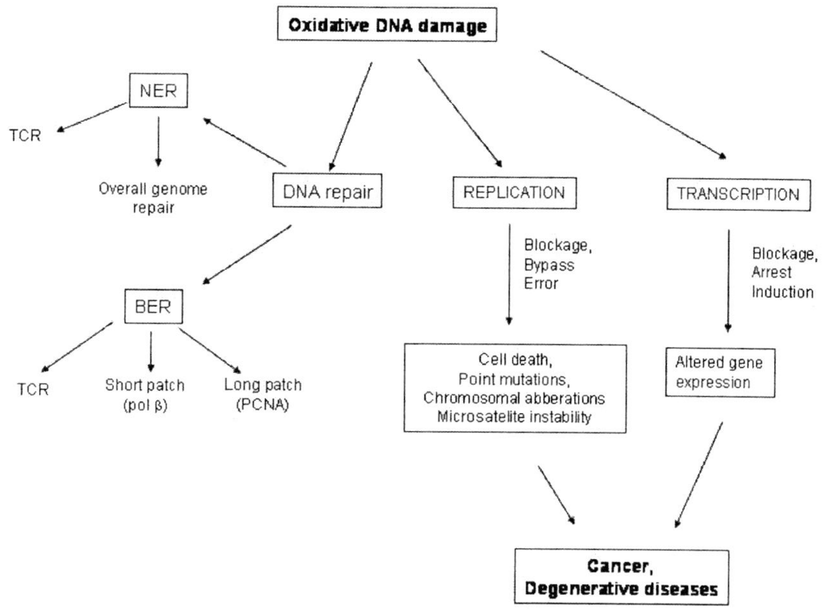

Figure 2. Consequences of oxidative DNA damage. If the DNA damage persists it causes multiple changes including mutations, chromosomal aberrations and microsatelite instability, known to be frequent phenotypes of cancer and other degenerative diseases. Oxidative DNA damage can be removed by several DNA repair mechanisms.

It has been demonstrated experimentally that most oxidative DNA lesions can be removed by repair systems with different, but in many cases overlapping, specificities. This overlap is very important as it means that inhibition or elimination of one repair activity does not necessarily render a particular base lesion irreparable. Some repair activities function as back-up mechanisms and are often less efficient than the primary repair pathway. The presence of multiple repair pathways actually underlines the seriousness of the threat to the genome and to biological function posed by oxidative damage.

Oxidized DNA bases are mostly repaired by the base-excision repair (BER) pathway, which is the main repair process in the removal of single lesions in DNA (Dianov et al., 1998). Nucleotide-excision repair (NER) plays a role in the repair of bulky adducts, such as exocyclic adducts induced by lipid peroxidation products and cross-links, which are not substrates for BER (Moller and Wallin, 1998), and is a back-up mechanism when BER mechanisms are compromised or saturated (Dianov et al., 1998; Evans et al., 2004).

BER, first discovered by Lindahl (Lindahl, 1974), is accomplished via two pathways that involve different subsets of repair enzymes and result in removal and replacement of one (short-patch BER) or several (long-patch BER) nucleotides (Figure 3) (Dianov and Lindahl, 1994; Matsumoto et al., 1994; Frosina et al., 1996; Klungland and Lindahl, 1997; Dianov et al., 1999; Zhang and Dianov, 2005). BER is initiated by a lesion-specific DNA glycosylase that recognizes and removes the damaged base by cleaving the *N*-glycosyl bond producing an abasic site (apurinic/apyrimidinic or AP site). An AP site is recognized and cleaved by AP-endonuclease (APE1), which incises the phosphodiester bond 5' to the AP site (Bohr, 2002), resulting in a 3'-hydroxyl group and 5-abasic deoxyribose phosphate (dRP). Removal of dRP is accomplished by the action of DNA polymerase ß (Pol ß) which catalyses ß-elimination of the 5'-sugar phosphate residue by associated AP lyase activity and adds one nucleotide to the 3'-OH end of the incised AP site (Matsumoto and Kim, 1995). DNA ligase completes the repair by sealing the DNA ends. These repair events create a single-nucleotide repair patch or short patch, where only one damaged nucleotide is removed and replaced by the corresponding normal nucleotide (Bohr, 2002; Dianov et al., 1992). When the 5'-sugar phosphate of AP site is modified (reduced or oxidized), the repair is switched to the long patch repair pathway, which involves DNA glycosylase, APE1, DNA polymerase δ or ε (Pol δ/ε), flap endonuclease (FEN-1), proliferating cell nuclear antigen (PCNA), and DNA ligase (Klungland and Lindahl, 1997; Dianov et al., 1999). This sub-pathway is PCNA dependent (Frosina et al., 1996), where DNA polymerase (δ or ε) adds several nucleotides to the repair gap, thus displacing the dRP as "flap" oligonucleotide. The resulting flap structure is recognized and excised by FEN-1 and the DNA is ligated by DNA ligase. These repair events result in a 2-6 nucleotide long repair patch (Dianov and Lindahl, 1994; Matsumoto et al., 1999). Approximately 90 % of 8-oxoG and Tg DNA lesions are repaired by single nucleotide BER, while the remainder, such as mismatch 8oxoG/A, are removed by long path BER (Bohr, 2002; Fortini at al., 2003).

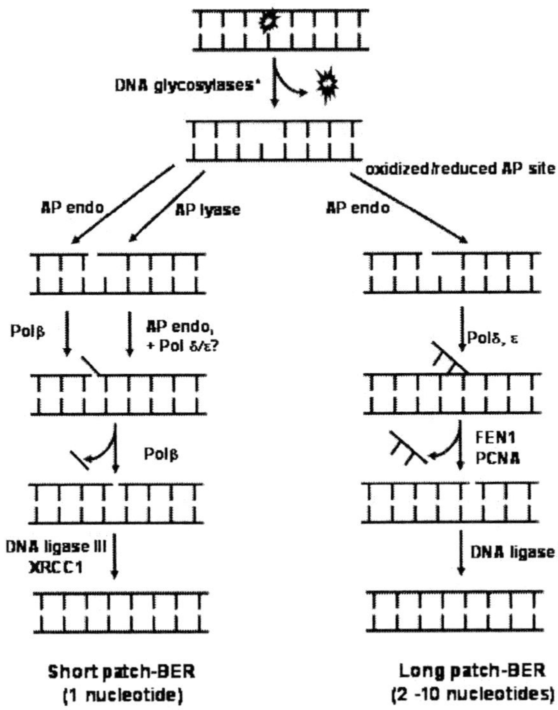

Figure 3. Alternative BER pathways. Short patch BER is initiated by bifunctional (i.e. OGG1) or monofinctional (UNG) glycosylase. Unmodified AP sites lead to one-nucleotide repair patch generated by Pol β and sealed in a common ligation step. Reduced or oxidized AP sites are processed via PCNA-dependent pathway, and involve cleavage of 2 to 10 nucleotides »flap« by FEN1. AP site, apyrimidic or apurinic site; endo, endonuclease; pol, polymerase; PCNA, proliferating cell nuclear antigen; FEN1, flap-endonuclease 1; XRCC1, X-ray cross-complementation protein.

A number of other DNA repair mechanisms, including post-replication repair, function as further lines of defense. Bacteria lacking one or more of the DNA repair mechanisms exhibit increased mutation frequencies. In humans there are several disease categories with functionally defective DNA repair systems (Digweed, 1993; Hansson, 1992). Some of these show increased cancer risk and increased levels of oxidative DNA damage.

Repair of Purine Derived Oxidative DNA Lesions ("GO-system")

8-oxoG is the most stable mutagenic lesion after oxidation of DNA, and can pair with cytosine and adenine nucleotides with almost equal efficiency, resulting in GC → TA transversions (Lindahl, 1993; Lindahl and Wood 1999). Furthermore, incorporation of 8-oxodGTP, generated by oxidation of dGTP, may lead to AT → CG and GC → TA transversions (Tajiri et al., 1995). The repair system that counteracts the deleterious effects of 8-oxoG was originally described as the "GO-system" in *Escherichia coli* and comprises the MutT, MutY and MutM proteins (Michaels and Miller, 1992; Fowler et al., 2003). In human cells, this system is more complex since, in addition to BER, it involves mismatch repair (MMR) and NER as well (Slupphaug et al., 2003). »GO system« relies on three lines of defense (Figure 4). The first is removal of 8-oxodGTP from the nucleotide pool by the nucleotidyl hydrolase hMTH1 (human MutT homologue 1) which efficiently degrades 8-oxodGTP (and also 8-oxodATP and 2-hydroxy-dATP) to their respective monophosphates (Sakumi et al., 1993; Fujikawa et al., 1999; Sakai et al., 2002). hMTH1 is localized in nuclei, cytosol and mitochondria (Kang et al., 1995). The importance of MTH1 in mutation avoidance is demonstrated by an increased rate of spontaneous tumorigenesis in *mth−/−* mice (Tsuzuki et al., 2001).

If 8-oxodGTP escapes sanitation of the nucleotide pool, it may be incorporated in DNA by polymerases. It has been shown that polymerase β discriminates poorly between dGTP and 8-oxodGTP (Kamath-Loeb et al., 1997; Miller et al., 2000) and that it exhibits a preference (24:1) for inserting 8-oxodGTP opposite a template dA over a dC (Miller et al., 2000).

Therefore the second line of defense is removal of 8-oxoG (either incorporated or formed by oxidation of DNA) opposite both bases. Several mammalian DNA glycosylases with distinct, but overlapping substrate specificities have been reported that remove 8-oxoG from DNA. The best studied is hOGG1 (also named hOGH1 and hMMH). hOGG1 is the functional counterpart of *E. coli* 8-oxoG glycosylase/AP-lyase MutM/Fpg (Aburatani et al., 1997; Boiteux and Radicella, 2000). It is double-strand specific and has a high specificity for 8-oxoG (or FaPyG) paired with C. It also removes 8-oxoG opposite A, but with a much lower specificity and without associated strand cleavage (Bjørås at al., 1997). Genes for mammalian or bacterial enzymes with high specificity for removal of 8-oxoG from 8-oxoG:A pairs have not yet been cloned. However, in yeast, yOGG2 protein has been shown to have a slight preference for 8-oxoG:A over 8-oxoG:C pairs (Bruner et al., 1998). An 8-oxoG:A-glycosylase, hOGG2, was however partially purified from HeLa cells (Hazra et al., 1998). The 36 kDa protein demonstrated selective removal of 8-oxoG from 8-oxoG:A and 8-oxodG:G pairs and was immunologically distinct from hOGG1. Mutations in the gene *hOGG1* (Audebert et al., 2000), as well as loss of heterozygosity (LOH) at this locus (Wikman et al., 2000), have been associated with kidney tumors and lung cancer. The available evidence from gene structure and expression analysis in different tissues implies that *OGG1* is a constitutively expressed housekeeping gene, although expression can vary with tissue source and between individuals (Radicella et al.,1997; Nishioka et al., 1999; Boiteux and Radicella, 2000; Hanaoka et al., 2000; Verjat et al., 2000).

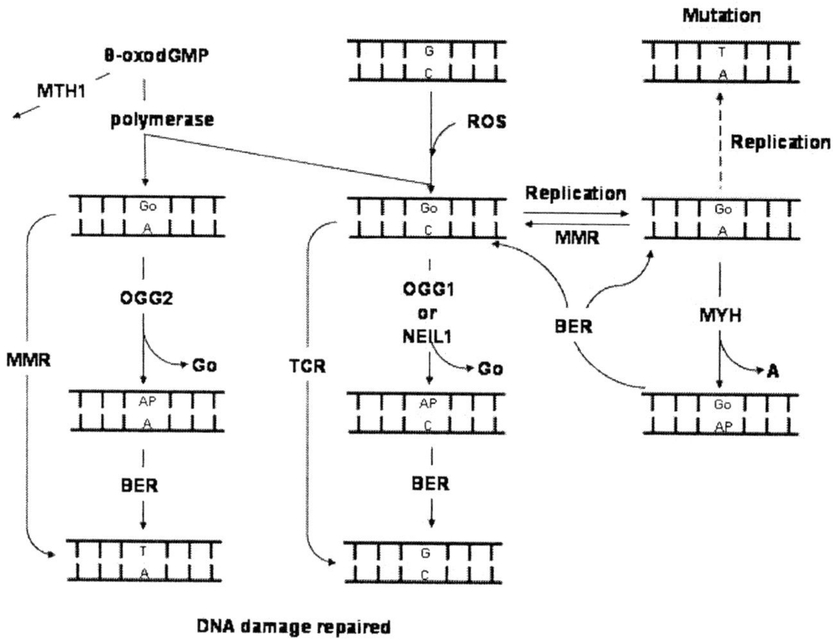

Figure 4. "GO-system" in mammalian cells. Misincorporation of 8-oxoG is reduced by hydrolysis of 8-oxodGTP by MTH1. Misincorporated 8-oxoG opposite A is repaired by OGG2-initiated BER or MMR. 8-oxoG formed in DNA by oxidation of G opposite C is repaired by OGG1-initiated BER or TCR. Misincorporated 8-oxoG opposite C is repaired by NEIL1-initiated replication-associated BER. If 8-oxoG:C pair escapes repair prior to the next round of replication, misincorporated A opposite 8-oxoG is removed by MYH-initiated replication-associated BER, MMR, or by a combination of proteins from both pathways (Adapted from Slupphaug et al., 2003).

NEIL is a mammalian ortholog belonging to the Fpg/Nei family (Hazra et al., 2002; Takao et al., 2002; Bandaru et al., 2002; Morland et al., 2002). The prime substrates of the recombinant hNEIL1 (Hazra et al., 2002; Bandaru et al., 2002; Morland et al., 2002) and hNEIL2 (Hazra et al., 2002a), as well as of mNEIL1 (Takao et al., 2002), are oxidized pyrimidines. However, hNEIL1 also showed substantial removal of 8-oxoG from 8-oxoG:C pairs (Hazra et al., 2002; Bandaru et al., 2002; Morland et al., 2002) and weak but significant activity against 8-oxoG:A pairs (Morland et al., 2002). NEIL1 plays a role in replication-associated repair of 8-oxoG:C (Hazra et al., 2002; Morland et al., 2002).

When repair of 8-oxoG:C-lesions is not completed before the next round of replication, dATP could be incorporated instead of dCTP by the replicative pol δ (Shibutani et al., 1991), causing GC → TA transversions. The third line of defense copes with such events by glycolytic removal of A, misincorporated opposite 8-oxoG. This is a function of hMYH (human MutY homologue), a double strand specific adenine glycosylase that targets A or 2-hydroxyadenine (2-OH-A) mismatched with either G or 8-oxoG (Slupska et al., 1996; Takao et al., 1999; Ohtsubo et al., 2000). hMYH is a simple glycosylase with no AP-lyase activity, similar to *E. coli* MutY protein (Slupska et al., 1999). The *in vivo* importance of such repair in humans is supported by the finding that inherited variants of hMYH at conserved amino acid residues, are associated with increased somatic GC → TA transversions in colorectal tumors (Al-Tassan et al., 2002). In human cells, different isoforms of the protein are located in nuclei and mitochondria (Ohtsubo et al., 2000). Nuclear hMYH1 is associated with immediate post-replicative BER, as observed with the nuclear hUNG2 (Otterlei et al., 1999), and probably proceeds via the long-patch, PCNA-dependent pathway.

Several other mechanisms have been found to contribute to removal of 8-oxoG from DNA. Transcription-coupled repair (TCR) of 8-oxoG is involved in removal of 8-oxoG from the transcribes (TS), but not the non-transcribed (NTS) strand (Le Page, et al., 2000). This may explain the lack of mutagenic phenotype of knockout mice deficient in OGG1 activity (Klungland et al., 1999). TCR of 8-oxoG has been demonstrated to require XPG, TFIIH and CSB proteins; defective TCR results in enhanced mutation frequencies at 8-oxoG.

Repair of Pyrimidine Derived Oxidative DNA Lesions

The bifunctional glycosylase hNTH1, a homologue of bacterial Nth, removes thymine glycol and some related bases, but not oxidized purines (Eide et al., 2001). As already discussed, NEIL1 (NEH1/FPG1) removes 8-oxoG and FaPy, but, it also removes thymine glycol, 5-hydroxyuracil, 5,6-dihydrouracil (5,6-DHU) and 5-hydroxycytosine (5-OHC) opposite different bases (Table 2). Oxidized pyrimidines appear to be the preferred substrates (Bandaru et al., 2002; Morland et al., 2002). NEIL1 has a broad substrate specificity and its presence explains the lack of obvious phenotype in knockout mice deficient in mNTH1, although thymine glycol produced in the mutant mouse liver DNA disappeared more slowly than in the wild-type mouse (Takao et al., 2002). NTH1 and NEIL1 show opposite preferences for removal of Tg opposite G (NEIL1) and A (NTH1). Other oxidized pyrimidines, such as 5-hydroxyuracil, 5,6-dihydrouracil and 5-hydroxycytosine are recognized by hNTH1, hNEIL1 and hNEIL2 (Table 2). The functional implications of this apparent redundancy may reflect distinct functions of the three enzymes with respect to nuclear, as opposed to mitochondrial, repair as well as in various stages of the cell cycle.

The uracil-DNA glycosylase SMUG1 is only present in higher eukaryotes. Human and murine SMUG1 efficiently remove U from single and double-stranded DNA, generally preferring U:G mispairs over U:A pairs (Nilsen et al., 2001; Kavli et al., 2002). SMUG1 is also the major activity for removal of 5-hydroxymethyluracil (HmU) from both HmU:A pairs and HmU:G mispairs (Kavli et al., 2002; Boorstein et al., 2001). This may be a major function of SMUG1 but, in addition, it may serve as an important backup enzyme for UNG2 in removing U from U:G mispairs. SMUG1 also removes 3,*N*4-ethenocytosine (εC) (Kavli et al., 2002).

Oxidized cytosines, such as isodialuric acid, 5-hydroxyuracil and alloxan, have structures resembling uracil and are in fact removed by human uracil-DNA glycosylase (UNG) (Dizdaroglu et al., 1996). Although the rate of removal of these derivatives is slow compared with the preferred substrate, uracil, it is approximately as fast as the removal of various oxidized substrates by other DNA glycosylases (Krokan et al., 1997; Dizdaroglu et al., 1996).

Repair of Exocyclic Adducts

All etheno adducts are potentially mutagenic *in vivo*, therefore, the repair processes eliminating exocyclic adducts from DNA should play a crucial role in maintaining the stability of genetic information. Like oxidized DNA bases, etheno adducts are also predominantly repaired by BER (Gros et al., 2002), and NER functions as a back-up when BER is saturated or does not function.

Several bacterial and mammalian glycosylases have been shown to be involved specifically in the repair of exocyclic adducts. The main substrates for *E. coli* AlkA protein are alkylated bases. However, AlkA is also involved in the repair of oxidative DNA damage, and it excises a broad range of modified bases including hypoxanthine (Saparbaev and Laval, 1994), 5-formyluracil (Bjelland et al., 1994) and fragmented thymine (Privezentzev et al., 2000), as well as ethenobases, N^2,3-ethenoguanine (Sodum and Chung, 1988; Habraken et al., 1991) and 1,N^6-ethenoadenine (Saparbaev et al., 1995). The *S. cerevisiae* MAG protein is the functional and structural homologue of AlkA, and similarly excises hypoxanthine and εA residues from DNA (Saparbaev et al., 1995; Saparbaev and Laval, 1994). The ANPG protein is the human counterpart of *E. coli* AlkA protein. This protein releases 3-methyladenine and 7-methylguanine from methylated DNA (O'Connor and Laval., 1991), but is also able to excise hypoxanthine (Saparbaev and Laval, 1994) and εA (Saparbaev et al., 1995; Singer et al., 1992) more efficiently than the AlkA protein. Recently, it has been shown that ANPG, but not AlkA, efficiently excises 1,N^2-εG from DNA (Saparbaev et al., 2002).

The *E. coli* mismatch-specific uracil-DNA glycosylase (MUG protein) is able to remove various ε-bases from DNA, including 3,N^4-ethenocytosine (Saparbaev and Laval, 1994), 8-(hydroxymethyl)-3,N^4-ethenocytosine (Hang et al., 2002) and 1,N^2-ethenoguanine (Saparbaev et al., 2002). The MUG protein was initially characterized as an enzyme excising uracil from G:U mismatch and it was proposed that it acts as a backup enzyme for UDG protein and may be important in avoiding mutations during the stationary phase of cell growth (Gallinari and Jiricny, 1996; Mokkapati et al., 2001). However, on comparing the catalytic efficiency of the MUG protein towards different substrates, εC was shown to be excised 50 times more efficiently than uracil from the G:U pair, suggesting that the primary *in vivo* function of MUG is the repair of εC (Saparbaev and Laval, 1994). Recently, it has been shown that MUG can excise, with very low efficiency, εA, hypoxanthine and 5-hydroxycytosine residues when present in DNA (O'Neill et al., 2003).

The human thymine-DNA glycosylase (hTDG) was initially found to be able to remove T mispaired with G (Neddermann and Jiricny, 1993). Its more complete characterization showed that hTDG exhibits wide substrate specificity, excising εC (Saparbaev and Laval, 1994; Hang et a., 1998) and mismatched uracil (Neddermann and Jiricny, 1994). The hTDG protein is the homologue of the *E. coli* MUG, indicating that εC-DNA glycosylase activity is conserved during evolution. Recently, in addition to hTDG, two human enzymes excising εC have been identified: methyl-CpG binding domain protein (MBD4/MED1) (Petronzelli et al.,

2000) and single-stranded monofunctional uracil-DNA glycosylase (SMUG1) (Kavli et al., 2002).

The main DNA adducts formed by vinyl carbamate (VCB) are 7-(2-oxoethyl)guanine and, to a lesser extent, N^2,3-ethenoguanine (N^2,3-εG) (Bolt , 1988; Fedtke et al., 1990). Low background levels of N^2,3-εG and other exocyclic DNA adducts have also been consistently detected in unexposed humans and animals (Chaudhary et al., 1994; Ham et al., 1999; Nath et al., 1994). It was established that the pure *E. coli* AlkA releases N^2,3-εG from chloroacetaldehyde-treated DNA in a concentration- and time-dependent manner (Matijasevic et al., 1992). In mammals, N^2,3-εG appears to be the substrate of mammalian alkyl-*N*-purine-DNA glycosylase (ANPG) (Singer et al., 1992).

The pro-mutagenic and genotoxic exocyclic DNA adduct 1,N^2- εG is a product formed in DNA exposed *in vitro* to lipid peroxidation-derived aldehydes such as 4-hydroxyalkenals and its peroxides (Morinello et al., 2001; Sodum and Chung, 1988; 1991). 1,N^2- εG is a lesion showing moderate DNA polymerase blocking activity *in vitro* (Langouet et al., 1997), and it causes misincorporations *in vitro* and in *E. coli* (Langouet et al., 1997, 1998). Furthermore, the presence of 1,N^2- εG in chromosomal DNA of Chinese hamster ovary cells induces deletions, rearrangements, double mutants, and base pair substitutions (Akasaka and Guengerich, 1999). The lesion is removed by two structurally unrelated proteins, the *E. coli* mismatch-specific uracil-DNA glycosylase (MUG) and the human alkyl-*N*-purine-DNA glycosylase (ANPG) (Saparbaev et al., 2002). Both the MUG and ANPG proteins preferentially excise 1,N^2- εG when it is opposite dC (Saparbaev et al., 2002).

Malondialdehyde, a mutagenic and carcinogenic compound, is generated endogenously as a product of lipid peroxidation and prostaglandin biosynthesis. The major adduct in DNA formed by malondialdehyde is a pyrimidopurinone, the pyrimido[1,2-a]purin-10(3H)-one (M_1G). Studies of site-specific mutagenesis indicate that in bacteria M_1G and PdG are repaired by the nucleotide excision repair (NER) pathway (Marnett, 1999; Niedernhofer et al., 2003; Fink and Marnett, 2001). Although it has been demonstrated that, in HeLa cell-free extract *in vitro*, PdG is a substrate for NER (Johnson et al., 1997), it has been proposed that *E. coli* MUG and human full-length ANPG proteins also play a role in the repair of six-membered exocyclic DNA adducts (Saparbaev et al., 2002).

Involvement of ROS in Metal Mediated Mutagenicity and Carcinogenicity

Humans have been in contact with metals since the beginning of our existence. Some metals are essential for the normal functioning of cells and the survival of organisms, others are not, but we cannot avoid exposure. Metals can contribute significantly to endogenous ROS formation. For example, although iron participates in a number of critical physiological processes, like oxygen transportation, xenobiotic metabolism and oxidative phosphorylation, and is a prosthetic group in many enzymes (Symons and Gutteridge, 1998), when present in excess of an organism's binding capacity it may be highly toxic, as indicated by iron overload-related diseases, including cancer development. The same considerations are valid for a number of other metals, some of which, like Zn, Cd and As are essentially not redox active.

Some metals, including chromium, nickel, arsenic, cadmium, and cobalt, have long been recognized as human and/or animal carcinogens (IARC, 1980, 1990, 1991, 1993). However, the underlying molecular mechanisms of metal carcinogenicity are still not well understood, especially since they are not mutagenic in bacterial test systems and only weakly mutagenic in cultured mammalian cells. Recent studies have identified several mechanisms of metal activity which may be relevant for the carcinogenic process, of which induction of oxidative DNA damage (Kasprzak, 1996) and interference with diverse DNA repair systems (Hartwig, 1998) lead to increased levels of mutation, thereby playing a crucial role in cancer initiation.

Various metals can induce ROS mediated DNA damage, and consequently mutations, through different mechanisms (Figure 5). Reactive oxygen species (ROS) like hydrogen peroxide (H_2O_2) and superoxide anions ($O_2{\bullet}^-$), that are continuously generated during oxygen metabolism, are converted into highly reactive hydroxyl radicals via Fenton- and Haber-Weiss type reactions in the presence of redox-active transition metals. Iron(II)/iron(III), nickel(II), and cobalt(II) can damage DNA in the presence of H_2O_2 *in vitro*. Chromium(VI) undergoes intracellular reduction to the stable chromium(III) species. In the course of reduction, reactive intermediates (ROS, sulfur-centered radicals, pentavalent and tetravalent chromium species) are generated, and different forms of structural DNA damage, including 8-oxo-guanine as an indicator of oxidative DNA damage, are induced (for review see Singh et al., 1998). Oxidative DNA damage might also occur via the inactivation of cellular defense systems, as has been postulated in the case of cadmium(II) (for review see Kasprzak, 1996).

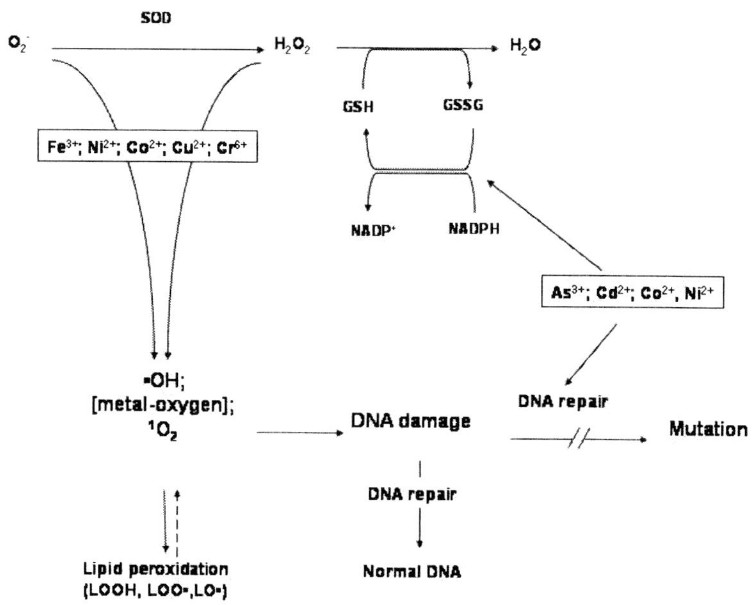

Figure 5. Shematic presentations of metal ROS mediated mutageneisis. Metal ions react with superoxide anion ($O2^-$) and H_2O_2 to produce highly reactive species such as hydroxyl free radical (•OH) and metal-oxygen complexes in biological systems, resulting in oxidative DNA damage. Certain metals inactivate protective mechanismes: antioxidative mechanisms and DNA repair systems.

Involvement of ROS in cadmium and arsenic mediated mutagenicity has been studied using the human-hamster hybrid A_L cell mutation assay in which both intragenic and multilocus deletions can readily be detected. In this test system both metals were shown to be strong mutagens that induced predominantly large deletion mutations (Filipič and Hei, 2004; Hei and Filipič, 2004). The formation of ROS within A_L cells exposed to cadmium or arsenic was confirmed by laser scanning confocal microscopy and the fluorescent probe 5-(and-6)-chloromethyl-2-,7-dichlorodihydrofluorescein diacetate. In addition, a dose dependent induction of 8-OHdG adducts in A_L cells exposed to cadmium and arsenic was demonstrated using immunoperoxidase staining coupled with a monoclonal antibody specific for 8-OHdG adducts. The involvement of ROS in the mutagenicity was confirmed by concurrent treatment of A_L cells with the oxyradical scavenger dimethyl sulfoxide, which significantly reduced the number of cadmium- or arsenic-induced mutations. In contrast, pre-treatment of cells with buthionine sulfoximine, that depletes intracellular glutathione, increased the cytotoxicity and mutagenicity of both metals (Filipič and Hei, 2004; Hei and Filipič, 2004).

The second important mechanism of metal induced mutagenicity is inhibition of DNA repair processes. As already discussed, DNA is continuously damaged by endogenous processes including ROS and by environmental agents, leading to several thousand lesions per cell per day (Lindahl, 1993). Efficient repair of these lesions is thus an important prerequisite to maintain DNA integrity. It has been shown that the repair of oxidative DNA base modifications is inhibited at low, non-cytotoxic concentrations of nickel(II) and cadmium(II), while (NER) is inhibited by nickel(II), cadmium(II), cobalt(II), and arsenic(III) (for recent review see Hartwig et al., 2002). In our studies we showed that cadmium inhibited repair of H_2O_2 induced 8-OHdG formation in A_L cells, which also explained the observed accumulation of 8-OHdG lesion after prolonged cell exposure to cadmium (Filipič and Hei, 2004). Furthermore we also showed that cadmium attenuated the repair of DNA single strand breaks induced by UV and methyl methane sulfonate (Fatur et al., 2003). One reason for repair inhibition appears to be the displacement of zinc(II) and magnesium(II). Potentially sensitive targets are the so-called zinc finger structures present in several DNA repair enzymes, such as the mammalian XPA protein and the bacterial formamidopyrimidine-DNA glycosylase (Fpg protein). Detailed studies revealed that each zinc finger protein exerts unique sensitivity towards toxic metal ions. Thus, toxic metal ions may lower genetic stability by inducing oxidative DNA damage and by decreasing the repair capacity towards DNA lesions induced by endogenous and exogenous mutagens, which in turn could increase the risk of tumor formation.

Conclusions

Cellular production of ROS and exposure to ROS are unavoidable, and oxidative damage to DNA is consequently abundant in human tissues. The most abundant oxidative DNA modification is considered to be 8-oxodG, which is also the most mutagenic one, resulting in the G→T transversions frequently found in oncogenes and tumor suppressor genes. However, there is also a plethora of other oxidative modifications of base and sugar residues and exocyclic adducts that occur frequently in DNA, but they are less well studied and their biological significance is less apparent.

Systems that defend the organism against oxidative damage to DNA are very complex and include enzymatic removal of reactive oxygen species, enzymatic nucleotide pool sanitation, and DNA repair. Nearly all known DNA repair pathways and trans-lesion synthesis, are involved in the repair of DNA oxidative damage. The important feature of the repair of oxidized purines and pyrimidines is that backup enzymes exist for removing these lesions in the case where BER pathways are overloaded or not functional. The network of interacting repair systems and backup enzymes within pathways also explains the lack of a serious phenotype in mice with targeted disruptions of single DNA glycosylases. The complexity of the defense mechanisms against oxidative DNA damage also indicates that oxidative stress constitutes a serious threat that requires massive efforts to avoid decay of the genome.

Numerous metal ions are critically involved in maintaining homeostasis of cellular redox levels. In addition, several nonessential heavy metals are toxic and carcinogenic. Despite the fact that the mechanism of their carcinogenicity is still not clear, there is strong evidence that all known carcinogenic metals induce ROS and oxidative DNA damage. Furthermore, most of them at the same time inhibit DNA repair pathways. In this context, toxic heavy metals constitute a group of environmental contaminants that may contribute more than is currently thought to the overall threat of cancer and other mutation mediated diseases.

The data from human studies obtained so far support the experimentally based indication of oxidative DNA damage as an important mutagenic and apparently carcinogenic factor. However, proof of a causal relationship in humans is still lacking. In the future the use of biomarkers may provide this evidence and allow further investigation of the qualitative and quantitative importance of oxidative DNA modification and carcinogenesis in humans, and also suggest possible preventive measures.

References

Aburatani H., Hippo Y., Ishida T., Takashima R., Matsuba C., Kodama T., Takao M., Yasui A., Yamamoto K., Asano M., Fukasawa K., Yoshinari T., Inoue H., Ohtsuka E., Nishimura S. (1997) Cloning and characterization of mammalian 8-hydroxyguanine-specific DNA glycosylase/apurinic, apyrimidinic lyase, a functional mutM homologue. *Cancer Res* 57, 2151–2156.

Akasaka S. and Guengerich F.P. (1999) Mutagenicity of site-specifically located 1,N2-ethenoguanine in Chinese hamster ovary cell chromosomal DNA. *Chem Res Toxicol* 12, 501–507.

Altman S.A., Zastawny T.H., Randers-Eichhorn L., Cacciuttolo M.A., Akman S.A., Dizdaroglu M., Rao G. (1995) Formation of DNA–protein cross-links in cultured mammalian cells upon treatment with iron ions. *Free Radical Biol Med* 19, 897–902.

Al-Tassan N., Chmiel N.H., Maynard J., Fleming N., Livingston A.L., Williams G.T., Hodges A.K., Davies D.R., David S.S., Sampson J.R., Cheadle J.P. (2002) Inherited variants of MYH associated with somatic G:C → T:A mutations in colorectal tumors. *Nat Genet* 30 227–232.

Audebert M., Radicella J.P., Dizdaroglu M. (2000) Effect of single mutations in the *OGG1* gene found in human tumors on the substrate specificity of the Ogg1 protein. *Nucleic Acids Res* 28, 2672–2678.

Bandaru V., Sunkara S., Wallace S.S., Bond J.P. (2002) A novel human DNA glycosylase that removes oxidative DNA damage and is homologous to *Escherichia coli* endonuclease VIII. *DNA Repair* 1, 517–529.

Bartsch H., Barbin A., Marion M.-J., Nair J., Guichard Y. (1994) Formation, detection, and role in carcinogenesis of ethenobases in DNA. *Drug Metab Revs* 26, 349–371.

Basu A.K., Loechler E.L., Leadon S.A., Essigmann J.M. (1989) Genetic effects of thymine glycol: site-specific mutagenesis and molecular modeling studies. *Proc Natl Acad Sci U S A* 86(20), 7677-7681.

Basu A.K., O'Hara S.M., Valladier P., Stone K., Mols O., Marnett L.J. (1988) Identification of adducts formed by reaction of guanine nucleotides with malondialdehyde and structurally related aldehydes. *Chem Res Toxicol* 1, 53–59.

Basu A.K., Wood M.L., Niedernhofer L.J., Ramos L.A., Essigmann J.M. (1993) Mutagenic and genotoxic effects of three vinyl chloride-induced DNA lesions: 1,N^6-ethenoadenine, 3,N^4-ethenocytosine, and 4-amino-5-(imidazol-2-yl)imidazole. *Biochemistry* 32, 12793–12801.

Basu A.K. and Marnett L.J. (1983) Unequivocal demonstration that malondialdehyde is a mutagen. *Carcinogenesis* 4, 331–333.

Beckman J.S. and Koppenol W.H. (1996) Nitric oxide, superoxide, and peroxynitrite: the good, the bad, and ugly. *Am J Physiol* 271(5 Pt 1), C1424-1437.

Bjelland S., Birkeland N.K., Benneche T., Volden G., Seeberg E. (1994) DNA glycosylase activities for thymine residues oxidized in the methyl group are functions of the AlkA enzyme in *Escherichia coli*. *J Biol Chem* 269, 30489–30495.

Bjørås M., Luna L., Johnson B., Hoff E., Haug T., Rognes T., Seeberg E. (1997) Opposite base-dependent reactions of a human base excision repair enzyme on DNA containing 7,8-dihydro-8-oxoguanine and abasic sites. *EMBO J* 16, 6314–6322.

Bohr V.A. (2002) Repair of oxidative DNA damage in nuclear and mitochondrial DNA, and some changes with aging in mammalian cells. *Free Radic Biol Med* 1, 32(9), 804-812.

Boiteux S. And Radicella J.P. (2000) The human OGG1 gene: structure, functions, and its implication in the process of carcinogenesis. *Arch Biochem Biophys* 377, 1–8.

Bolt H.M. (1988) Roles of etheno-DNA adducts in tumorigenicity of olefins. *Crit Rev Toxicol* 18, 299–309.

Bolton J.L., Trush M.A., Penning T.M., Dryhurst G., Monks T.J. (2000) Role of quinones in toxicology. *Chem Res Toxicol* 13, 135-160.

Boorstein R.J., Cummings A., Marenstein D.R., Chan M.K., Ma Y.L., Neubert T.A., Brown S.M., Teebor G.W. (2001) Definitive identification of mammalian 5-hydroxymethyluracil DNA *N*-glycosylase activity as SMUG1. *J Biol Chem* 276, 41991–41997.

Borges F., Fernandes E., Roleira F. (2002) Progress towards the discovery of xanthine oxidase inhibitors. *Curr Med Chem* 9, 195–217.

Bos J.L. (1988) The ras gene family and human carcinogenesis. *Mutat Res* 195(3), 255-271.

Breen A.P. and Murphy A. (1995) Reactions of oxyl radical with DNA. *Free Radic Biol Med* 18, 1033–1077.

Bruner S.D., Nash H.M., Lane W.S., Verdine G.L. (1998) Repair of oxidatively damaged guanine in *Saccharomyces cerevisiae* by an alternative pathway. *Curr Biol* 8, 393–403.

Burcham P.C. and Marnett L.J. (1994) Site-specific mutagenesis by propanodeoxyguanosine adduct carried on an M13 genome. *J Biol Chem* 269, 28844–28850.

Butler J. and Hoey B.M. (1993) The one-electron reduction potential of several substrates can be related to their reduction rates by cytochrome-P-450 reductase. *Biochim Biophys Acta* 1161, 73–78.

Cadet J., Berger M., Douki T., Ravanat J.-L. (1997) Oxidative damage to DNA: formation, measurement, and biological significance. *Rev Physiol Biochem Pharmacol* 131, 1–87.

Cadet J., Bourdat A.G., D'Ham C., Duarte V., Gasparutto D., Romieu A., Ravanat J.L. (2000) Oxidative base damage to DNA: specificity of base excision repair enzymes. *Mutat Res.* 462(2-3), 121-128.

Cadet J., Delatour T., Douki T., Gasparutto D., Pouget J.P., Ravanat J.L., Sauvaigo S. (1999) Hydroxyl radicals and DNA base damage. *Mutat Res* 8, 424(1-2), 9-21.

Chaudhary A.K., Nokubo M., Marnett L.J., Blair I.A. (1994) Analysis of the malondialdehyde-2′-deoxyguanosine adduct in rat liver DNA by gas chromatography/electron capture negative chemical ionization mass spectrometry. *Biol Mass Spectrom* 23, 457–464.

Cheng K.C., Cahill D.S., Kasai H., Nishimura S., Loeb L.A. (1992) 8-Hydroxyguanine, an abundant form of oxidative DNA damage, causes G-T and A-C substitutions. *J Biol Chem* 5, 267(1), 166-172.

Chung F., Young R., Hecht S.S. (1984) Formation of cyclic 1,N^2-propanodeoxyguanosine adducts in DNA upon reaction with acrolein and crotonaldehyde. *Cancer Res* 44, 990–995.

Chung F.-L., Chen H.C., Nath R.G. (1996) Lipid peroxidation as a potential endogeneous source for the formation of exocyclic DNA adducts. *Carcinogenesis* 17, 2105–2111.

Cooke M.S., Evans M.D., Dizdaroglu M., Lunec J. (2003) Oxidative DNA damage: mechanisms, mutation, and disease. *FASEB J* 17(10), 1195-1214.

Delaney M.O., Wiederholt C.J., Greenberg M.M. (2002) Fapy.dA induces nucleotide misincorporation translesionally by a DNA polymerase. *Angew Chem Int Ed Engl* 1, 41(5), 771-773.

Dianov G. and Lindahl T. (1994) Reconstitution of the DNA base excision-repair pathway. *Curr Biol* 4, 1069–1076.

Dianov G. and Lindahl T. (1994) Reconstitution of the DNA base excision-repair pathway. *Curr Biol* 1, 4(12), 1069-1076.

Dianov G., Bischoff C., Piotrowski J., Bohr V.A. (1998) Repair pathways for processing of 8-oxoguanine in DNA by mammalian cell extracts. *J Biol Chem* 11, 273(50), 33811-33816.

Dianov G., Price A., Lindahl T. (1992) Generation of single-nucleotide repair patches following excision of uracil residues from DNA. *Mol Cell Biol* 12(4), 1605-1612.

Dianov G.L., Prasad R., Wilson S.H., Bohr V.A. (1999) Role of DNA polymerase beta in the excision step of long patch mammalian base excision repair. *J Biol Chem* 14, 274(20), 13741-13743.

Digweed M. (1993) Human genetic instability syndromes: single gene defects with increased risk of cancer. *Toxicol Lett* 67, 259–281.

Dirksen M.L., Blakely W.F., Holwitt E., Dizdaroglu M. (1988) Effect of DNA conformation on the hydroxyl radical-induced formation of 8,5′-cyclopurine 2′-deoxyribonucleoside residues in DNA. *Int J Radiat Biol* 54, 195–204.

Dizdaroglu M. (1991) Chemical determination of free radical induced damage to DNA. *Free Radic Biol Med* 10, 225–242.

Dizdaroglu M. (1994) Chemical determination of oxidative DNA damage by gas chromatography–mass spectrometry. *Meth Enzymol* 234, 3–16.

Dizdaroglu M. (1986) Free-radical-induced formation of an 8,5′-cyclo-2′-deoxyguanosine moiety in deoxyribonucleic acid. *Biochem J* 238, 247–254.

Dizdaroglu M. (1992) Oxidative damage to DNA in mammalian chromatin. *Mutat Res* 275, 331–342.

Dizdaroglu M. *Mechanisms of free radical damage to DNA*. In: *DNA & Free Radicals: Techniques, Mechanisms & Applications*; O.I. Aruoma (Ed.), OICA International, Saint Lucia, 1998; 3–26.

Dizdaroglu M., Dirksen M.L., Jiang H.X., Robbins J.H. (1987) Ionising-radiation-induced damage in the DNA of cultured human cells. Identification of 8,5-cyclo-2-deoxyguanosine. *Biochem J* 241, 929–932.

Dizdaroglu M., Gajewski E., Reddy P., Margolis S.A. (1989) Structure of a hydroxyl radical induced DNA–protein cross-link involving thymine and tyrosine in nucleohistone. *Biochemistry* 28, 3625–3628.

Dizdaroglu M., Karakaya A., Jaruga P., Slupphaug G., Krokan H.E. (1996) Novel activities of human uracil DNA *N*-glycosylase for cytosine-derived products of oxidative DNA damage. *Nucleic Acids Res* 24, 418–422.

Douki T., Delatour T., Paganon F., Cadet J. (1996) Measurement of oxidative damage at pyrimidine bases in γ-irradiated DNA. *Chem Res Toxicol* 9, 1145–1151.

Du M.Q., Carmichael P.L., Phillips D.H. (1994) Induction of activating mutations in the human c-Ha-ras-1 proto-oncogene by oxygen free radicals. *Mol Carcinog* 11(3), 170-175.

Eide L., Luna L., Gustad E.C., Henderson P.T., Essigmann J.M., Demple B., Seeberg E. (2001) Human endonuclease III acts preferentially on DNA damage opposite guanine residues in DNA. *Biochemistry* 40, 6653–6659.

el Ghissassi F., Barbin A., Nair J., Bartsch H. (1995) Formation of 1,N6-ethenoadenine and 3,N4-ethenocytosine by lipid peroxidaton products and nucleic acid bases. *Chem Res Toxicol* 8, 278–283.

Evans J., Maccabee M., Hatahet Z., Courcelle J., Bockrath R., Ide H., Wallace S. (1993) Thymine ring saturation and fragmentation products: lesion bypass, misinsertion and implications for mutagenesis. *Mutat Res* 299(3-4), 147-156.

Evans M.D., Dizdaroglu M., Cooke M.S. (2004) Oxidative DNA damage and disease: induction, repair and significance. *Mutat Res* 567(1), 1-61.

Fatur T., Lah T.T., Filipič M. (2003) Cadmium inhibits repair of UV-, methyl methanesulfonate- and N-methyl-N-nitrosourea-induced DNA damage in Chinese hamster ovary cells. *Mutat Res* 529, 109-116.

Fedtke N., Boucheron J.A., Walker V.E., Swenberg J.A. (1990) Vinyl chloride-induced DNA adducts. II: Formation and persistence of 7-(2’-oxoethyl)guanine and N2,3-ethenoguanine in rat tissue DNA. *Carcinogenesis* 11, 1287–1292.

Feig D.I., Sowers L.C., Loeb L.A. (1994) Reverse chemical mutagenesis: identification of the mutagenic lesions resulting from reactive oxygen species-mediated damage to DNA. *Proc Natl Acad Sci USA* 5, 91(14), 6609-6613.

Filipič M. and Hei T.K. (2004) Mutagenicity of cadmium in mammalian cells: implication of oxidative DNA damage. *Mutat Res* 546, 81-91.

Fink S.P. and Marnett L.J. (2001) The relative contribution of adduct blockage and DNA repair on template utilization during replication of 1,N2-propanodeoxyguanosine and pyrimido. *Mutat Res* 485, 209–218.

Fink S.P., Reddy G.R., Marnett L.J. (1997) Mutagenicity in *Escherichia coli* of the major DNA adduct derived from the endonenous mutagen malondialdehyde. *Proc Natl Acad Sci USA* 94, 8652–8657.

Floyd R.A. (1990) The role of 8-hydroxyguanine in carcinogenesis. *Carcinogenesis* 11(9), 1447-1450.

Fortini P., Pascucci B., Parlanti E., D'Errico M., Simonelli V., Dogliotti E. (2003) 8-Oxoguanine DNA damage: at the crossroad of alternative repair pathways. *Mutat Res* 29, 531(1-2), 127-139.

Fowler R., White S., Koyama C., Moore S., Dunn R., Schaaper R. (2003) Interactions among the *Escherichia coli* mutT, mutM, and mutY damage prevention pathways. *DNA Repair* 2, 159–173.

Fridovich I. (1986) Biological effects of the superoxide radical. *Arch Biochem Biophys* 247, 1–11.

Frosina G., Fortini P., Rossi O., Carrozzino F., Raspaglio G., Cox L.S., Lane D.P., Abbondandolo A., Dogliotti E. (1996) Two pathways for base excision repair in mammalian cells. *J Biol Chem* 271, 9573–9578.

Fujikawa K., Kamiya H., Yakushiji H., Fujii Y., Nakabeppu Y., Kasai H. (1999) The oxidized forms of dATP are substrates for the human MutT homologue, the hMTH1 protein. *J Biol Chem* 274, 18201–18205.

Gallinari P. and Jiricny J. (1996) A new class of uracil-DNA glycosylases related to human thymine-DNA glycosylase. *Nature* 383, 735–738.

Graziewicz M.A., Zastawny T.H., Olinski R., Speina E., Siedlecki J., Tudek B. (2000) Fapyadenine is a moderately efficient chain terminator for prokaryotic DNA polymerases. *Free Radic Biol Med* 1, 28(1), 75-83.

Grollman A.P. and Moriya M. (1993) Mutagenesis by 8-oxoguanine: an enemy within. *Trends Genet* 9(7), 246-249.

Gros L., Saparbaev M.K., Laval J. (2002) Enzymology of the repair of free radicals-induced DNA damage. *Oncogene* 21, 8905–8925.

Habraken Y., Carter C.A., Sekiguchi M., Ludlum D.B. (1991) Release of N2,3-ethanoguanine from haloethylnitrosourea-treated DNA by *Escherichia coli* 3-methyladenine DNA glycosylase II. *Carcinogenesis* 12, 1971–1973.

Halliwell B. and Gutteridge J.M.C. *Free Radicals in Biology and Medicine*. Claredon Press, Oxford, 1989.

Ham A.J., Ranasinghe A., Morinello E.J., Nakamura J., Upton P.B., Johnson F., Swenberg J.A. (1999) Immunoaffinity/gas chromatography/high-resolution mass spectrometry method for the detection of N(2), 3-ethenoguanine. *Chem Res Toxicol* 12, 1240–1246.

Hanaoka T., Yamano Y., Hashimoto H., Kagawa J., Tsugane S. (2000) A preliminary evaluation of intra- and interindividual variations of hOGG1 messenger RNA levels in peripheral blood cells as determined by a real-time polymerase chain reaction technique. *Cancer Epidemiol Biomark Prev* 9, 1255–1258.

Hang B., Downing G., Guliaev A.B., Singer B. (2002) Novel activity of *Escherichia coli* mismatch uracil-DNA glycosylase (Mug) excising 8-(hydroxymethyl)-3,N4-ethenocytosine, a potential product resulting from glycidaldehyde reaction. *Biochemistry* 41, 2158–2165.

Hang B., Medina M., Fraenkel-Conrat H., Singer B. (1998) A 55-kDa protein isolated from human cells shows DNA glycosylase activity toward 3,N4-ethenocytosine and the G/T mismatch. *Proc Natl Acad Sci USA* 95, 13561–13566.

Hansson J. (1992) Inherited defects in DNA repair and susceptibility to DNA-damaging agents. *Toxicol Lett* 64–65, 141–148.

Harrison R. (2002) Structure and function of xanthine oxidoreductase: Where are we now? *Free Rad Biol Med* 33, 774–797.

Hartwig A. (1998) Carcinogenicity of metal compounds: possible role of DNA repair inhibition. *Toxicol Lett* 102–103, 235–239.

Hartwig A. and Schwerdtle T. (2002) Interactions by carcinogenic metal compounds with DNA repair processes: toxicological implications. *Toxicol Lett* 127, 47-54.

Hazra I.K., Izumi T., Maidt L., Floyd R.A., Mitra S. (1998) The presence of two distinct 8-oxoguanine repair enzymes in human cells: their potential complementary roles in preventing mutation. *Nucleic Acids Res* 26, 5116– 5122.

Hazra T.K., Izumi T., Boldogh I., Imhoff B., Kow Y.W., Jaruga P., Dizdaroglu M., Mitra S. (2002) Identification and characterization of a human DNA glycosylase for repair of modified bases in oxidatively damaged DNA. *Proc Natl Acad Sci USA* 99, 3523–3528.

Hazra T.K., Kow Y.W., Hatahet Z., Imhoff B., Boldogh I., Mokkapati S.K., Mitra S., Izumi T. (2002a) Identification and characterization of a novel human DNA glycosylase for repair of cytosine-derived lesions. *J Biol Chem* 277, 30417–30420.

Hei T.K. and Filipič M. (2004) Role of oxidative damage in the genotoxicity of arsenic. *Free Rad Med Biol* 37, 574-581.

Henle E.S., Luo Y., Gassmann W., Linn S. (1996) Oxidative damage to DNA constituents by iron-mediated Fenton reactions. The deoxyguanosine family. *J Biol Chem* 271, 21177–21186.

Henner W.D., Rodriguez L.O., Hecht S.M., Haseltine W.A. (1983) □-ray induced deoxyribonucleic acid strand breaks. *J Biol Chem* 258, 711–713.

Hollstein M., Shomer B., Greenblatt M., Soussi T., Hovig E., Montesano R., Harris C.C. (1996) Somatic point mutations in the p53 gene of human tumors and cell lines: updated compilation. *Nucleic Acids Res* 1, 24(1), 141-146.

Hollstein M., Sidransky D., Vogelstein B., Harris C.C. (1991) p53 mutations in human cancers. *Science* 5, 253(5015), 49-53.

Hsu I.C., Metcalf R.A., Sun T., Welsh J.A., Wang N.J., Harris C.C. (1991) Mutational hotspot in the p53 gene in human hepatocellular carcinomas. *Nature* 4, 350(6317), 427-428.

Hussain S.P., Aguilar F., Amstad P., Cerutti P. (1994) Oxy-radical induced mutagenesis of hotspot codons 248 and 249 of the human p53 gene. *Oncogene* 9(8), 2277-2781.

IARC. *Monographs on the Evaluation of Carcinogenic Risks to Humans*, vol. 49, IARC, Lyon (1990).

IARC. *Monographs on the Evaluation of Carcinogenic Risks to Humans*, vol. 52, IARC, Lyon (1991).
IARC. *Monographs on the Evaluation of Carcinogenic Risks to Humans*, vol. 58, IARC, Lyon (1993).
IARC. *Monographs on the Evaluation of the Carcinogenic Risk of Chemicals to Humans*, vol. 23, IARC, Lyon (1980).
Ide H., Tedzuka K., Shimzu H., Kimura Y., Purmal A.A., Wallace S.S., Kow Y.W. (1994) Alpha-deoxyadenosine, a major anoxic radiolysis product of adenine in DNA, is a substrate for *Escherichia coli* endonuclease IV. *Biochemistry* 28, 33(25), 7842-7847.
Johnson K.A., Fink S.P., Marnett L.J. (1997) Repair of propanodeoxyguanosine by nucleotide excision repair *in vivo* and *in vitro*. *J Biol Chem* 272, 11434–11438.
Kamath-Loeb A.S., Hizi A., Kasai H., Loeb L.A. (1997) Incorporation of the guanosine triphosphate analogs 8-oxodGTP and 8-NH2-dGTP by reverse transcriptases and mammalian DNA polymerases. *J Biol Chem* 272, 5892–5898.
Kamiya H. and Kasai H. (1997) Substitution and deletion mutations induced by 2-hydroxyadenine in *Escherichia coli*: effects of sequence contexts in leading and lagging strands. *Nucleic Acids Res* 15, 25(2), 304-311.
Kamiya H., Miura H., Murata-Kamiya N., Ishikawa H., Sakaguchi T., Inoue H., Sasaki T., Masutani C., Hanaoka F., Nishimura S. (1995) 8-Hydroxyadenine (7,8-dihydro-8-oxoadenine) induces misincorporation in *in vitro* DNA synthesis and mutations in NIH 3T3 cells. *Nucleic Acids Res* 11, 23(15), 2893-2899.
Kang D., Nishida J., Iyama A., Nakabeppu Y., Furuichi M., Fujiwara T., Sekiguchi M., Takeshige K. (1995) Intracellular localization of 8-oxo-dGTPase in human cells, with special reference to the role of the enzyme in mitochondria. *J Biol Chem* 270, 14659–14665.
Kasai H., Hayami H., Yamaizumi Z., Saito H., Nishimura S. (1984) Detection and identification of mutagens and carcinogens as their adducts with guanosine derivates. *Nucleic Acids Res* 12, 2127–2136.
Kasprzak K.S. and Buzard G.S.: *The role of metals in oxidative damage and redox cell signaling derangement.* In: *Molecular Biology and Toxicology of Metals*; J. Koropatnick and R. Zalups (Ed.), Taylor and Francis, London 2000; pp. 477–527.
Kasprzak K.S. In: *Toxicology of Metals*; L.W. Chang (Ed.), Lewis Publishers, Boca Raton, New York, London, Tokyo, 1996; pp. 299–320.
Kavli B., Sundheim O., Akbari M., Otterlei M., Nilsen H., Skorpen F., Aas P.A., Hagen L., Krokan H.E., Slupphaug G. (2002) HUNG2 is the major repair enzyme for removal of uracil from U:A matches, U:G mismatches, and U in single-stranded DNA, with hSMUG1 as a broad specificity backup. *J Biol Chem* 277, 39926–39936.
Kawanishi M., Matsuda T., Nakayama A., Takebe H., Matsui S., Yagi T. (1998) Molecular analysis of mutations induced by acrolein in human fibroblast cells using supf shuttle vector plasmids. *Mut Res Gen Toxicol Environ Mutagen* 417, 65–73.
Kawanishi S., Hiraku Y., Murata M., Oikawa S. (2002) The role of metals in site specific DNA damage with reference to carcinogens. *Free Rad Med Biol* 32, 822–832.
Keck K. (1968) Formation of cyclonucleotides during irradiation of aqueous solutions of purine nucleotides. *Z Naturforsch* B 23, 1034–1043.
Kehrer J.P. (1993) Free-radicals as mediators of tissue-injury and disease. *Crit Rev Toxicol* 23, 21–48.

Klungland A. and Lindahl T. (1997) Second pathway for completion of human DNA base excision-repair: reconstitution with purified proteins and requirement for DNase IV (FEN1). *EMBO J* 16, 3341–3348.

Klungland A. and Lindahl T. (1997) Second pathway for completion of human DNA base excision-repair: reconstitution with purified proteins and requirement for DNase IV (FEN1). *EMBO J* 2, 16(11), 3341-3348.

Klungland A., Rosewell I., Hollenbach S., Larsen E., Daly G., Epe B., Seeberg E., Lindahl T., Barnes D.E. (1999) Accumulation of premutagenic DNA lesions in mice defective in removal of oxidative base damage. *Proc Natl Acad Sci USA* 96, 13300–13305.

Kovacic P. and Wakelin L.P.G. D (2001) NA molecular electrostatic potential: novel perspectives for the mechanism of action of anticancer drugs involving electron transfer and oxidative stress. *Anti-Cancer Drug Des* 16, 175–184.

Krokan H.E., Standal R., Slupphaug G. (1997) DNA glycosylases in the base excision repair of DNA. *Biochem J* 325, 1–16.

Langouet S., Mican A.N., Muller M., Fink S.P., Marnett L.J., Muhle S.A., Guengerich F.P. (1998) Misincorporation of nucleotides opposite five-membered exocyclic ring guanine derivatives by *Escherichia coli* polymerases *in vitro* and *in vivo*: 1,N2-ethenoguanine, 5,6,7,9-tetrahydro-9-oxoimidazo[1, 2-a]purine, and 5,6,7,9-tetrahydro-7-hydroxy-9-oxoimidazo[1, 2-a]purine. *Biochemistry* 37, 5184–5193

Langouet S., Muller M., Guengerich F.P. (1997) Misincorporation of dNTPs opposite 1,N2-ethenoguanine and 5,6,7,9-tetrahydro-7-hydroxy-9-oxoimidazo[1,2-a]purine in oligonucleotides by *Escherichia coli* polymerases I exo- and II exo-, T7 polymerase exo-, human immunodeficiency virus-1 reverse transcriptase, and rat polymerase beta. *Biochemistry* 36, 6069–6079.

Le Page F., Klungland A., Barnes D.E., Sarasin A., Boiteux S. (2000) Transcription coupled repair of 8-oxoguanine in murine cells: the Ogg1 protein is required for repair in nontranscribed sequences but not in transcribed sequences. *Proc Natl Acad Sci USA* 97, 8397–8402.

Lindahl T. (1974) An N-glycosidase from Escherichia coli that releases free uracil from DNA containing deaminated cytosine residues. *Proc Natl Acad Sci USA* 71(9), 3649-3653.

Lindahl T. (1993) Instability and decay of the primary structure of DNA. *Nature* 22, 362 (6422), 709-715.

Lindahl T. and Wood R.D. (1999) Quality control by DNA repair. *Science* 286, 1897–1905.

Luo Y., Henle E.S., Linn S. (1996) Oxidative damage to DNA constituents by iron-mediated Fenton reactions. *J Biol Chem* 271, 21167–21176.

Maccabee M., Evans J.S., Glackin M.P., Hatahet Z., Wallace S.S. (1994) Pyrimidine ring fragmentation products. Effects of lesion structure and sequence context on mutagenesis. *J Mol Biol* 18, 236(2), 514-530.

Marnett L.J. (1999) Lipid peroxidation-DNA damage by malondialdehyde. *Mutat Res* 424, 83–95.

Marnett L.J. and Burcham P.C. (1993) Endogenous DNA adducts: Potential and paradox. *Chem Res Toxicol* 6, 771–785.

Mates M. (2000) Effects of antioxidant enzymes in the molecular control of reactive oxygen species toxicology. *Toxicology* 16, 153(1-3), 83-104.

Matijasevic Z., Sekiguchi M., Ludlum D.B. (1992) Release of N2,3-ethenoguanine from chloroacetaldehyde-treated DNA by *Escherichia coli* 3-methyladenine DNA glycosylase II. *Proc Natl Acad Sci USA* 89, 9331–9334.

Matsumoto Y. and Kim K. (1995) Excision of deoxyribose phosphate residues by DNA polymerase beta during DNA repair. *Science* 4, 269(5224), 699-702.

Matsumoto Y., Kim K., Bogenhagen D.F. (1994) Proliferating cell nuclear antigen-dependent abasic site repair in *Xenopus laevis* oocytes: an alternative pathway of base excision DNA repair. *Mol Cell Biol* 14, 6187–6197.

Matsumoto Y., Kim K., Hurwitz J., Gary R., Levin D.S., Tomkinson A.E., Park M.S. (1999) Reconstitution of proliferating cell nuclear antigen-dependent repair of apurinic/apyrimidinic sites with purified human proteins. *J Biol Chem* 19, 274(47), 33703-33708.

McBride T.J., Preston B.D., Loeb L.A. (1991) Mutagenic spectrum resulting from DNA damage by oxygen radicals. *Biochemistry* 8, 30(1), 207-213.

Michaels M.L. and Miller J.H. (1992) The GO system protects organisms from the mutagenic effect of the spontaneous lesion 8-hydroxyguanine (7,8-dihydro-8-oxoguanine). *J Bacteriol* 174, 6321–6325.

Miller H., Prasad R., Wilson S.H., Johnson F., Grollman A.P. (2000) 8-oxodGTP incorporation by DNA polymerase beta is modified by active-site residue Asn279. *Biochemistry* 39, 1029–1033.

Mokkapati S.K., Fernandez de Henestrosa A.R., Bhagwat A.S. (2001) *Escherichia coli* DNA glycosylase Mug: a growth-regulated enzyme required for mutation avoidance in stationary-phase cells. *Mol Microbiol* 41, 1101–1111.

Moller P. and Wallin H. (1998) Adduct formation, mutagenesis and nucleotide excision repair of DNA damage produced by reactive oxygen species and lipid peroxidation product. *Mutat Res* 410(3), 271-290.

Morinello E.J., Ham A.J., Ranasinghe A., Sangaiah R., Swenberg J.A. (2001) Simultaneous quantitation of N(2),3-ethenoguanine and 1,N(2)-ethenoguanine with an immunoaffinity/gas chromatography/high-resolution mass spectrometry assay. *Chem Res Toxicol* 14, 327–334.

Moriya M. (1993) Single-stranded shuttle phagemid for mutagenesis studies in mammalian cells: 8-oxoguanine in DNA induces targeted G.C-->T.A transversions in simian kidney cells. *Proc Natl Acad Sci USA* 1, 90(3), 1122-1126.

Moriya M., Zhang W., Johnson F., Grollman A.P. (1994) Mutagenic potency of exocyclic DNA adducts: marked differences between *Escheriachia coli* and simian kidney cells. *Proc Natl Acad Sci USA* 91, 11899–11903.

Morland I., Rolseth V., Luna L., Rognes T., Bjørås M., Seeberg E. (2002) Human DNA glycosylases of the bacterial Fpg/MutM superfamily: an alternative pathway for the repair of 8-oxoguanine and other oxidation products in DNA. *Nucleic Acids Res* 30, 4926–4936.

Murata-Kamiya N., Kamiya H., Iwamoto N., Kasai H. (1995) Formation of a mutagen, glyoxal, from DNA treated with oxygen free radicals. *Carcinogenesis* 16, 2251–2253.

Nackerdien Z., Rao G., Cacciuttolo M.A., Gajewski E., Dizdaroglu M. (1991) Chemical nature of DNA–protein cross-links produced in mammalian chromatin by hydrogen peroxide in the presence of iron or copper ions. *Biochemistry* 30, 4873–4879.

Nath R.G. and Chung F. (1994) Detection of exocyclic 1,N^2-propanodeoxyguanosine adducts as common DNA lesions in rodents and humans. *Proc Natl Acad Sci USA* 91, 7491–7495.

Nath R.G., Chen H.J., Nishikawa A., Young-Sciame R., Chung F.L. (1994) A 32P-postlabeling method for simultaneous detection and quantification of exocyclic etheno and propano adducts in DNA. *Carcinogenesis* 15, 979–984.

Nath R.G., Ocando J.E., Chung F. (1996) Detection of 1,N^2-propanodeoxyguanosine adducts as potential endogenous lesions in rodent and human tissues. *Cancer Res* 56, 452–456.

Neddermann P. and Jiricny J. (1993) The purification of a mismatch-specific thymine-DNA glycosylase from HeLa cells. *J Biol Chem* 268, 21218–21224.

Neddermann P. and Jiricny J. (1994) Efficient removal of uracil from G:U mispairs by the mismatch-specific thymine DNA glycosylase from HeLa cells. *Proc Natl Acad Sci USA* 91, 1642–1646.

Niedernhofer L.J., Daniels J.S., Rouzer C.A., Greene R.E., Marnett L.J. (2003) Malondialdehyde, a product of lipid peroxidation, is mutagenic in human cells. *J Biol Chem* 278, 31426–31433.

Nilsen H., Haushalter K.A., Robins P., Barnes D.E., Verdine G.L., Lindahl T. (2001) Excision of deaminated cytosine from the vertebrate genome: role of the SMUG1 uracil-DNA glycosylase. *EMBO J* 20, 4278–4286.

Nishioka K., Ohtsubo T., Oda H., Fujiwara T., Kang D., Sugimachi K., Nakabeppu Y. (1999) Expression and differential intracellular localization of two major forms of human 8-oxoguanine DNA glycosylase encoded by alternatively spliced OGG1 mRNAs. *Mol Biol Cell* 10, 1637–1652.

Nishiyama T., Hagiwara Y., Hagiwara H., Shibamoto T. (1994) Inhibitory effect of 2″-*o*-glycosyl isovitexin and α-tocopherol on genotoxic glyoxal formation in a lipid peroxidation system. *Fd Chem Toxic* 32, 1047–1051.

Nordberg J. and Arner E.S. (2001) Reactive oxygen species, antioxidants, and the mammalian thioredoxin system. *Free Radic Biol Med* 1, 31(11), 1287-1312.

O'Connor T.R. and Laval J. (1991) Human cDNA expressing a functional DNA glycosylase excising 3-methyladenine and 7-methylguanine. *Biochem Biophys Res Commun* 176, 1170–1177.

O'Neill R.J., Vorob'eva O.V., Shahbakhti H., Zmuda E., Bhagwat A.S., Baldwin G.S. (2003) Mismatch Uracil Glycosylase from *Escherichia coli*: a general mismatch or a specific dna glycosylase? *J Biol Chem* 278, 20526–20532.

Ohtsubo T., Nishioka K., Imaiso Y., Iwai S., Shimokawa H., Oda H., Fujiwara T., Nakabeppu Y. (2000) Identification of human MutY homolog (hMYH) as a repair enzyme for 2-hydroxyadenine in DNA and detection of multiple forms of hMYH located in nuclei and mitochondria. *Nucleic Acids Res* 28, 1355–1364.

Olinski R., Gackowski D., Foksinski M., Rozalski R., Roszkowski K., Jaruga P. (2002) Oxidative DNA damage: assessment of the role in carcinogenesis, atherosclerosis, and acquired immunodeficiency syndrome. *Free Radic Biol Med* 15, 33(2), 192-200.

Olinski R., Nackerdien Z., Dizdaroglu M. (1992) DNA–protein cross-linking between thymine and tyrosine in chromatin of gamma-irradiated or H_2O_2-treated cultured human cells. *Arch Biochem Biophys* 297, 139–143.

Otterlei M., Warbrick E., Nagelhus T.A., Haug T., Slupphaug G., Akbari M., Aas P.A., Steinsbekk K., Bakke O., Krokan H.E. (1999) Post-replicative base excision repair in replication foci. *EMBO J* 18, 3834–3844.

Palejwala V.A., Simha D., Humayun M.Z. (1991) Mechanisms of mutagenesis by exocyclic DNA adducts. Transfection of M13 viral DNA bearing a site-specific adduct shows that ethenocytosine is a highly efficient RecA-independent mutagenic noninstructional lesion. *Biochemistry* 30, 8736–8743.

Pandya G.A. and Moriya M. (1996) 1,*N*^{6}-Ethenodeoxyadenosine, a DNA adduct highly mutagenic in mammalian cells. *Biochemistry* 35, 11487–11492.

Petronzelli F., Riccio A., Markham G.D., Seeholzer S.H., Genuardi M., Karbowski M., Yeung A.T., Matsumoto Y., Bellacosa A. (2000) Investigation of the substrate spectrum of the human mismatch-specific DNA N-glycosylase MED1 (MBD4): fundamental role of the catalytic domain. *J Cell Physiol* 185, 473–480.

Privezentzev C.V., Saparbaev M., Sambandam A., Greenberg M.M., Laval J. (2000) AlkA protein is the third *Escherichia coli* DNA repair protein excising a ring fragmentation product of thymine. *Biochemistry* 39, 14263–14268.

Proctor PH and Reynolds ES. (1984) Free Radicals and Disease in Man. *Physiological Chemistry and Physics and Medical NMR*, 16 175-195.

Radicella J.P., Dherin C., Desmaze C., Fox M.S., Boiteux S. (1997) Cloning and characterization of hOGG1, a human homolog of the OGG1 gene of *Saccharomyces cerevisiae*. *Proc Natl Acad Sci USA* 94, 8010–8015.

Raleigh J.A., Kremers W., Whitehouse R. (1976) Radiation chemistry of nucleotides: 8,5′-cyclonucleotide formation and phosphate release initiated by hydroxyl radical attack on adenosine monophosphates. *Radiat Res* 65, 422–441.

Reardon J.T., Bessho T., Kung H.C., Bolton P.H., Sancar A. (1997) *In vitro* repair of oxidative DNA damage by human nucleotide excision repair system: possible explanation for neurodegeneration in xeroderma pigmentosum patients. *Proc Natl Acad Sci USA* 19, 94(17), 9463-9468.

Richeson C.E., Mulder P., Bowry V.W., Ingold K.U. (1998) The complex chemistry of peroxynitrite decomposition: new insights. *J Am Chem Soc* 120, 7211-7219.

Rossman T.G. and Goncharova E.I. (1998) Spontaneous mutagenesis in mammalian cells is caused mainly by oxidative events and can be blocked by antioxidants and metallothionein. *Mutat Res* 18, 402(1-2), 103-110.

Sakai Y., Furuichi M., Takahashi M., Mishima M., Iwai S., Shirakawa M., Nakabeppu Y. (2002) A molecular basis for the selective recognition of 2-hydroxy-dATP and 8-oxo-dGTP by human MTH1. *J Biol Chem* 277, 8579–8587.

Sakumi K., Furuichi M., Tsuzuki T., Kakuman T., Kawabata S., Maki H., Sekiguchi M. (1993) Cloning and expression of a cDNA for a human enzyme that hydrolyzes 8-oxo-dGTP, a mutagenic substrate for DNA synthesis. *J Biol Chem* 268, 23524–23530.

Salles B., Sattler U., Bozzato C., Calsou P. (1999) Repair of oxidative DNA damage in vitro: a tool for screening antioxidative compounds. *Food Chem Toxicol* 37(9-10), 1009-1014.

Saparbaev M. and J. Laval (1994) Excision of hypoxanthine from DNA containing dIMP residues by the *Escherichia coli*, yeast, rat, and human alkylpurine DNA glycosylases. *Proc Natl Acad Sci USA* 91, 5873–5877.

Saparbaev M., Kleibl K., Laval J. (1995) *Escherichia coli*, *Saccharomyces cerevisiae*, rat and human 3-methyladenine DNA glycosylases repair 1,N6-ethenoadenine when present in DNA. *Nucl Acids Res* 23, 3750–3755.

Saparbaev M., Langouet S., Privezentzev C.V., Guengerich F.P., Cai H., Elder R.H., Laval J. (2002) 1,N(2)-ethenoguanine, a mutagenic DNA adduct, is a primary substrate of

Escherichia coli mismatch-specific uracil-DNA glycosylase and human alkylpurine-DNA-N-glycosylase. *J Biol Chem* 277, 26987–26993.

Seto H., Okuda T., Takesue T., Ikemura T. (1983) Reaction of malondialdehyde with nucleic acid: I. Formation of fluorescent pyrimido[1,2-α]purin-10(3*H*)-one nucleotides. *Bull Chem Soc Jpn* 56, 1799–1802.

Shapiro R. and Hachmann J. (1966) The reaction of guanine derivates with 1,2-dicarbonyl compounds. *Biochemistry* 5, 2799–2807.

Shibutani S., Takeshita M., Grollman A.P. (1991) Insertion of specific bases during DNA synthesis past the oxidation damaged base 8-oxodG. *Nature* 349, 431–434.

Sies H. (1991) Oxidative stress: from basic research to clinical application. *Am J Med* 30, 91(3C), 31S-38S.

Singer B., Antoccia A., Basu A.K., Dosanjh M.K., Fraenkel-Conrat H., Gallagher P.E., Kusmierek J.T., Qiu Z.H., Rydberg B. (1992) Both purified human 1,N6-ethenoadenine-binding protein and purified human 3-methyladenine-DNA glycosylase act on 1,N6-ethenoadenine and 3-methyladenine. *Proc Natl Acad Sci USA* 89, 9386–9390.

Singh J., Carlisle D.L., Pritchard D.E., Patierno S.R. (1998) Chromium-induced genotoxicity and apoptosis: relationship to chromium carcinogenesis. *Oncology Reports* 5, 1307–1318.

Slupphaug G., Kavli B., Krokan H.E. (2003) The interacting pathways for prevention and repair of oxidative DNA damage. *Mutat Res* 29, 531(1-2), 231-251.

Slupska M.M., Baikalov C., Luther W.M., Chiang J.H., Wei Y.F., Miller J.H. (1996) Cloning and sequencing a human homolog (hMYH) of the *Escherichia coli mutY* gene whose function is required for the repair of oxidative DNA damage. *J Bacteriol* 178, 3885–3892.

Slupska M.M., Luther W.M., Chiang J.H., Yang H.J., Miller J.H. (1999) Functional expression of hMYH, a human homolog of the *Escherichia coli* MutY protein. *J Bacteriol* 181, 6210–6213.

Sodum R.S. and Chung F.L. (1988) 1,N2-ethenodeoxyguanosine as a potential marker for DNA adduct formation by trans-4-hydroxy-2-nonenal. *Cancer Res* 48,. 320–323.

Sodum R.S. and Chung F.L. (1991) Stereoselective formation of in vitro nucleic acid adducts by 2,3-epoxy-4-hydroxynonanal. *Cancer Res* 51, 137–143.

Steinberg D. (1997) Low density lipoprotein oxidation and its pathobiological significance. *J BiolChem* 272, 20963–20966.

Symons M.C.R. and Gutteridge J.M.C. *Free radicals and iron: chemistry, biology, and medicine*. Oxford: Oxford Science Publications, 1998.

Tajiri T., Maki H., Sakumi K., Sekiguchi M. (1995) Functional cooperation of MutT, MutM and MutY proteins in preventing mutations caused by spontaneous oxidation of guanine nucleotide in *Escherichia coli*. *Mutat Res* 336, 257–267.

Tajiri T., Maki H., Sekiguchi M. (1995) Functional cooperation of MutT, MutM and MutY proteins in preventing mutations caused by spontaneous oxidation of guanine nucleotide in Escherichia coli. *Mutat Res* 336(3), 257-267.

Takao M., Kanno S., Kobayashi K., Zhang Q.M., Yonei S., van der Horst G.T.J., Yasui A. (2002) A back-up glycosylase in Nth1 knock-out mice is a functional Nei (endonuclease VIII) homologue. *J Biol Chem* 277, 42205–42213.

Takao M., Kanno S., Shiromoto T., Hasegawa R., Ide H., Ikeda S., Sarker A.H., Seki S., Xing J.Z., Le X.C., Weinfeld M., Kobayashi K., Miyazaki J., Muijtjens M., Hoeijmakers J.H.J., van der Horst G., Yasui A. (2002) Novel nuclear and mitochondrial glycosylases revealed by disruption of the mouse *Nth1* gene encoding an endonuclease III homolog for repair of thymine glycols. *EMBO J* 21, 3486–3493.

Takao M., Zhang Q.M., Yonei S., Yasui A. (1999) Differential subcellular localization of human MutY homolog (hMYH) and the functional activity of adenine:8-oxoguanine DNA glycosylase. *Nucleic Acids Res* 27, 3638–3644.

Tan X., Grollman A.P., Shibutani S. (1999) Comparison of the mutagenic properties of 8-oxo-7,8-dihydro-2'-deoxyadenosine and 8-oxo-7,8-dihydro-2'-deoxyguanosine DNA lesions in mammalian cells. *Carcinogenesis* 20(12), 2287-2292.

Teebor G.W., Boorstein R.J., Cadet J. (1988) The repairability of oxidative free radical mediated damage to DNA: a review. *Int J Radiat Biol* 54, 131–150.

Téoule R., Bert C., Bonicel A. (1977) Thymine fragment damage retained in the DNA polynucleotide chain after gamma irradiation in aerated solutions (II). *Radiat Res* 72, 190–200.

Toyokuni S., Mori T., Hiai H., Dizdaroglu M. (1995) Treatment of Wistar rats with a renal carcinogen, ferric nitrilotriacetate, causes DNA–protein cross-linking between thymine and tyrosine in their renal chromatin. *Int J Cancer* 62, 309–313.

Tremblay S., Douki T., Cadet J., Wagner J.R. (1999) 2′-Deoxycytidine glycols, a missing link in the free radical-mediated oxidation of DNA. *J Biol Chem* 274, 20833–20838.

Tsuzuki T., Egashira A., Igarashi H., Iwakuma T., Nakatsura Y., Tominaga Y., Kawate H., Nakao K., Nakamura K., Ide F., Kura S., Nakabeppu Y., Katsuki M., Ishikawa T., Sekiguchi M. (2001) Spontaneous tumorigenesis in mice defective in the *MTH1* gene encoding 8-oxo-dGTPase. *Proc Natl Acad Sci USA* 98, 11456–11461.

Van Heerebeek L., Meischl C., Stooker W., Meijer C.L.J.M., Niessen H.W.M., Roos D. (2002) NADPH oxidase(s): New source(s) of reactive oxygen species in the vascular system? *J Clin Pathol* 55, 561–568.

Verjat T., Dhenaut A., Radicella J.P., Araneda S. (2000) Detection of 8-oxoG DNA glycosylase activity and OGG1 transcripts in the rat CNS. *Mutat Res* 460, 127–138.

Vignais P.V. (2002) The superoxide-generating NADPH oxidase: Structural aspects and activation mechanism. *Cell Mol Life Sci* 59, 1428–1459.

von Sonntag C. *The Chemical Basis of Radiation Biology*, Taylor and Francis, New York, 1987.

Wagner J.R., van Lier J.E., Berger M., Cadet J. (1994) Thymidine hydroperoxides: structural assignment, conformational features, and thermal decomposition in water. *J Am Chem Soc* 116, 2235–2242.

Wang D., Kreutzer D.A., Essigmann J.M. (1998) Mutagenicity and repair of oxidative DNA damage: insights from studies using defined lesions. *Mutat Res* 25, 400(1-2), 99-115.

Wiederholt C.J. and Greenberg M.M. (2002) Fapy.dG instructs Klenow exo(-) to misincorporate deoxyadenosine. *J Am Chem Soc* 26, 124(25), 7278-7279.

Wikman H., Risch A., Klimek F., Schmezer P., Spiegelhalder B., Dienemann H., Kayser K., Schulz V., Drings P., Bartsch H. (2000) hOGG1 polymorphism and loss of heterozygosity (LOH): significance for lung cancer susceptibility in a Caucasian population. *Int J Cancer* 88, 932–937.

Winter C.K., Segall H.J., Haddon W.F. (1986) Formation of cyclic adducts of deoxyguanosine with the aldehydes *trans*-4-hydroxy-2-hexanal and *trans*-4-hydroxynonenal *in vitro*. *Cancer Res* 46, 5682–5686.

Yau T.M. (1979) Mutagenicity and cytotoxicity of malonaldehyde in mammalian-cells. *Mech Ageing Dev* 11, 137–144.

Yoshida M., Makino K., Morita H., Terato H., Ohyama Y., Ide H. (1997) Substrate and mispairing properties of 5-formyl-2'-deoxyuridine 5'-triphosphate assessed by *in vitro* DNA polymerase reactions. *Nucleic Acids Res* 15, 25(8), 1570-1577.

Zhang Q.M. and Dianov G.L. (2005) DNA repair fidelity of base excision repair pathways in human cell extracts. *DNA Repair (Amst)* 3, 4(2), 263-270.

Zollner H., Esterbauer H., Schauenstein E. (1975) Relationship between chemical constitution and therapeutic activity of alpha,beta-unsaturated aldehydes in Ehrlich-solid-tumor of mouse. *Zeit Krebs Klin Onkol* 83, 27–30.

In: New Developments in Mutation Research
Editor: Charles L. Valon, pp. 133-148
ISBN 1-59454-664-9

Chapter 6

TRANSLESION DNA SYNTHESIS ACROSS MONO ADP-RIBOSYLATED DEOXYGUANOSINE BY Y-FAMILY DNA POLYMERASES

***Masanobu Kawanishi*[1], *Kazuki Matsukawa*[1], *Eiji Ohashi*[2], *Takeji Takamura*[3], *Yukari Totsuka*[3], *Masahiko Watanabe*[3], *Takashi Sugimura*[3], *Keiji Wakabayashi*[3], *Fumio Hanaoka*[4], *Haruo Ohmori*[2] and *Takashi Yagi*[1*]**

[1]Research Institute for Advanced Science and Technology, Osaka Prefecture University, Sakai, Osaka, Japan

[2]Institute for Virus Research, Kyoto University, Shogoin, Sakyo-ku, Kyoto, Kyoto, Japan

[3]Cancer Prevention Basic Research Project, National Cancer Center Research Institute, Tokyo, Japan

[4]Graduate School of Frontier Bioscience, Osaka University, and CREST, Japan Science and Technology Corporation, Osaka, Japan

ABSTRACT

Pierisin-1 is a potent apoptosis-inducing protein found in a pupal extract of the cabbage white butterfly, *Pieris rapae*. Pierisin-1 catalyzes mono (ADP-ribosyl)ation of the 2'-deoxyguanosine (dG) residue and produces a bulky adduct, N^2-(ADP-ribos-1-yl)-2'-deoxyguanosine N^2-ADPR-dG in DNA, which leads to the apoptosis of tumor cells. N^2-ADPR-dG generated G to C and T transversion mutations in mammalian cells when assayed with the endogenous *HPRT* gene and the *supF* gene on a shuttle vector as the marker gene. As mutations are often generated during bypassing of damaged nucleotides in DNA, here we explored the translesion DNA synthesis across N^2-ADPR-dG conducted by mammalian Y-family DNA polymerases *in vitro*. A site-specifically modified

[*] To whom correspondence should be addressed: Takashi Yagi, Research Institute for Advanced Science and Technology, Osaka Prefecture University, 1-2 Gakuen-cho, Sakai-city, Osaka, 599-8570 Japan
Phone: +81(0)72-254-9862, Fax: +81(0)72-254-9938, E-mail: yagi-t@riast.osakafu-u.ac.jp

oligodeoxynucleotide containing a single N^2-ADPR-dG was used as a template for ^{32}P-labelled-primer extension reactions catalyzed by DNA polymerase (Pol) η, ι and κ. The results showed that Pol κ bypassed through N^2-ADPR-dG by inserting dCMP preferentially opposite the lesion. Pol κ also extended DNA synthesis from a mismatched primer terminus where dG, dA or dT had been placed opposite the lesion. Pol ι incorporated dCMP and dTMP opposite N^2-ADPR-dG, but could not extend the DNA strand past the lesion. Pol η inserted dCMP and extended the DNA strand poorly past the lesion. These results could not clarify completely which polymerase generates G to C and T transversions as observed *in vivo* mutation analyses, but indicated that Pol κ could extend DNA synthesis from the mismatch terminus once a misincorporation of non-complementary nucleotide occurred. Therefore, Pol κ may play an important role in both the error-free and error-prone bypass of N^2-ADPR-dG.

INTRODUCTION

Pierisin-1 is a 98-kDa protein, which is abundant in the fifth instar larvae and the early phase pupae of the cabbage white butterfly, *Pieris rapae* [1-4]. This protein exhibits potent cytotoxic effects against various human cell lines [1, 2]. Human uterine cervical carcinoma HeLa cells are most sensitive to pierisin-1 with an IC_{50} value of 0.043 ng/mL, whereas human leukemia U937 cells are least sensitive with an IC_{50} value of 150 ng/mL [5-7]. Pierisin-1 induces apoptotic cell death in most cancer cell lines, resulting in characteristic molecular features such as cleavage of DNA to nucleosome units and of poly(ADP-ribose) polymerase [2, 5, 8].

The molecular cloning of a complementary DNA (cDNA) of pierisin-1 from *Pieris rapae* revealed that the protein consists of 850 amino acids and its 27-kDa N-terminal portion shares sequence homology with the enzyme units of ADP-ribosylating toxins, including the A-subunit of cholera toxin [9]. Pierisin-1 also has a glutamic acid residue at the putative NAD-binding site, conserved in all toxins, and its substitution caused a loss of cytotoxic activity [9, 10]. Moreover, inhibitors of ADP-ribosylating enzymes reduce pierisin-1-induced apoptosis. These results suggest an essential role for ADP-ribosylating activity in exerting the cytotoxicity of pierisin-1 [9]. Similarly, pierisin-2 from another cabbage butterfly *Pieris brassicae*, the deduced amino acid sequence of which is 91% identical with that of pierisin-1, has been suggested to exert its action through ADP-ribosyltransferase [7, 11].

The C-terminal region of pierisin-1 shares sequence similarity with HA-33 (or HA1), a subcomponent of hemagglutinin of botulinum toxin [12, 13]. An *in vitro* expressed protein consisting of only the N-terminal region exhibited cytotoxicity and apoptosis-inducing activity when it was incorporated by electroporation [6]. However, the N-terminal truncated protein alone could not be incorporated into the cells. The remaining 71-kDa C-terminal region, which consists of possible lectin domains of a ricin B-chain, binds to receptors on the surface of target cells and incorporates the protein into cells [6]. Two neutral glycosphingolipids, globotriaosylceramide (Gb3) and globotetraosylceramide (Gb4) have been identified as receptor molecules [14]. Pierisin-1-insensitive mouse melanoma MEB4 cells (IC_{50}: 270 ng/mL, [6]) lacks Gb3 and Gb4, whereas pierisin-1-sensitive cells such as HeLa and human gastric carcinoma TMK-1 cells (IC_{50}: 0.75 ng/mL, [5]) have large amounts of Gb3 and Gb4 receptors [14]. The presence of pierisin-1 receptors on the cells is an important factor affecting sensitivity to pierisin-1 [14].

As described above, pierisin-1 binds to Gb3 and Gb4 receptors in the C-terminal region in a manner similar to ricin, and then it is incorporated into cells and exhibits cytotoxicity by ADP-ribosylation catalyzed by the N-terminal region [14]. Remarkably, the target molecule for ADP-ribosylation by pierisin-1 is DNA, not proteins, thus providing a sharp contrast to bacteria-derived ADP-ribosylating toxins such as cholera toxin, pertussis toxin and diphtheria toxin [15]. Pierisin-1 efficiently catalyzes mono (ADP-ribosyl)ation of double-stranded DNA with β-NAD [15, 16]. The ADP-ribose moiety of β-NAD is transferred by pierisin-1 to the amino group at the N^2 position of deoxyguanosine (dG) and a bulky adduct, N^2-(ADP-ribos-1-yl)-2'-deoxyguanosine (N^2-ADPR-dG: shown in Figure 1), is produced [15]. Using a ^{32}P-postlabeling method, N^2-ADPR-dG was detected in DNA from pierisin-1-treated HeLa cells [15]. Pierisin-1 is, therefore, a very unique ADP-ribosyltransferase targeting the N^2 amino group of dG residues of DNA, and the resulting product N^2-ADPR-dG is a novel DNA adduct.

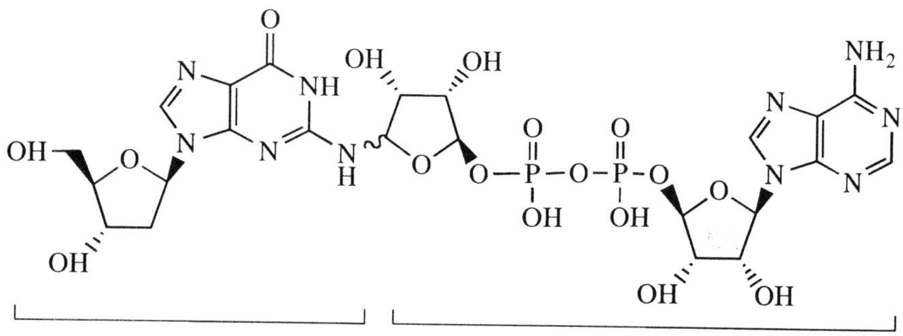

Figure 1. Chemical structure of N^2-(ADP-ribos-1-yl)-2'-deoxyguanosine (N^2-ADPR-dG).

DNA lesions frequently lead to mutagenesis and carcinogenesis. Metabolites of benzo[*a*]pyrene (B[*a*]P) and tamoxifen (TAM) react also with the exocyclic N^2-amino group of dG residues to form N^2-BPDE(B[*a*]P diol epoxides)-dG and N^2-TAM-dG, respectively [17-20]. These chemicals predominantly induce G to T transversions [21-24]. Generation of N^2-ADPR-dG by pierisin-1 also enhances mutations. Mutational analyses with the hypoxanthine-guanine phosphoribosyltransferase (*HPRT*) locus in Chinese hamster lung (CHL) cells or the *supF* gene in the shuttle vector plasmid pMY189 as the target marker gene revealed that pierisin-1 enhanced G:C to C:G and G:C to T:A base substitutions [25].

Most of DNA lesions are removed by multiple DNA repair mechanisms, but some of them may escape from the repair and block the progression of the replication fork. The ordinary replicative DNA polymerases (Pol) with high-fidelity due to concomitant proofreading 3'-5' exonuclease activity, such as mammalian Pol δ and Pol ε and bacterial Pol I and Pol III, are unable to replicate damaged DNA [26]. Therefore, organisms have highly specialized DNA polymerases that can accommodate DNA lesions in their active site and carry out aberrant DNA synthesis, which is characterized by poor replication accuracy, the favored formation of non-Watson-Crick base pairs, efficient mismatch extension, and most importantly, an ability to replicate through DNA damage [27]. The synthesis of DNA across lesions is generally called "translesion DNA synthesis" (TLS) and usually requires one or more members of the recently discovered Y-family specialized DNA polymerases suited for lesion bypass [26]. These polymerases are found in prokaryotes, archaea and eukaryotes, which include organisms from unicellular life forms to humans. The Y-family consists of, for example, eukaryotic Pol η, Pol ι, Pol κ and Rev1, as well as prokaryotic Pol IV and Pol V [28]. In addition, the B-family polymerase Pol ζ working in cooperation with some of the Y-family polymerases is also able to replicate damaged templates [29-32]. In human beings, a defect in Pol η causes xeroderma pigmentosum variant (XPV), a hereditary disease with sunlight-sensitive and cancer prone symptoms [33-35]. Biochemical studies on their properties and crystallographic data have deepened our understanding of the question how the Y-family polymerases cope with damaged DNA templates [27, 36-41].

In this study, we explored the TLS across the novel DNA adduct N^2-ADPR-dG conducted by human Y-family DNA polymerases *in vitro*. Using a site-specifically modified oligodeoxynucleotide containing a single N^2-ADPR-dG as a template, ^{32}P-labelled-primer extension reactions were performed with human DNA polymerases η, ι and κ.

Materials and Methods

Materials

DNA polymerase η, ι and κ were purified as described [33, 42, 43]. The Pol κ protein used in this study is a truncated form, which is a catalytically active fragment lacking the C-terminal 310 amino acid residues [43]. β-NAD was obtained from Sigma-Aldrich Japan (Tokyo, Japan). Pronase (1,000 units/mg) was purchased from Kaken Seiyaku (Tokyo, Japan). [γ-^{32}P]ATP (3,000 Ci/mmol) was obtained from Amersham Biosciences Ltd. (Buckinghamshire, U. K.), and T4 polynucleotide kinase and Klenow fragment were purchased from Toyobo Co. Ltd., (Osaka, Japan). Micrococcal nuclease and phosphodiesterase II (bovine spleen) were obtained from Worthington Biochemical Co. (New Jersey, USA), and calf intestine alkaline phosphatase was from Takara Bio Co. (Kyoto, Japan). All oligonucleotides used in this study (Table 1) were of HPLC grade and purchased from Sigma Genosys Japan (Tokyo, Japan).

Table 1. Sequences of oligodeoxynucleotides

Name	Sequence
template	5'-CCTTCCXTCTCCCTCTCCTCATTCCACTAC-3', 30mer
-3 primer	5'-AATGAGGAGAGGGA-3', 14mer
-1 primer	5'-TGAGGAGAGGGAGA-3', 14mer
A-primer	5'-GAGGAGAGGGAGAA-3', 14mer
T-primer	5'-GAGGAGAGGGAGAT-3', 14mer
G-primer	5'-GAGGAGAGGGAGAG-3', 14mer
C-primer	5'-GAGGAGAGGGAGAC-3', 14mer

X = dG or N^2-ADPR-dG

Preparation of Mono-ADP ribosylated Oligonucleotides

Pierisin-1 was purified from the pupae of *P. rapae*, as described [2], and treated with Pronase to cleave the protein into an N-terminal region with the ADP-ribosyltransferase domain and a C-terminal region with the receptor-binding domain [9]. The Pronase-treated pierisin-1 has more than 10-fold the activity of untreated pierisin-1, and therefore was used for the modification of oligonucleotides. For site-specific modifications, 50 μg of 30mer oligonucleotide (5'-CCTTCCGTCTCCCTCTCCTCATTCCACTAC, the G underlined is a target of rybosylation) was mono-ADP ribosylated in 100 μL of buffer containing 10 μg of Pronase-treated pierisin-1, 1 mM of β-NAD, 2 mM of EDTA (pH 7.5), 50 mM of Tris-HCl (pH 7.5) and 50 μg of 7mer oligo (5'-AGACGGA) at 37 °C for 16 hours [15, 44]. Then the oligonucleotides were recovered by ethanol precipitation and redissolved in 100 μL of water. The ADP-ribosylated 30mer oligonucleotide was separated using an Agilent 1100 series HPLC system equipped with an Agilent G1315B photodiode array detector (Agilent Technologies, Waldbronn, Germany) and a column of TOSOH TSK-Gel OligoDNA RP (ϕ4.6 mm x 150 mm; TOSOH Co., Tokyo, Japan) at a flow rate of 1 ml/min. The solvent system was as follows: a linear gradient from 7.5 % acetonitrile to 12.5 % acetonitrile over 20 min in 100 mM ammonium acetate (pH 7.0) [15, 44]. In these conditions, α and β isomers of ADP-ribosylated 30 mer oligonucleotide, were eluted at 16.9 min and 17.9 min, and the unmodified 30mer oligonucleotide elutes at 19.6 min. These α- and β-isoforms are easily epimerized under aqueous conditions [15]. β-NAD and the presumed ADP-ribosylated 7mer oligonucleotide eluted before 10 min. The amount of modified 30 mer oligonucleotide was estimated from the absorbance at 260 nm. Fractions containing the modified oligonucleotides were collected and the nucleotides were concentrated by ethanol precipitation. As contamination by 7mer nucleotide or unmodified 30mer oligonucleotide of the modified nucleotide could be critical in following experiments, a portion of the nucleotide was labeled with [γ-^{32}P]ATP, and then its purity was checked by denaturing 20% polyacrylamide gel electrophoresis (PAGE). Bands were identified by autoradiography. The modified oligonucleotide migrated slower than the unmodified oligonucleotide.

In order to confirm that the adduct had formed on the oligonucleotide, 1 nmol of ADP-ribosylated oligonucleotide was digested to mononucleotides by 22 units of micrococcal nuclease and 0.1 unit of phosphodiesterase II for 4 hours at 37 °C in 17 mM sodium succinate buffer (pH 6.0) containing 8 mM $CaCl_2$, then to nucleosides by 10 units of calf intestine alkaline phosphatase for 4 hours at 37 °C in 50 mM Tris-HCl (pH 8.0) [19]. The digest was separated by HPLC with a column of COSMOSIL 5C18-MS (ϕ4.6 mm x 250 mm; Nacarai Tesque, Kyoto) at a flow rate of 1 ml/min. The solvent system was as follows: isoclatic 4 % acetonitrile in 50 mM triethylamine acetate (TEAA) (pH 7.0) for 20 minutes then a linear gradient from 4 % to 15 % acetonitrile over 15 min in 50 mM TEAA (pH 7.0). In these conditions, normal nucleotides eluted in 16 min and both isomers of N^2-(ADP-ribos-1-yl)-2'-deoxyguanosine eluted at 37 min. The retention time and on-line UV spectrum of the isomers were identical with the authentic specimen obtained previously [15].

^{32}P-labeling at 5'-terminus of Primers

Six kinds of primer were used for the experiment: 5'-AATGAGGAGAGGGA (-3 primer), 5'-TGAGGAGAGGGAGA (-1 primer), 5'-GAGGAGAGGGAGAA (A primer), 5'-GAGGAGAGGGAGAT (T primer), 5'-GAGGAGAGGGAGAG (G primer), and 5'-GAGGAGAGGGAGAC (C primer). Briefly, 20 pmol of each primer was incubated at 37 °C for 30 min in 100 μL of buffer containing 2.22 MBq of [γ-^{32}P]ATP, 1 unit of T4 PNK, 50 mM Tris-HCl (pH 8.0), 10 mM $MgCl_2$, and 10 mM 2-mercaptoethanol. The labeled primers were recovered by ethanol precipitation and resuspended in 10 μL of water.

Preparation of Primer/template Mixture

Eight pmol of ^{32}P-labeled 14mer primer was annealed to 10 pmol of the modified 30mer oligonucleotide in 25 μL of buffer containing 10 mM Tris-HCl (pH 8.0), 1 mM EDTA, and 100 mM NaCl by heating to 70 °C for 5 min then cooling to room temperature slowly [34]. This mixture therefore contains the template at 0.4 pmol/μL.

DNA Polymerase Assays

For running start assays, 1 μL of a mixture of "-3 primer" (5'-^{32}P-AATGAGGAGAGGGA) with the template (5'-CCTTCCXTCTCCCTCTCCTCATTCC-ACTAC, X = dG or N^2-ADPR-dG) was used for the polymerization reaction (10 μL) catalyzed by a DNA polymerase. The reaction was carried out at 37 °C for 5 min in 10 μM each of four dNTPs, 50 mM Tris-HCl (pH 8.0), 8 mM $MgCl_2$ and 5 mM 2-mercaptoethanol for Klenow fragment (3' to 5' exo +), for 15 min in 100 μM each of four dNTPs, 40 mM Tris-HCl (pH 8.0), 5 mM $MgCl_2$, 10 mM dithiothreitol (DTT), 250 μg/ml BSA, and 60 mM KCl for Pol η and Pol κ, and for 15 min in 100 μM each of four dNTPs, 40 mM Tris-HCl (pH 8.0), 5 mM $MgCl_2$, 10 mM 2-mercaptoethanol, and 250 μg/ml BSA for Pol ι. Reactions were stopped by adding 10 μL of formamide dye and heating to 90 °C for 2 min [34].

For insertion assays, the reaction was carried out as above, except for that "-1 primer" (5'-^{32}P-TGAGGAGAGGGAGA) was used instead of the "-3 primer" and 300 μM of each dNTP was used instead of 100 μM each of four dNTPs. For extension assays, one of "A primer" (5'-^{32}P-GAGGAGAGGGAGAA); "T primer" (5'-^{32}P-GAGGAGAGGGAGAT); "G primer (5'-^{32}P-GAGGAGAGGGAGAG); or "C primer" (5'-^{32}P-GAGGAGAGGGAGAC) was used as the primer and the reaction was carried out in the presence of 300 μM dGTP.

Polyacrylamide gel Electrophoresis and Autoradiography

Two microliter of a reaction mixture with formamide dye was subjected to electrophoresis on a 20% polyacrylamide gel containing 7M urea. Bands were identified by autoradiography, and the radioactivity was measured with a FLA-3000 image analyzer (Fuji photofilm Co. Ltd., Kanagawa, Japan). The amount of extended product was calculated as:

Extended product [%] = 100 x (Radioactivity of extended primer [photo-stimulated luminescence: psl] - radioactivity of background [psl]) / (radioactivity of total primers [psl] - radioactivity of background [psl]).

RESULTS

Running Start Experiment with the Klenow Fragment

For *in vitro* TLS experiments, it is essential to use highly purified preparation of damaged DNA template that is free from contaminated with unmodified DNA. To confirm the feasibility of the experimental system, primer extension reactions catalyzed by the Klenow fragment (3'→5' exo$^+$) of *Escherichia coli* DNA polymerase I, one of the high fidelity polymerases, were carried out in the presence of four dNTPs using "-3 primer" with unmodified or N^2-ADPR-dG-modified 30 mer templates. As shown in Figure 2, primer extension on an unmodified template produced a fully extended 23 mer product with 0.0008 or 0.008 units of the Klenow fragment. In the reaction using the N^2-ADPR-dG-modified template, the primer extension reaction stalled one or two bases before the adduct, and even with 0.008 units of the Klenow enzyme, no fully extended 23 mer product was observed. Instead, there was a weak band at the position of 22 mer, implying that the Klenow enzyme skipped over the lesion at a low frequency to one nucleotide deletion product. These results indicate that N^2-ADPR-dG strongly blocks normal DNA replication by the Klenow fragment and that the N^2-ADPR-dG-modified template used in this study is not contaminated with unmodified template.

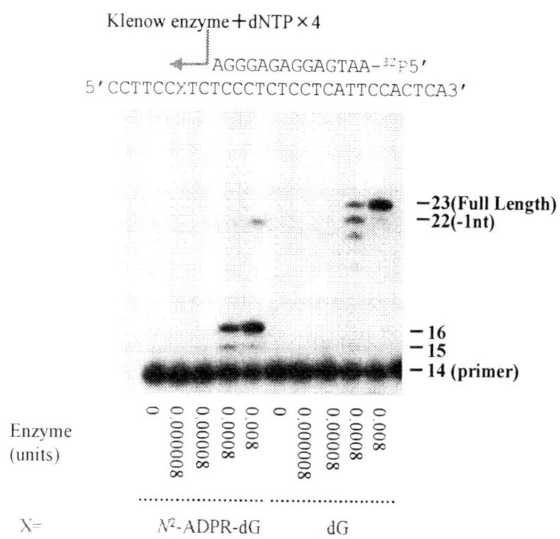

Figure 2. Translesion DNA synthesis (running start) catalyzed by the Klenow (exo$^+$) enzyme. Various amounts of the Klenow enzyme were incubated with a 30mer template containing either a N^2-ADPR-dG or a dG at the position indicated by X annealed to a ^{32}P-5'-end-labeled 14mer primer (-3 primer)(shown on top) with all four dNTPs. The primer allowed the addition of two nucleotides before encountering the adduct.

Running Start Experiment with Y-family Polymerases

Pol η, Pol ι and Pol κ were assayed for polymerase activity on both unmodified and N^2-ADPR-dG-modified templates annealed with "-3 primer". Pol η and Pol κ were capable of carrying out DNA synthesis beyond N^2-ADPR-dG, but Pol η appears to pause one base before the lesion more frequently than Pol κ (Figure 3). Pol ι added elongated one or two nucleotides from the 3'-terminal of the primer on both unmodified and modified templates and stopped further elongation, probably due to its very low processibility (Figure 3).

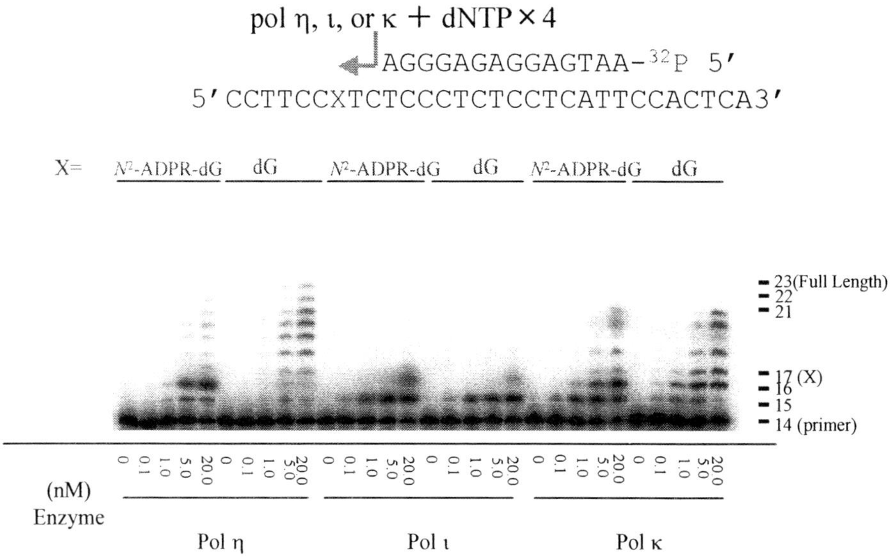

Figure 3. Translesion DNA synthesis (running start) catalyzed by Pol η, Pol ι and Pol κ. Various amounts of Pol η, Pol ι or Pol κ were incubated with a 30 mer template containing either a N^2-ADPR-dG or a dG at the position indicated by X annealed to a ^{32}P-5'-end-labeled 14mer primer (-3 primer)(shown on top) with all four dNTPs. The primer allowed the addition of two nucleotides before encountering the lesion.

Insertion Experiment with Y-family Polymerases

To examine which base was inserted opposite N^2-ADPR-dG by Pol η, Pol ι and Pol κ (20 nM each), the "-1 primer" terminating one base before the lesion site was annealed to the template and polymerization reactions were performed with the template-primer substrates in the presence of a single dNTP. When the reaction products were analyzed, very similar patterns were observed with damaged and undamaged DNA templates, as shown in Figure 4. Pol κ inserted almost exclusively dCMP opposite N^2-ADPR-dG and dG. Pol ι inserted dCMP most preferentially, dTMP to a less extent, and dGMP and dAMP to much less extents, opposite the lesion and dG. Pol η also inserted dCMP preferentially and three others to lesser extents, opposite both.

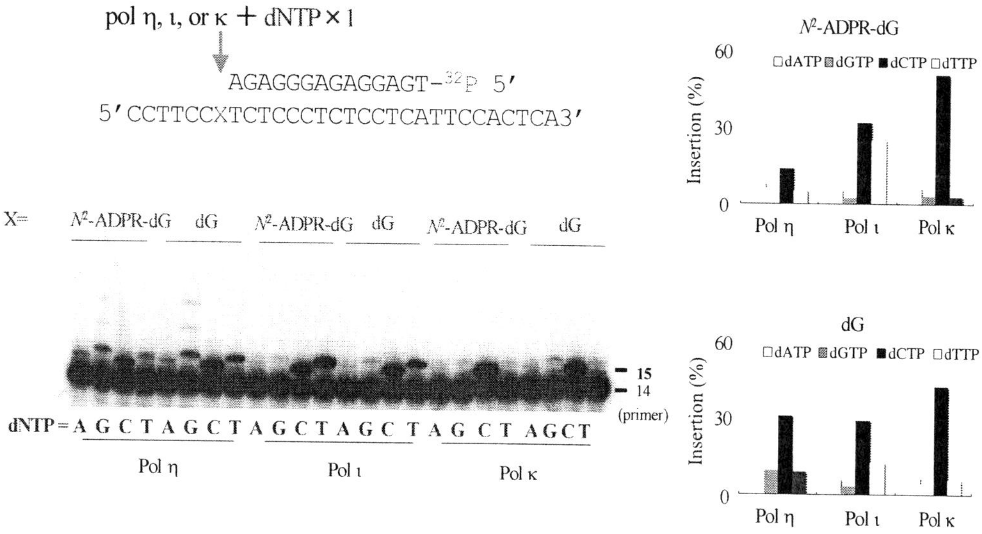

Figure 4. Insertion of nucleotide opposite N^2-ADPR-dG by Pol η, Pol ι and Pol κ. ^{32}P-5'-end-labeled 14 mer primer (-1 primer) which terminates just before the lesion site was annealed to the template containing either a N^2-ADPR-dG or a dG at the position indicated by X (shown on top). Insertion reactions by 20 nM of Pol η, Pol ι or Pol κ were performed with the template-primer substrates in the presence of a single dNTP indicated below the gel image. The amount of extended product was calculated as described in MATERIALS AND METHODS and is shown in the graphs (right).

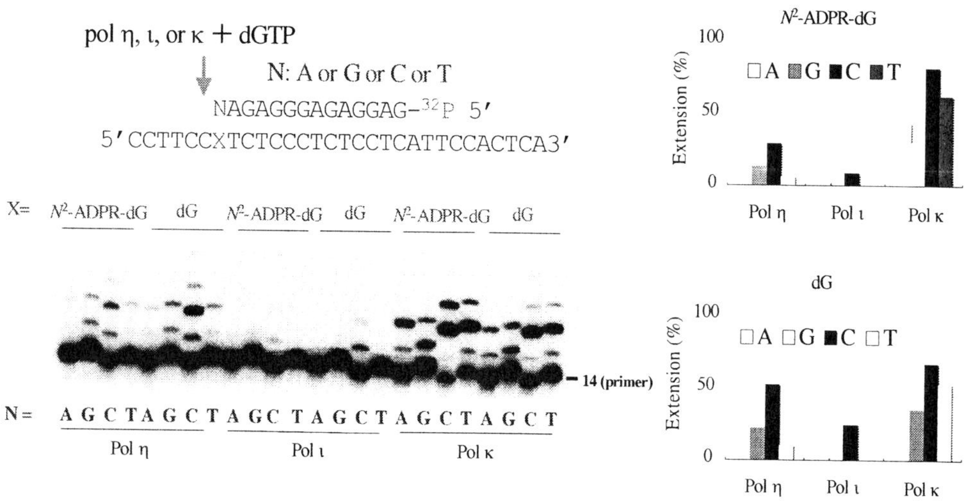

Figure 5. Extension from a mismatch terminus of primers opposite N^2-ADPR-dG.

Four kinds of primer (A-primer, T-primer, G-primer, or C-primer) were annealed with unmodified or N^2-ADPR-dG-modified templates (shown on top). In these template-primer substrates, each of four bases (indicated as N) in the 3'-terminus of the primers correctly or incorrectly paired with N^2-ADPR-dG or dG on the templates. Extension reactions by 20 nM of Pol η, Pol ι or Pol κ were carried out in the presence of dGTP. The amount of all extended bands was calculated and is shown in the graphs (right).

Extension Experiment with Y-family Polymerases

Subsequently, the ability of the polymerases to extend a mismatched or matched base placed opposite N^2-ADPR-dG was examined. For this end, each of the primers having A, G, C or T at the 3'-terminus of the primers correctly or incorrectly paired with N^2-ADPR-dG or dG on the templates was incubated with Pol η, Pol ι and Pol κ (20 nM each). The results are shown in Figure 5. Pol κ extended the DNA chain efficiently from the 3'-matched primer (C: N^2-ADPR-dG), but also from the three mismatched pairs significantly, on both of the damaged and undamaged templates. Pol η also catalyzed extensions from C: N^2-ADPR-dG pair and also from dG: N^2-ADPR-dG pair to some extent, although less efficiently compared with Pol κ. Pol ι slightly extended the DNA strand only from the correctly base-paired primer.

DISCUSSION

DNA lesions can often block DNA replication, therefore cells possess specialized low-fidelity, and often error-prone, DNA polymerases that can bypass such lesions and promote replication of damaged DNA. Since the adduct N^2-ADPR-dG induces G to C or T transversion mutations in mammalian cells [25] and this lesion inhibits DNA replication by Pol α and Pol β *in vitro* [44] , we explored the TLS across N^2-ADPR-dG conducted by Y-family DNA polymerases Pol η, Pol ι and Pol κ *in vitro*. The results showed that Pol κ catalyzed an efficient TLS past N^2-ADPR-dG by inserting dC in preference to other bases opposite the lesion. Pol κ extended the DNA strand also from a mismatch terminus where dG, dA or dT had been incorporated opposite the lesion. Pol ι incorporated dCMP and dTMP opposite the lesion, but did not extend beyond it. Pol η inserted dCMP preferentially and three others to lesser extents opposite N^2-ADPR-dG, and extended further, however, with low efficiency. These results indicate that Pol κ plays a crucial role in bypassing N^2-ADPR-dG. Consistent with this, we have recently obtained the results that Pol κ-deficient mouse embryonic fibroblast (MEF) cells are more sensitive to pierisin-1 than the wild-type MEF cells (M. Ishiguro, E. O. and H.O. unpublished observation).

Similar *in vitro* studies using site-specifically modified oligodeoxynucleotides showed that Pol κ accurately bypasses similar DNA lesions with adducts at the N^2 position of the dG residue, such as N^2-BPDE-dG and the estrogen-derived adduct N^2-[3-methoxyestra-1,3,5(10)-trien-6-yl]-2'-deoxyguanosine (45, 46). However, Pol κ is highly inaccurate during TLS across the 2-acetylaminofluorene-derived dG adduct *N*-(deoxyguanosin-8-yl)-2-acetylaminofluorene, in which position C8 of the dG residue is the adducted site (47). Thus the position of adducts in the bases may influence the accuracy of the TLS by Pol κ (48).

Pol κ can also bypass 8-oxo-guanine, abasic site and thymine-glycol by inserting dA (43, 49, 50). However, Pol κ cannot bypass UV-induced T-T cyclobutane pyrimidine dimer and 6-4 photoproduct (43, 49). The promoter regions of the mouse and human genes coding for Pol κ contain XRE (xenobiotic responsive elements)-like sequence to which ligand-activated arylhydrocarbon receptor (AhR) with AhR nuclear translocator (Arnt) binds, and expression of Pol κ gene is stimulated by 3-methychoranthrene, a polycyclic aromatic hydrocarbon (PAH) similar to B[*a*]P (51). Furthermore, Pol κ is highly expressed in the adrenal cortex of embryonic mice, which is a site of active steroid biosynthesis (52). These findings may implicate that Pol κ has evolved to cope with some endogenous DNA damages including estrogen-modified dG (51), and therefore can bypass structurally similar bulky DNA adducts such as xenobiotic PAHs and mono ADP ribose adducts.

In our experiments, we were not able to clarify which polymerase generates G to C and T base substitutions by itself. Johnson *et al.* [31] proposed a model for mutagenic bypass of DNA lesions, in which the two DNA polymerases act sequentially. According to this model, one polymerase incorporates deoxynucleotides opposite DNA lesions, and then the other extends the DNA strand from such deoxynucleotides. Pol κ has an ability to extend the DNA strand from primer-terminal mispairs opposite damaged DNA templates [53, 54]. In fact, Pol κ extended the DNA strand from a mismatch terminus opposite the N^2-ADPR-dG. Pol κ by itself may not achieve the mutagenic TLS across the N^2-ADPR-dG, but it may do so together with another polymerase, acting as a promiscuous extender of primer-terminal mispairs. In the experiments described above, Pol ι inserted dT opposite N^2-ADPR-dG, but not extending further. On the other hand, Pol κ extended the DNA strand from the mismatch terminus. These results suggest that Pol ι specifically incorporates dT opposite the lesion, and then Pol κ readily extends the mispair. In fact, N^2-ADPR-dG induces G to A mutations not predominantly but moderately in the *supF* gene [25]. Pol ι may act in cooperation with Pol κ to induce the mutation *in vivo*.

In general, G to C transversions are thought to be rare events in both spontaneous and chemically induced mutations [55, 56]. B[*a*]P and TAM generate adducts at N^2 position of dG residue and predominantly induce G to T transversions in various DNA targets including *HPRT* and *supF* genes and site-specifically modified oligodeoxynucleotides [21-24]. It has been shown that in the TLS across N^2-BPDE-dG, Pol η incorporates dA opposite the lesion and subsequently Pol κ extends the DNA chain [53, 57, 58]. In our experiment Pol η inserted any base opposite N^2-ADPR-dG with low efficiency, therefore Pol η might be responsible for the G to T and C mutations. Examining the spectrum of pierisin-1-induced mutations in XPV cells lacking Pol η and Pol κ-knockout mouse cells should help us to understand the involvement of Pol η and Pol κ, respectively, in the mutagenesis.

Conclusion

TLS across the novel DNA lesion N^2-ADPR-dG produced by pierisin-1 was investigated using site-specifically modified oligodeoxynucleotides and the human Y-family polymerases η, ι and κ *in vitro*. Results of ^{32}P-labeled-primer extension experiments showed that Pol κ inserted dCMP correctly opposite the N^2-ADPR-dG and efficiently extended the DNA chain past the lesion. On the other hand, Pol κ extended the DNA strand from a mismatch terminus opposite the lesion as well. Therefore, Pol κ may play an important role in both the error-free and error-prone bypass of N^2-ADPR-dG.

Acknowledgements

This work was supported in part by a Grant-in-Aid for Cancer Research from the Ministry of Health, Labor, and Welfare of Japan and Grants-in-aid for Scientific Research from the Ministry of Education, Culture, Sports, Science and Technology.

References

[1] Koyama, K., Wakabayashi, K., Masutani, M., Koiwai, K., Watanabe, M., Yamazaki, S., Kono, T., Miki, K., & Sugimura, T. (1996). Presence in *Pieris rapae* of cytotoxic activity against human carcinoma cells. *Jpn. J. Cancer Res.*, 87(12), 1259-1262.

[2] Watanabe, M., Kono, T., Koyama, K., Sugimura, T., & Wakabayashi, K. (1998). Purification of pierisin, an inducer of apoptosis in human gastric carcinoma cells, from cabbage butterfly, *Pieris rapae. Jpn. J. Cancer Res.*, 89(5), 556-561.

[3] Watanabe, M., Nakano, T., Shiotani, B., Matsushima-Hibiya, Y., Kiuchi, M., Yukuhiro, F., Kanazawa, T., Koyama, K., Sugimura, T., & Wakabayashi, K. (2004). Developmental stage-specific expression and tissue distribution of pierisin-1, a guanine-specific ADP-ribosylating toxin, in *Pieris rapae*. Comp. Biochem. Physiol.: *A Mol. Integr. Physiol.*, 139(2), 125-131.

[4] Sugimura, T. (2002). Serendipitous discoveries from sudden inspirations and the joy of being a scientist. *Biochem. Biophys. Res. Commun.*, 296(5), 1037-1038.

[5] Kono, T., Watanabe, M., Koyama, K., Kishimoto, T., Fukushima, S., Sugimura, T., & Wakabayashi, K. (1999). Cytotoxic activity of pierisin, from the cabbage butterfly, *Pieris rapae*, in various human cancer cell lines. *Cancer Lett.*, 137(1), 75-81.

[6] Kanazawa, T., Watanabe, M., Matsushima-Hibiya, Y., Kono, T., Tanaka, N., Koyama, K., Sugimura, T., & Wakabayashi, K. (2001). Distinct roles for the N- and C-terminal regions in the cytotoxicity of pierisin-1, a putative ADP-ribosylating toxin from cabbage butterfly, against mammalian cells. *Proc. Natl. Acad. Sci. U. S. A.*, 98(5), 2226-2231.

[7] Matsushima-Hibiya, Y., Watanabe, M., Kono, T., Kanazawa, T., Koyama, K., Sugimura, T., & Wakabayashi, K. (2000). Purification and cloning of pierisin-2, an apoptosis-inducing protein from the cabbage butterfly, *Pieris brassicae. Eur. J. Biochem.*, 267(18), 5742-5750.

[8] Kanazawa, T., Kono, T., Watanabe, M., Matsushima-Hibiya, Y., Nakano, T., Koyama, K., Tanaka, N., Sugimura, T., & Wakabayashi, K. (2002). Bcl-2 blocks apoptosis caused by pierisin-1, a guanine-specific ADP-ribosylating toxin from the cabbage butterfly. Biochem. *Biophys. Res. Commun.*, 296(1), 20-25.

[9] Watanabe, M., Kono, T., Matsushima-Hibiya, Y., Kanazawa, T., Nishisaka, N., Kishimoto, T., Koyama, K., Sugimura, T., & Wakabayashi, K. (1999). Molecular cloning of an apoptosis-inducing protein, pierisin, from cabbage butterfly: possible involvement of ADP-ribosylation in its activity. *Proc. Natl. Acad. Sci. U. S. A.*, 96(19), 10608-10613.

[10] Watanabe, M., Enomoto, S., Takamura-Enya, T., Nakano, T., Koyama, K., Sugimura, T., & Wakabayashi, K. (2004). Enzymatic properties of pierisin-1 and its N-terminal

domain, a guanine-specific ADP-ribosyltransferase from the cabbage butterfly. *J. Biochem.* (Tokyo), 135(4), 471-477.

[11] Takamura-Enya, T., Watanabe, M., Koyama, K., Sugimura, T., & Wakabayashi, K. (2004). Mono(ADP-ribosyl)ation of the N^2 amino groups of guanine residues in DNA by pierisin-2, from the cabbage butterfly, *Pieris brassicae*. *Biochem. Biophys. Res. Commun.*, 323(2), 579-582.

[12] Tsuzuki, K., Kimura, K., Fujii, N., Yokosawa, N., Indoh, T., Murakami, T., & Oguma, K. (1990). Cloning and complete nucleotide sequence of the gene for the main component of hemagglutinin produced by Clostridium botulinum type C. *Infect. Immun.*, 58(10), 3173-3177

[13] Inoue, K., Fujinaga, Y., Watanabe, T., Ohyama, T., Takeshi, K., Moriishi, K., Nakajima, H., Inoue, K., & Oguma, K. (1996). Molecular composition of Clostridium botulinum type A progenitor toxins. *Infect. Immun.*, 64(5), 1589-1594

[14] Matsushima-Hibiya, Y., Watanabe, M., Hidari, K. I., Miyamoto, D., Suzuki, Y., Kasama, T., Koyama, K., Sugimura, T., & Wakabayashi, K. (2003). Identification of glycosphingolipid receptors for pierisin-1, a guanine-specific ADP-ribosylating toxin from the cabbage butterfly. *J. Biol. Chem.*, 278(11), 9972-9978.

[15] Takamura-Enya, T., Watanabe, M., Totsuka, Y., Kanazawa, T., Matsushima-Hibiya, Y., Koyama, K., Sugimura, T., & Wakabayashi, K.. (2001). Mono(ADP-ribosyl)ation of 2'-deoxyguanosine residue in DNA by an apoptosis-inducing protein, pierisin-1, from cabbage butterfly. *Proc. Natl. Acad. Sci. U. S. A.*, 98(22), 12414-12419.

[16] Watanabe, M., Takamura-Enya, T., Kanazawa, T., Totsuka, Y., Matsushima-Hibiya, Y., Koyama, K., Sugimura, T., & Wakabayashi, K. (2002). Mono(ADP-ribosyl)ation of DNA by apoptosis-inducing protein, pierisin. *Nucleic Acids Res. Suppl.*, 2002(2), 243-244.

[17] Meehan, T., and Straub, K. (1979). Double stranded DNA stereoselectively binds benzo[*a*]pyrene diol epoxides. *Nature*, 277, 410-412.

[18] Cheng, S. C., Hilton, B. D., Roman, J. M., & Dipple, A. (1989). DNA adducts from carcinogenic and noncarcinogenic enantiomers of benzo[*a*]pyrene dihydrodiol epoxide. *Chem. Res. Toxicol.*, 2(5), 334-340.

[19] Dasaradhi, L., & Shibutani, S. (1997). Identification of tamoxifen-DNA adducts formed by R-sulfate tamoxifen and R-acetoxytamoxifen. *Chem. Res. Toxicol.*, 10(2), 189-196.

[20] Osborne, M. R., Hewer, A., Hardcastle, I. R., Carmichael, P. L., & Phillips, D. H. (1996). Identification of the major tamoxifendeoxyguanosine adduct formed in the liver DNA of rats treated with tamoxifen. *Cancer Res.*, 56(1), 66-71.

[21] Yang, J. L., Chen, R. H., Maher, V. M., and McCormick, J. J. (1991). Kinds and location of mutations induced by (±)-7β, 8α-dihydroxy-9α,10α-epoxy-7,8,9,10-tetrahydrobenzo[*a*]-pyrene in the coding region of the hypoxanthine (guanine) phosphoribosyltransferase gene in diploid human fibroblasts. *Carcinogenesis,* 12(1), 71-75.

[22] Yang, J. L., Maher, V. M., and McCormick, J. J. (1987). Kinds of mutations formed when a shuttle vector containing adducts of (±)-7β, 8α-dihydroxy-9α,10α-epoxy-7,8,9,10-tetrahydrobenzo[*a*]-pyrene replicates in human cells. *Proc. Natl. Acad. Sci. U.S.A.*, 84(11), 3787-3791.

[23] Mazur, M., and Glickman, B. W. (1988). Sequence specificity of mutations induced by benzo[*a*]pyrene-7,8-diol-9,10-epoxide at endogenous *aprt* gene in CHO cells. Somat. Cell. *Mol. Genet.,* 14(4), 393-400.

[24] Terashima, I., Suzuki, N., and Shibutani, S. (1999). Mutagenic potential of α-(N^2-deoxyguanosinyl)tamoxifen lesions, the major DNA adducts detected in endometrial tissues of patients treated with tamoxifen. *Cancer Res.,* 59(9), 2091-2095.

[25] Totsuka, Y., Kawanishi, M., Nishigaki, R., Matsukawa, K., Yagi, T., Takamura-Enya, T., Watanabe, M., Sugimura, T., & Wakabayashi, K. (2003). Analysis of *HPRT* and *supF* mutations caused by pierisin-1, a guanine specific ADP-ribosylating toxin derived from the cabbage butterfly. *Chem. Res. Toxicol.,* 16(8), 945-952.

[26] Plosky, B. S., &Woodgate, R. (2004). Switching from high-fidelity replicases to low-fidelity lesion-bypass polymerases. Curr. Opin. *Genet. Dev.,* 14(2), 113-119.

[27] Goodman, M. F. (2002). Error-prone repair DNA polymerases in prokaryotes and eukaryotes. *Annu. Rev. Biochem*., 71, 17-50.

[28] Ohmori, H., Friedberg, E. C., Fuchs, R. P., Goodman, M. F., Hanaoka, F., Hinkle, D., Kunkel, T. A., Lawrence, CW, Livneh, Z., Nohmi, T., Prakash, L., Prakash, S., Todo, T., Walker, G. C., Wang, Z., & Woodgate, R. (2001). The Y-family of DNA polymerases. *Mol. Cell*, 8(1), 7-8.

[29] Morrison, A., Christensen, R. B., Alley, J., Beck, A. K., Bernstine, E.G., Lemontt, J. F., & Lawrence, C. W. (1989). REV3, a *Saccharomyces cerevisiae* gene whose function is required for induced mutagenesis, is predicted to encode a nonessential DNA polymerase. *J. Bacteriol*., 171(10), 5659-5667.

[30] Nelson, J. R., Lawrence, C. W., & Hinkle, D. C. (1996). Thymine-thymine dimer bypass by yeast DNA polymerase ζ. *Science*, 272(5268), 1646-1649.

[31] Johnson, R. E., Washington, M. T., Haracska, L., Prakash, S., & Prakash, L. (2000). Eukaryotic polymerases ι and ζ act sequentially to bypass DNA lesions. *Nature*, 406(6799), 1015-1019.

[32] Woodgate, R. (2001). Evolution of the two-step model for UV-mutagenesis. *Mutat. Res.*, 485(1), 83-92.

[33] Masutani, C., Kusumoto, R., Yamada, A., Dohmae, N., Yokoi, M., Yuasa, M., Araki, M., Iwai, S., Takio, K., & Hanaoka, F. (1999). The XPV (xeroderma pigmentosum variant) gene encodes human DNA polymerase η. *Nature*, 399(6737), 700-704.

[34] Masutani, C., Araki, M., Yamada, A., Kusumoto, R., Nogimori, T., Maekawa, T., Iwai, S., & Hanaoka, F. (1999). Xeroderma pigmentosum variant (XP-V) correcting protein from HeLa cells has a thymine dimer bypass DNA polymerase activity. *EMBO J.,* 18(12), 3491-3501.

[35] Johnson, R. E., Kondratick, C. M., Prakash, S., & Prakash, L. (1999). hRAD30 mutations in the variant form of xeroderma pigmentosum. *Science,* 285(5425), 263-265.

[36] Trincao, J., Johnson, R. E., Escalante, C. R., Prakash, S., Prakash, L., & Aggarwal, A. K. (2001). Structure of the catalytic core of *S. cerevisiae* DNA polymerase η: implications for translesion DNA synthesis. *Mol. Cell*, 8(2), 417-26.

[37] Zhou, B. L., Pata, J. D., & Steitz, T. A. (2001). Crystal structure of a DinB lesion bypass DNA polymerase catalytic fragment reveals a classic polymerase catalytic domain. *Mol. Cell*, 8(2), 427-37.

[38] Ling, H., Boudsocq, F., Woodgate, R., & Yang, W. (2001). Crystal structure of a Y-family DNA polymerase in action: a mechanism for error-prone and lesion-bypass replication. *Cell*, 107(1), 91-102.

[39] Silvian, L. F., Toth, E. A., Pham, P., Goodman, M. F., & Ellenberger, T. (2001). Crystal structure of a DinB family error-prone DNA polymerase from *Sulfolobus solfataricus*. *Nat. Struct. Biol.*, 8(11), 984-989.

[40] Ling, H., Boudsocq, F., Plosky, B. S., Woodgate, R., & Yang, W. (2003). Replication of a *cis-syn* thymine dimer at atomic resolution. *Nature*, 424(6952), 1083-1087.

[41] Jansen, J. G., & de Wind, N. (2003). Biological functions of translesion synthesis proteins in vertebrates. *DNA Repair (Amst).*, 2(10), 1075-1085.

[42] Tissier, A., Frank, E. G., McDonald, J. P., Iwai, S., Hanaoka, F., & Woodgate, R. (2000). Misinsertion and bypass of thymine-thymine dimers by human DNA polymerase ι. *Genes Dev.*, 14(13), 1642-1650.

[43] Ohashi, E., Ogi, T., Kusumoto, R., Iwai, S., Masutani, C., Hanaoka, F., & Ohmori, H. (2000). Error-prone bypass of certain DNA lesions by the human DNA polymerase κ. *Genes Dev.*, 14(13), 1589-1594.

[44] Watanabe, M., Kanazawa, T., Totsuka, Y., Shiotani, B., Hibiya, Y., Koyama K, Sugimura T, & Wakabayashi K. (2002). ADP-rybosylated dG adduct formation by pierisin-1, from *Pieris rapae,*.and translesion DNA synthesis across the adduct. Jpn. *J. Cancer Res.,* 93 suppl., 49.

[45] Suzuki, N., Ohashi, E., Kolbanovskiy, A., Geacintov, N.E., Grollman, A.P., Ohmori, H., & Shibutani, S. (2002). Translesion synthesis by human DNA polymerase κ on a DNA template containing a single stereoisomer of dG-(+)- or dG-(-)-*anti*-N^2-BPDE(7,8-dihydroxy-anti-9,10-epoxy-7,8,9,10-tetrahydrobenzo[*a*]pyrene). *Biochemistry*, 41(19), 6100-6106.

[46] Suzuki, N., Itoh, S., Poon, K., Masutani, C., Hanaoka, F., Ohmori, H., Yoshizawa, I., & Shibutani, S. (2004). Translesion synthesis past estrogen-derived DNA adducts by human DNA polymerases η and κ. *Biochemistry*, 43(20), 6304-6311.

[47] Suzuki, N., Ohashi, E., Hayashi, K., Ohmori, H., Grollman, A.P., & Shibutani, S. (2001). Translesional synthesis past acetylaminofluorene-derived DNA adducts catalyzed by human DNA polymerase κ and *Escherichia coli* DNA polymerase IV. *Biochemistry*, 40(50), 15176-15183.

[48] (48) Yasui, M., Dong, H., Bonala, R. R., Suzuki, N., Ohmori, H., Hanaoka, F., Johnson, F., Grollman, A. P., & Shibutani, S. (2004). Mutagenic properties of 3-(deoxyguanosin-N^2-yl)-2-acetylaminofluorene, a persistent acetylaminofluorene-derived DNA adduct in mammalian cells. *Biochemistry*, 43(47), 15005-15013.

[49] Zhang, Y., Yuan, F., Wu, X., Wang, M., Rechkoblit, O., Taylor, J.S., Geacintov, N.E., & Wang, Z. (2000). Error-free and error-prone lesion bypass by human DNA polymerase κ *in vitro*. *Nucleic Acids Res.*, 28(21), 4138-4146.

[50] Fischhaber, P.L., Gerlach, V.L., Feaver, W.J., Hatahet, Z., Wallace, S.S., & Friedberg, E.C. (2002). Human DNA polymerase κ bypasses and extends beyond thymine glycols during translesion synthesis *in vitro*, preferentially incorporating correct nucleotides. *J. Biol. Chem.*, 277(40), 37604-37611.

[51] Ohmori, H., Ohashi, E., & Ogi, T. (2004). Mammalian pol κ: Regulation of its expression and lesion substrates. *Adv. Protein Chem.*, 69, 265-278.

[52] Velasco-Miguel, S., Richardson, J.A., Gerlach, V.L., Lai, W.C., Gao, T., Russell, L.D., Hladik, C.L., White, C.L., & Friedberg, E.C. (2003). Constitutive and regulated expression of the mouse *Dinb* (*Polκ*) gene encoding DNA polymerase kappa. *DNA Repair,* 2(1), 91-106.

[53] Zhang, Y., Wu, X., Guo, D., Rechkoblit, O., Geacintov, N. E., & Wang, Z. (2002). Two-step error-prone bypass of the (+)- and (-)-*trans-anti*-BPDE-N^2-dG adducts by human DNA polymerases η and κ. *Mutat. Res.*, 510(1-2), 23-35.

[54] Washington, M.T., Johnson, R. E., Prakash, L., & Prakash, S. (2002). Human DINB1-encoded DNA polymerase κ is a promiscuous extender of mispaired primer termini. *Proc. Natl. Acad. Sci. U. S. A.,* 99(4), 1910-1914.

[55] Zhang, L.-H., Vrieling, H., Zeeland, A. A. V., & Jenssen, D. (1992). Spectrum of spontaneously occurring mutations in the *hprt* gene of V79 chinese hamster cells. *J. Mol. Biol.,* 223(3), 627-635.

[56] Lewis, P. D., Harvey, J. S., Waters, E. M., Skibinski, D. O. F., & Parry, J. M. (2001). Spontaneous mutation spectra in *supF* comparative analysis of mammalian cell line base substitution spectra. *Mutagenesis*, 16(6), 503-515.

[57] Chiapperino, D., Kroth, H,. Kramarczuk, I. H., Sayer, J. M., Masutani, C., Hanaoka, F., Jerina, D. M., & Cheh, A. M. (2002). Preferential misincorporation of purine nucleotides by human DNA polymerase η opposite benzo[*a*]pyrene 7,8-diol 9,10-epoxide deoxyguanosine adducts. *J. Biol. Chem.*, 277(14), 11765-11771.

[58] Rechkoblit, O., Zhang, Y., Guo, D., Wang, Z., Amin, S., Krzeminsky, J., Louneva, N., & Geacintov, N.E. (2002). Trans-Lesion synthesis past bulky benzo[a]pyrene diol epoxide N^2-dG and N^6-dA lesions catalyzed by DNA bypass polymerases. *J. Biol. Chem.,* 277(34), 30488-30494.

In: New Developments in Mutation Research
Editor: Charles L. Valon, pp. 149-188
ISBN 1-59454-664-9

Chapter 7

DNA NON-HOMOLOGOUS END-JOINING IN UNICELLULAR ORGANISMS

Miroslav Chovanec[1],* Boris Kysela[2],*
Zuzana Dudášová[1] and Aidan J. Doherty[3]
[1]Laboratory of Molecular Genetics, Cancer Research Institute, Bratislava, Slovak Republic
[2]Division of Reproductive and Child Health, Section of Medical and Molecular Genetics, The Medical School, University of Birmingham, Birmingham, United Kingdom
[3]Genome Damage and Stability Centre, University of Sussex, Brighton, United Kingdom

ABSTRACT

DNA double-strand breaks (DSB) are considered to be one of the most serious forms of DNA damage, because if left unrepaired, they can cause cell death and, if misrepaired, they can lead to genomic instability and, ultimately, the development of cancer in multicellular organisms. Efficient DSB repair is therefore critical to preserve genomic stability and cell viability. There are two major pathways for the repair of DSB in a cell: homologous recombination (HR) and non-homologous end-joining (NHEJ).

In mammalian cells, DSB are repaired primarily through the NHEJ pathway. The essential mammalian NHEJ factors are the DNA-dependent protein kinase (DNA-PK), consisting of the catalytic subunit (DNA-PKcs) and the two DNA end-binding components (the KU70 and KU80 proteins forming a stable complex), the DNA ligase IV/XRCC4 complex and the ARTEMIS protein. In contrast to mammalian cells, the budding yeast *Saccharomyces cerevisiae* repairs DSB mainly by HR, although NHEJ pathway is also of notable relevance. This organism possesses the KU70, KU80, DNA ligase IV and XRCC4 homologues (YKu70, YKu80, Lig4 and Lif1, respectively), while it lacks clear functional homologues of either ARTEMIS or DNA-PKcs. However, *S. cerevisiae* cells possess an additional regulatory NHEJ factor, the Nej1 protein, which has not yet been found in any other organisms. Furthermore, there is evidence in the budding

* To whom correspondence may be addressed. E-mails: miroslav.chovanec@savba.sk (M. Chovanec) or b.kysela@bham.ac.uk (B. Kysela).

yeast for a requirement for the Mre11/Rad50/Xrs2 complex for NHEJ, which does not appear to be the case in mammals.

DSB by NHEJ had previously been thought to be restricted to the eukarya. However, functionally homologous end-bridging and ligation activities have now been identified in prokarya and recent papers have demonstrated that bacterial Ku and ligase proteins form a *bona fide* NHEJ system that encodes all the recognition, processing and ligation activities required for DSB repair. These findings provide new evolutionary insights into the basic biochemistry of this repair pathway, and suggest that the basic DSB repair process is evolutionary much older than previously thought.

Here we review the current knowledge on yeast and bacterial NHEJ. We will investigate the possible origins of this NHEJ repair system, examining the insights it has given us into the biochemistry underlying the DSB repair process and how it may also play a role in stationary phase mutagenesis leading to genome diversification under conditions of stress.

Introduction

DNA double-strand breaks (DSB), which are one of the most severe types of DNA damage, can be induced by a number of different agents. Endogenous DSB arise by cellular processes such as transformation, conjugation, and transduction in bacteria; mating-type switching and transposition of certain elements in yeast; and V(D)J recombination in mammalian cells. Furthermore, DSB are created endogenously by action of free radicals, as intermediates of mitotic and meiotic recombination and during DNA replication when replication machinery encounters single-stranded break. DSB are also produced exogenously by exposure to ionizing radiation (IR) and a wide range of chemical compounds, such as radiomimetic drug bleomycin. Since unrepaired or misrepaired DSB can initiate processes leading to mutagenesis, tumorigenesis and cell death, repair of DSB is essential to maintain genome stability and cell viability. Cells have evolved two major pathways for the repair of DSB, homologous recombination (HR) and non-homologous end-joining (NHEJ) (for reviews, see [1-6]).

NHEJ was originally discovered in mammalian cells, where the DNA-dependent protein kinase (DNA-PK) consisting of the catalytic (DNA-PKcs) and DNA end-binding (KU70/80) subunits, the DNA ligase IV/XRCC4 complex, and the ARTEMIS protein, are the core NHEJ factors (recently reviewed in [4,7-15]). According to current models of mammalian NHEJ, the KU70/80 heterodimer is thought to be the first player in the process, binding with high affinity to the broken DNA ends. Once bound, KU70/80 recruits the end-bridging, nucleolytic, polymerization, and ligation factors [16]. Recruitment of DNA-PKcs results in the formation of the DNA-PK complex, a likely mechanistic function of which is to maintain the proximity of the two broken DNA ends and hence to serve as an end-to-end synapsis factor [9,10,17]. Upon DNA-PK complex formation, the kinase activity of the DNA-PKcs, being responsible for several subsequent phosphorylation events, is activated (for reviews, see [8,10,18]). Autophosphorylation of DNA-PK remodels DNA-PK-DNA end complex, thereby allowing access to DNA ends for further processesing [17]. Nucleolytic factor acting on a subset of DSB repaired by NHEJ is the ARTEMIS protein [16,19], a significant fraction of which is in a complex with DNA-PKcs. ARTEMIS protein alone displays 5' to 3' exonuclease activity, however upon complex formation with, and phosphorylation by, DNA-PKcs, it acquires endonucleolytic activity on 5' and 3' overhangs, as well as hairpins [20]. The polymerization step during NHEJ is mediated by DNA polymerases μ and λ (POL μ and λ), as well as terminal deoxynucleotidyl transferase (TdT) [16,21-23]. NHEJ is completed by the ligation reaction carried out by the DNA ligase IV/XRCC4 complex. DNA ligase IV is responsible for the enzymatic part of the reaction as it is a DSB repair ligase, while XRCC4 serves to modulate its stability and activity and to target it to the site of DSB (reviewed in [4,7-9,11-15]).

The core NHEJ components have also been discovered in other organisms including bacteria and budding yeast [24-38], suggesting a high degree of evolutionary conservation of this process. Therefore, investigating NHEJ in these two unicellular organisms, both of which have already proven to be the powerful model systems for studying DNA repair processes, can contribute to addressing the fundamental questions about this process. The present review is focused on the current knowledge of bacterial and yeast NHEJ. We discuss the possible origins of the NHEJ repair pathway, examining the insights it has given us into the biochemistry underlying the DSB repair process and how it may also function in stationary phase mutagenesis leading to genome diversification under conditions of stress.

NON-HOMOLOGOUS END-JOINING IN *SACCHAROMYCES CEREVISIAE*

The Synapsis Factors of the *S. cerevisiae* Non-Homologous End-Joining

The Yku70/80 Complex

KU was identified as an autoimmune antigen in sera from patients with polymyositis-scleroderma overlap syndrome [39]. It is composed of the two subunits (69.842 and 82.713 kDa), which form an extremely stable complex, the KU70/80 heterodimer. The complex binds tightly and specifically to the free DNA ends in a sequence-independent manner. The crystal structure of KU70/80, that has recently been reported, showed that KU70/80 heterodimer forms a dyad-symmetrical molecule resembling a basket-like shape, with a preformed ring that encircles duplex DNA. The reader is referred to the original paper by Walker *et al.* [40] for more detailed information on KU structure.

KU70/80 functions in a number of fundamental cellular processes, with its role in NHEJ being one of the most important. In NHEJ, KU70/80 binds to the broken DNA ends and serves as an alignment, recruitment and stimulating factor (reviewed in [41-45]). Recently, Ma and colleagues [16] reconstituted the NHEJ reactions *in vitro*, using mammalian purified proteins, and showed that KU70/80 has only a small effect in the reaction that contains only the core components, a fact that is even inconsistent with KU70/80 being a synapsis factor [46]. Interestingly, in this experimental system, KU70/80 has a more substantial effect on NHEJ when POL μ or λ is present. This observation is in line with the findings reporting that KU70/80 also facilitates recruitment of POL μ and λ [16] as well as DNA-PKcs [47-50] and the DNA ligase IV/XRCC4 complex [51,52]. Active recruitment of repair factors constitutes a functional role (rather than merely a physical role) for KU70/80 in mammalian NHEJ for the joining of incompatible DNA ends.

A yeast homologue of KU70/80, Yku70/80 (also referred to as Hdf1/Hdf2), has been identified [24,26,27,53]. In a similar manner to KU70/80, it binds to DNA ends, irrespective of sequence context [53,54]. Yku70/80 is composed of 70.647 and 71.240 kDa subunits, both of which display a high degree of homology with their human counterparts [24,24].

One of the most common assays to monitor the operation of NHEJ in *S. cerevisiae* is based on examining its ability to repair DSB in a plasmid sequence. In this assay, which is called the transformation-based plasmid repair assay, a *S. cerevisiae* strain is transformed with a plasmid that has been linearized by action of a restriction enzyme. A supercoiled version of the same plasmid is transformed into the yeast strain in parallel in order to normalize for differences in transformation efficiency between strains and between repeats of the same experiment. Since the plasmid can only be propagated in yeast when recircularized, the number of transformants obtained with the linear plasmid normalized to the number obtained with the supercoiled plasmid provides a quantitation of the ability of the yeast strain to repair DSB. To prevent the DSB from becoming repaired by HR, the sites for restriction enzyme cleavage of the plasmid are within regions bearing no homology to yeast chromosomal sequences.

In wild type yeast cells, the plasmid with the restriction enzyme-generated cohesive DNA ends is repaired with high efficiency, which is in sharp contrast to a situation in Yku70/80-deficient cells. Based on this, Yku70/80 has been considered as an essential factor of the *S. cerevisiae* NHEJ [25,26,53,55]. Moreover, in Yku70/80-deficient cells, plasmids are not repaired accurately. Instead, molecules that underwent losses of up to several hundred base pairs at the joining site, with junctional overlapping sequences of 3-15 base pairs (bp), are yielded. Unexpectedly, Yku70/80 does not seem to play a role in the rejoining of plasmid with blunt DNA ends [26].

Yku70/80-deficient cells show moderate sensitivity to bleomycin [27,28] and methyl methanesulfonate (MMS) [27,53], but no sensitivity to IR [54], ultraviolet (UV) light [26] and hydroxyurea (HU) [54]. Moreover, they display strongly reduced frequency of mating-type switching and spontaneous mitotic recombination [28], altered subnuclear organization of telomeres [56], severely diminished repression of the transcription of RNA polymerase II genes in close proximity to telomeres – a process called telomere position effect (TPE) [57-59], a temperature-sensitive phenotype for growth at restrictive temperature [24-26,54,60], and greatly shortened telomeres even at the permissive temperature [26,55,57,61]. In haploid state, Yku70/80-defective cells also exhibit a reduction of replication-independent mutation frequency (for further details, see text below) [62].

Since Yku70/80 interacts with the Tel1 protein [27,57,61], that shares homology with phosphatidyl-inositol-3 kinase-related kinase (PIKK) family, Tel1 was assumed to represent a DNA-PKcs equivalent in *S. cerevisiae* [63]. In recent years, however, detailed study of the *tel1* mutant phenotype [57] has shown that Tel1 and Yku70/80 operate independently from one another. Therefore, Tel1 does not seem to be an essential component of the *S. cerevisiae* NHEJ and a functional homologue of the DNA-PKcs. Instead, Tel1 possibly represents the *S. cerevisiae* ATM (Ataxia telangiectasia mutated) homologue [64] and *S. cerevisiae* cells do not have a clear DNA-PKcs homologue [65].

The Mre11/Rad50/Xrs2 Complex

Various genetic studies, such as integration [66] and deletion [67,68] frequency measurements, HO endonuclease-induced chromosomal DSB assays [53,57,69], and transformation-based plasmid repair assays (see text above) [53,55,57], carried out in the Mre11/Rad50/Xrs2-deficient cells, as well as biochemical results obtained using purified proteins [70], provided clear evidence for the involvement of the Mre11/Rad50/Xrs2 complex in the *S. cerevisiae* NHEJ. This is in contrast to the situation in mammalian cells, where satisfactory evidence for the involvement of the MRE11/RAD50/NBS1 complex, a mammalian homologue of Mre11/Rad50/Xrs2, in NHEJ is still absent [53,57,67,69,70].

The Mre11/Rad50/Xrs2 complex consists of 692, 1312 and 854 amino acid subunits, with molecular weights of 77.650, 152.568 and 96.364 kDa, respectively [71-74]. Interestingly, whereas NBS1 appears to be only functional rather than real structural homologue of Xrs2, both MRE11 and RAD50 represent clear structural and functional homologues of their yeast counterparts [75-79]. The Mre11/Rad50/Xrs2 complex, a stoichiometry of which has been shown to be 2:2:1 [70], possesses Mn^{2+}-dependent 3' to 5' single-stranded and double-stranded DNA (ssDNA and dsDNA, respectively) endonuclease, ssDNA exonuclease and hairpin cleavage activities that are specified by four phosphoesterase motifs residing in the N-terminal part of Mre11 [55,70,80-83]. Rad50 subunit of the complex binds and hydrolyzes ATP [84,85], which enhances nuclease activity of Mre11 *in vitro* [83]. This observation is in agreement with *in vivo* findings that demonstrated a null phenotype with respect to DNA repair conferred by the *RAD50* mutant allele that is incapable of ATP binding [85]. As expected, Mre11/Rad50/Xrs2 possesses a DNA-binding activity, which can be exhibited by all the individual subunits separately [80,82,83,86]. The crystal structure of Rad50/Mre11 has recently been solved. For further details on this structure, please see the original papers by Hopfner *et al.* [87-89].

As well as showing a generalized NHEJ defect, *in vivo* assays mentioned above showed that Mre11/Rad50/Xrs2 functions epistatically in NHEJ with Yku70/80 and Lig4/Lif1 [55,57] and that virtually all NHEJ products arising in Mre11/Rad50/Xrs2-debilitated cells were joined accurately [57]. Furthermore, they suggested that nuclease activity of Mre11/Rad50/Xrs2 may not play a role in NHEJ and that the function of the complex may rather be to tether together and stabilize broken DNA ends and/or to recruit other NHEJ factors [57,67-69]. Consistent with this observation, it has been shown that nuclease-defective Mre11 protein did not confer a mutant phenotype in the plasmid repair assay [57,80] and that over-expression of Exo1 (5' to 3' exonuclease that also exhibits flap endonuclease activity) in the *mre11* mutant failed to rescue its NHEJ deficiency [90]. Furthermore, this notion has been strengthened by recent biochemical data [70] demonstrating that the Mre11/Rad50/Xrs2 complex functions in NHEJ to end-bridge the broken DNA ends and to provide a scaffold upon which the NHEJ machinery is assembled. Nuclease activity of Mre11/Rad50/Xrs2 is not required for this function. Study by Chen *et al.* [70] also showed that Mre11/Rad50/Xrs2 and Lig4/Lif1 interact directly *via* Xrs2 and Lif1 subunits, respectively, and that this interaction specifically stimulates intermolecular DNA end-joining activity of Lig4/Lif1. Furthermore, the study highlighted the specific functional interplay among the Yku70/80, Mre11/Rad50/Xrs2, and Lig4/Lif1 complexes, as it demonstrated that DNA end-joining by Lig4/Lif1 in the presence of Mre11/Rad50/Xrs2 is further enhanced by Yku70/80.

In addition to their abnormal phenotype for NHEJ, Mre11/Rad50/Xrs2-deficient cells display dramatic telomeric attrition (losing approximately 65% of the terminal repeats at permissive temperature) but not TPE. Moreover, they show defects in HR and meiotic recombination, premature senescence, suppression of gross chromosomal rearrangements, elevated rates of spontaneous mitotic recombination, and delayed mating-type switching

[55,57,71,80-82,91-95], suggesting a more complex involvement of Mre11/Rad50/Xrs2 in maintaining of genome stability.

The End-Processing Factors of the *S. cerevisiae* Non-Homologous End-Joining

Although the NHEJ pathway repairs DSB in a template-independent fashion through simple ligation of the broken DNA ends (recently reviewed in [11,14]), DNA end-processing is often required prior to ligation in cases where broken DNA ends are neither complementary nor ligatable due to chemical modifications or sequence incompatibilities. A favored mode of DNA end-processing involves short tracts of DNA sequence homology, so-called microhomologies, situated close to the break site, that presumably facilitate alignment of the DNA ends [69,96,97]. After initial alignment, the processing of DNA ends by nucleases and DNA polymerases (to remove single-stranded flaps and fill-in gaps) is likely to be necessary to generate ligatable DNA termini [98,99]. However, very little is known to date on the identity of enzymatic activities that contribute to the end-processing reactions during NHEJ.

One of these non-core NHEJ components required for rejoining of not fully compatible DNA ends in *S. cerevisiae* is a DNA polymerase β-like enzyme, the product of the *POL4* gene [100,101]. Although Pol4 was previously regarded to be a DNA polymerase β homologue [100-103], nowadays it is rather classified as a homologue of POL λ [104]. The Pol4 DNA polymerase is the sole *S. cerevisiae* member of the functionally diverse Pol X family of nucleotidyl transferases, a family that in mammalian cells includes POL β, λ, μ and σ, as well as TdT [104-106]. Notably, DNA polymerases μ and λ possess the same domain structure as Pol4 [107] and function in the gap-filling in mammalian NHEJ [21-23,108]. Therefore, Pol4 and DNA polymerase μ or λ could have similar functions in NHEJ in both *S. cerevisiae* and mammalian cells.

The *pol4* mutant displays no altered sensitivity to the toxic effects of a variety of DNA damaging agents, including UV light, IR, MMS, ethyl methanesulfonate (EMS), *N*-methyl-*N*′-nitro-*N*-nitrosoguanidine and hydrogen peroxide (H_2O_2) [100,109], although study by Leem *et al.* [101] demonstrated its weak MMS and IR sensitivity. Moreover, the mutant does not show any defect in UV-induced recombination and mutagenesis, mitotic growth [101], spontaneous mitotic recombination, chromosome loss, sporulation and spore viability [100], and process of stationary phase mutagenesis (more extensively discussed below) [110]. On the other hand, it exhibits abnormally high levels of intragenic meiotic recombination [101] and an increased frequency of illegitimate mating [100,101,103,109], the latter suggesting an impairement in NHEJ. Nevertheless, the *pol4* mutant does not have a generalized NHEJ defect compared to the core NHEJ mutants: it only shows a defect in rejoining of one type of junction, mixed joins, requiring removal of a 5′- or 3′-terminal mismatch. Consistently with this, Pol4 appears to be required for NHEJ only if efficient gap-filling polymerization at mixed joins in conjunction with the removal of 5′- and 3′-terminal mismatches is needed [99]. However, although Pol4 fills gaps with displacement of mismatched DNA termini, it lacks both 5′- and 3′-terminal nuclease activities and therefore it does not seem to function beyond gap-filling.

Recently, it has been demonstrated that the N-terminal BRCT domain of Pol4 physically interacts with Lig4 subunit of the Lig4/Lif1 complex. This interaction specifically stimulates the DNA synthesis and ligation activities of Pol4 and Lig4/Lif1, respectively [107], and thus couples the two steps of the microhomology-dependent NHEJ, gap-filling and ligation, in order to allow ligation to occur immediately on completion of the minimum synthesis required to form a ligatable substrate. Taken together, it appears that Pol4 can efficiently fill-in short gaps formed by the alignment of complementary single strands at the ends of duplex DNA and that this reaction is specifically stimulated by the Lig4/Lif1 complex.

The *rad27* mutant cells are highly sensitive to killing by MMS, EMS, methyl nitrosourea (MNU), ethyl nitrosourea (ENU), bleomycin, but only moderately sensitive to UV light, H_2O_2, IR and nitrogen mustard [111-119]. Moreover, these cells display increased levels of spontaneous and UV-induced mutagenesis, mitotic HR rates, and plasmid loss events [111,113-115]. Interestingly, the *rad27* mutant is viable only if HR is functional [114,115,120], suggesting that the DNA lesions accumulating in *rad27* strain are substrates for HR. Alternatively, this may also indicate that Rad27 is a component of NHEJ. In line with this, Rad27 appears to play a role in completion of DSB repair [121] as well as in other processes that are connected to DSB, such as short-sequence recombination [122-130], stability of telomeric repeats [131], Ty1 mobility [132] and gross chromosomal rearrangements [95]. In addition, Rad27 affects premature ageing [133] and temperature-sensitivity [120]. Nevertheless, the most important function of Rad27 resides in the processing of the Okazaki fragments during normal DNA replication (reviewed in [134,135]).

The Rad27 protein, the yeast FEN-1 homologue, is a structure-specific nuclease possessing flap endonuclease and 5' to 3' exonuclease activities [136], the two activities that may have the use in NHEJ. Indeed, it was demonstrated that Rad27 is effectively involved in the end-processing of a subset of 5' flaps during NHEJ [98]. Recently, a function of Rad27 in NHEJ has more precisely been delineated: Rad27 physically and functionally interacts with both Pol4 and Lig4/Lif1 and all these proteins coordinately process and join DNA molecules with incompatible 5' ends [137].

The Ligation Factors of the *S. cerevisiae* Non-Homologous End-Joining

Since Lig4 is an ATP-dependent DSB repair DNA ligase and Lif1 is its stimulating, stabilizing and targeting co-factor, these two proteins are considered to be the *S. cerevisiae* homologues of the mammalian DNA ligase IV and XRCC4, respectively. Lig4 and Lif1 form a highly stable complex consisting of 944 and 421 amino acid subunits, a molecular weight of which is 108.514 and 48.259 kDa [29-33,138]. While there is no obvious domain structure within Lif1, Lig4 contains an ATP-dependent DNA ligase domain (amino acid residues 253-466) as well as two BRCT domains (amino acid residues 683-770 and 836-928) [http://smart.embl-heidelberg.de]. It has been revealed that a region between the two BRCT domains in Lig4, rather than these domains *per se*, is required for the interaction with the central part of Lif1 [32,138]. The yeast complex is very likely to have a 1:2 (Lig4:Lif1) stoichiometry, as judged from the crystal structure of the mammalian complex [139], according to which, the two helical tails of the XRCC4 dimer are in contact with each other over their entire length. In two particular regions they form a left-handed, parallel coiled coil, a platform to which DNA ligase IV adheres. For more detailed information on DNA ligase

IV/XRCC4 and, by inference, Lig4/Lif1 structure, the reader is referred to the papers by Junop *et al.* [140] and Sibanda *et al.* [139].

The most striking phenotype of cells lacking Lig4/Lif1 is observed in transformation-based plasmid repair assays [29-33], which parallels the situation in other NHEJ mutants (see text above). In this assay, a dramatic reduction in the ability to repair restriction enzyme-generated cohesive as well as blunt-ended DSB in plasmid DNA was observed in Lig4/Lif1-defective cells, a result that predicted an essential role for the Lig4/Lif1 complex in the *S. cerevisiae* NHEJ. Since correct religation of the plasmid in Lig4/Lif1-debilitated cells represented only a small fraction of events, these plasmid-based DSB seem to be repaired by an error-prone mechanism that is independent of Lig4/Lif1 [29,30].

Cells lacking functional Lig4/Lif1 complex do not display increased sensitivity to DSB-inducing agents, if they are dividing. In non-dividing stage, they are slightly sensitive toward these agents. This could be due to increased requirement for the NHEJ pathway in cells lacking a homologous sequence, in order to carry out homology-driven repair. This is consistent with findings showing that efficiency of NHEJ is increased in stationary [141] and haploid [142] cells compared to exponentially growing and diploid cells, respectively, as a consequence of the action of additional NHEJ components (see text below). In addition, Lig4/Lif1-defective cells are not UV sensitive [29-31,33] and do not show any obvious defects in telomere length maintenance [30,32,57] and in growth at 30°C. Growth at 37°C, however, requires re-examination, because growth defects at 37°C were observed uniquely in a single study [29-33]. Regarding meiotic defects, the Tomkinson [33] and Lindahl groups [29] demonstrated that Lig4/Lif1-defective cells have no impairment in sporulation or that sporulation in these cells is less efficient compared to the wild type, respectively. Therefore, meiosis is another phenotype that needs re-examination. Similar to the situation in Yku70/80-defective haploid cells, a reduction of replication-independent mutation frequency was observed in Lig4/Lif1 lacking haploid cells (for further details, see one of the sections below) [62].

Additional Non-Homologous End-Joining Factors of *S. cerevisiae*

Until recently, the Sir2, Sir3 and Sir4 proteins, forming the Sir2/3/4 (silent information regulator) complex (see, for example, [143]), had been believed to be members of the NHEJ machinery in *S. cerevisiae*. The linkage between Sir2/3/4 and NHEJ was originally discovered through searching for the Yku70-interacting proteins, which revealed the Yku70-Sir4 interaction. Subsequent phenotypic characterization of the *sir2*, *sir3* and *sir4* mutants showed that the corresponding proteins are required for NHEJ and deletion formation by illegitimate recombination. Moreover, mutations of the *SIR2*, *SIR3* and *SIR4* genes conferred increased sensitivity to IR in a genetic background with a mutation in the *RAD52* gene [144]. Taken together, these results strongly suggested a function of the Sir2/3/4 protein complex in DSB repair by NHEJ. Subsequently, however, impact of Sir2/3/4 on NHEJ was shown to be attributable to transcriptional derepression of the α and a genes resident at the *HML* and *HMR* silent mating-type loci, respectively. New findings showed no NHEJ impairment in Sir2/3/4-defective cells that retained a haploid mating-type (*sir* mutants in which all mating-type genes were inactivated) and decreased NHEJ in a *MAT*a/*MAT*α diploid cell relative to a diploid cell expressing only one *MAT* allele [142,145]. These results therefore provided clear evidence for the control of NHEJ by mating-type heterozygosity and for the existence of NHEJ component that is controlled by a Mata1-Matα2 repressor. Indeed, an additional regulatory NHEJ factor, called Nej1 (NHEJ regulator 1) [146-148] or Lif2 (Lif1-interacting factor) [149] (from now on referred to as only Nej1), has shortly afterwards been identified and shown to be encoded by a haploid-specific gene, expression of which is virtually undetectable in the presence of the Mata1-Matα2 repressor because of a consensus-binding site for this repressor in the promoter region of the *NEJ1* gene [146,148]. Therefore, it appears that Nej1 functions to downregulate NHEJ in diploid cells that contain a homologous donor sequence allowing to carry out HR. In line with this, Nej1 could be one of several factors that can regulate utilization of the two DSB repair pathways, NHEJ and HR, in *S. cerevisiae* (see also text below) [11,14].

The *NEJ1* gene product consists of 342 amino acid residues and has a molecular weight of 39.115 kDa [149]. There is no obvious domain structure within Nej1, except for the two putative transmembrane helices in the N-terminus of the protein [146,148, http://www.yeastgenome.org]. Hence, it is not surprising that database searches have so far failed to identify Nej1 homologues in other organisms at the sequence level. However, considering the low level of primary sequence conservation between Lif1 and XRCC4, it cannot be ruled out that there might be a functional counterpart of Nej1, perhaps also highly diverged in sequence, in mammals [146,150]. To date, no apparent biochemical function has been assigned to Nej1. The Nej1 protein has been demonstrated to interact with Lif1 [146-149,151], but not with any other NHEJ factor [146]. This interaction is mediated by the N- and C-terminal parts of Lif1 and Nej1, respectively [146,149], and presumably targets Lif1 into the nucleus to interact with Lig4 [148], although this would require re-examination since the Åström group [146] reported that Nej1 does not regulate the stability or localization of Lif1.

Nej1 clearly functions in DSB repair by NHEJ because the *nej1* mutant is defective in plasmid repair assays [146-148], HO endonuclease-induced chromosomal DSB assays [146,149] and in assays measuring transposition of maize Ac/Ds transposable elements in *S. cerevisiae* [152]. Furthermore, it is partially defective in suicide deletion screen [153]. In contrast, the mutant does not display sensitivity to MMS and UV [148], growth defect at 30°C and 37°C [149] and impairment in postdiauxic/stationary-phase stimulation of NHEJ (postdiauxic/stationary-phase stimulation of NHEJ means that NHEJ efficiency is consistently induced several fold when cells left the exponential phase of fermentative growth and made the transition to respiratory metabolism and stationary phase), the last one indicating that mating-type and growth-phase regulation of NHEJ are separable phenomena [153]. Another role of Nej1, antithetical to one in promoting end fusion at the site of DSB, has recently been revealed: in the absence of telomerase, Nej1 protects telomeres from NHEJ-dependent end fusions that result in dicentric chromosomes and classic breakage-fusion-breakage cycles [154], demonstrating a possible role of Nej1 in preserving genomic stability in the budding yeast [155].

Although it seems that there are unlikely to be any other major NHEJ factors in *S. cerevisiae* [147,153], because under normal conditions, Yku70/80, Mre11/Rad50/Xrs2 and Lig4/Lif1 may provide all of the essential functions for NHEJ, 16 additional minor regulatory factors [153] apart from Nej1 have recently been identified. Cells mutated in these factors display partial defects in NHEJ. Three of these mutants, *doa1*, *mck1* and *fyv6*, having the largest consistent defect in suicide deletion assay, are impaired in postdiauxic/stationary-phase stimulation of NHEJ, suggesting partial dependency of NHEJ on the nutritional status of the culture. Therefore, there are at least two regulatory inputs into the regulation of NHEJ in *S. cerevisiae*, dependent either on mating-type or nutrition [153], which can increase NHEJ efficiency under certain circumstances [11].

Current Model of Non-Homologous End-Joining in *S. cerevisiae*

Based on data mentioned above, we present the following model for *S. cerevisiae* NHEJ. In the initiating step, the Yku70/80 heterodimer binds to the broken DNA ends. Afterwards, it recruits further NHEJ component, the Mre11/Rad50/Xrs2 complex. Mre11/Rad50/Xrs2 serves as a DNA end-bridging as well as an end-alignment factor that, in cooperation with Yku70/80, targets the Lig4/Lif1/Nej1 complex to the break site. Furthermore, interplay between the Yku70/80 and Mre11/Rad50/Xrs2 complexes stimulates DNA ligase activity of Lig4/Lif1/Nej1, which is required for the completion of NHEJ. In cases where the broken DNA ends are not fully compatible and/or are damaged, a situation that prevents rejoining by simple religation, substantial DNA end-processing occurs prior to ligation. This requires nucleolytic and gap-filling factors, which are likely to be the Rad27 and Pol4 proteins, respectively. Rad27 and Pol4 are presumably brought to the broken DNA ends along with the Lig4/Lif1/Nej1 complex through their interaction with the Lig4 subunit. As a result of action of Rad27 and Pol4, ligatable DNA ends are generated. The NHEJ process is then completed by ligation reaction that is carried out by the Lig4/Lif1/Nej1 complex, as stated above (Figure 1).

Interplay between Homologous Recombination and Non-Homologous End-Joining in *Saccharomyces cerevisiae*

Eukaryotic cells can repair DSB by HR and NHEJ (reviewed in [1-3,6,156-159]). The process of HR involves a resection of the 5' ends at a DSB to create 3' nucleoprotein filaments, followed by strand exchange with a homologous donor duplex, synthesis from the broken 3' termini, and ultimately resolution of the extended D-loop [6,160]. In contrast, DSB repair by NHEJ represents a mechanism that does not require extensive sequence homology (for recent reviews, see [4,8,9,11,13-15,161]).

The choice of DSB repair mechanism depends on several factors. Since in *S. cerevisiae* NHEJ is efficient at rejoining DNA ends with cohesive overhangs, but not blunt ends or ends with non-cohesive overhangs that are repaired by HR [25,99,145,149,162,163], the nature of the DSB is one of these factors. Moreover, the utilization of either NHEJ or HR is dependent also on cell type, because NHEJ pathway in diploid cells is suppressed by a repression of Nej1 as a consequence of the *MAT*a/*MAT*α mating-type [142,145,146,148,149,153], thereby encouraging activity of HR. Furthermore, diploid cells committed to HR possess additional repair pathway choices (for instance, the choice of partner), which are under the control of DNA checkpoint proteins, supporting the notion that cells at different physiological stages possess a different constellation of regulators influencing the balance of repair events (for a review, see [164]). Additionally, helicase activity, which is under mating-type control, also appears to be important in regulating the balance of these events.

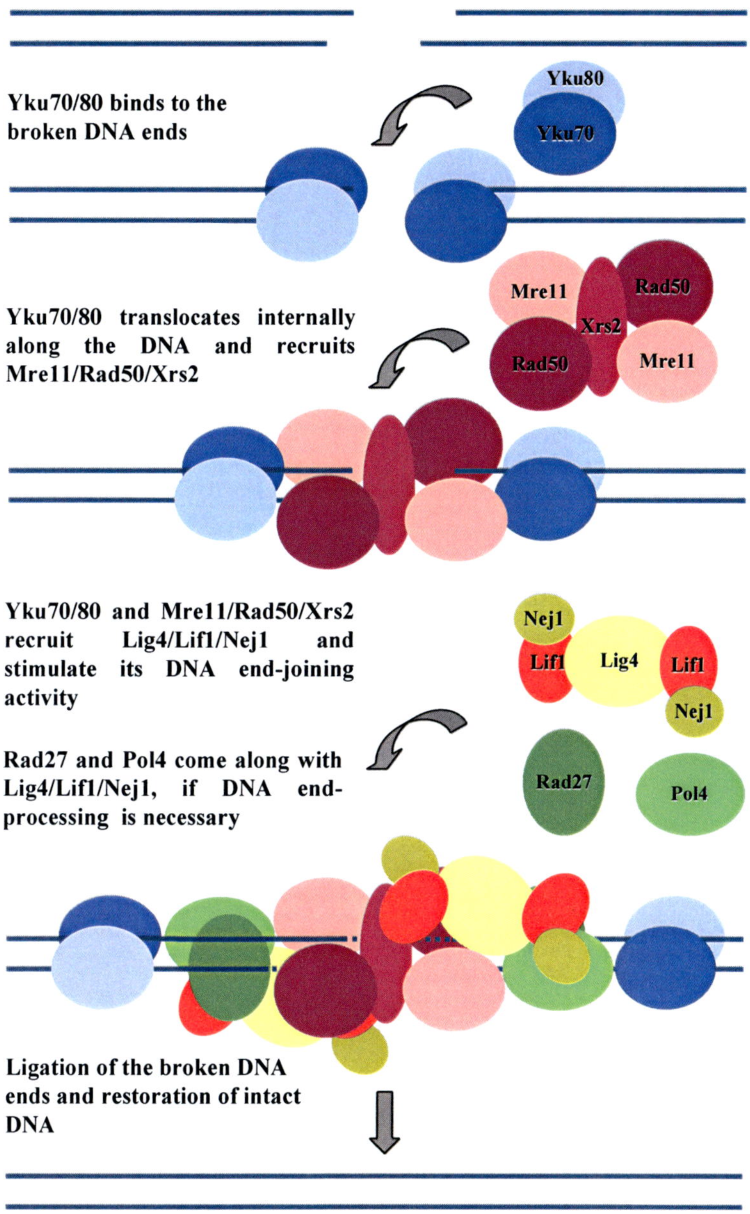

Figure 1. Model for *Saccharomyces cerevisiae* non-homologous end-joining
Yku70/80 is the first component that locates DSB. Following its binding to the broken DNA ends, Yku70/80 recruits the Mre11/Rad50/Xrs2 complex that serves as an end-bridging and alignment factor. Consequently, Yku70/80, in conjunction with Mre11/Rad50/Xrs2, recruits the Lig4/Lif1/Nej1 complex that catalyze the final step of NHEJ, ligation. In case that DNA ends are not complementary or/and are damaged, the DNA end-processing activities that generate ligatable DNA termini take part prior to ligation. These are nucleolytic and gap-filling factors, which are the *RAD27* and *POL4* gene products, respectively.

The cell cycle and growth phase can also influence DSB repair, favoring NHEJ in G_1 and postdiauxic/stationary phase in *S. cerevisiae* [141]. This is in line with data obtained for vertebrate cells [165], suggesting that favoring NHEJ and HR in cells lacking and possessing a homologous donor sequence to repair DSB, respectively, is conserved from unicellular to multicellular organisms.

Originally, it has been suggested that the choice between HR and NHEJ is based on simple competition model [145,166]. However, it has been shown that the defect in either pathway not only did not enhance the remaining pathway, but in fact led to its mild decrease [141,167]. This is in accordance with observation by the Kupiec group [168] showing that in yeast strains, eradicating HR does not increase the absolute frequency of end-joining events, suggesting that a simple competition model [145,166] between NHEJ and HR does not seem to adequately account for DSB repair pathway utilization [141,168]. Another recent report showed that proteins involved in HR (Rad52) and NHEJ (Yku70/80) do not compete for identical broken end substrates, since they bind to different DNA structures [169].

In this context, Frank-Vaillant and Marcand [170] found that an unprocessed DNA ends generated by the HO endonuclease in the budding yeast are stable for a significant period and subsequently the 5' strands of DNA ends are progressively degraded, producing long 3' ssDNA tails [171,172]. *In vivo*, intact DNA ends interact with Yku70/80, but not with Rad52 protein, which interacts only with processed DNA ends. Thus, once 5' to 3' processing commences, NHEJ pathway is suppressed [170] and the cells become committed to repair by HR that is initiated by arising 3' ssDNA tails [156]. In this connection, the term commitment to gene conversion was established as step in the repair process at which the cells no longer execute NHEJ and complete DSB repair by HR [168]. Ultimately, it implies that NHEJ precedes HR temporally, and that the availability of substrate dictates the particular pathway used [170]. The principle that NHEJ acts before HR might be conserved throughout evolution. However, since NHEJ in mammals accommodates also non-cohesive ends and thus could compete with HR on partially resected ends [173], the situation in higher eukaryotes is presumably more complex and the transition from NHEJ to HR may not be as sharp as in unicellular organisms [170].

Moreover, because the chromatin surrounding DSB is hypoacetylated, histone modifications seem to be a key in regulating DSB repair [174] (for more details, see text below). In addition to the extensive chromatin modifications, the *S. cerevisiae* Sir2, Sir3 and Sir4 proteins, typically involved in silencing at the telomeres and at the silent mating-type loci [175], dissociate from telomeres following DNA damage and relocate in complexes at the site of the damage in a Mec1-dependent manner [176-178]. These complex changes at the site of DSB strengthen the notion that the accessibility of DSB and their repair is affected by chromatin structure, as discussed below.

CHROMATIN STRUCTURE ANDDNA DOUBLE-STRAND BREAK REPAIR BY NON-HOMOLOGOUS END-JOINING IN *SACCHAROMYCES CEREVISIAE*

The timely and accurate repair of eukaryotic DNA damage is also dependent on efficient localization of DNA lesions within chromatin. In eukaryotic cells, DNA is tightly assembled with histone and non-histone proteins to form chromatin. The compaction of DNA into chromatin affects many, if not all, nuclear processes including DNA replication, repair, recombination and transcription. The fundamental subunit of chromatin is the nucleosome, in which DNA is wrapped around an octamer core composed of two of each of the core histone proteins. In addition, eukaryotic chromosomal DNA contains linker histones, such as histone H1. The structural unit encompassing histone H1 and the core histones associate with around 166 bp and has been termed the chromatosome. The complex structural packaging of DNA raises the question how it might impact upon the efficiency of NHEJ and what additional factors may be required for an efficient DSB repair in the context of chromatin.

The efficiency of NHEJ in various yeast mutants is often assayed by transforming yeast with linearized plasmids harboring a selectable marker. Transformation efficiency is directly related to religation of the plasmid required for its maintenance. Disruption of genes encoding the known components of NHEJ machinery leads to a severe defect in NHEJ (see text above) [25]. This assay is unlikely to provide a detailed insight into contributions of chromatin components machinery, since the introduced DNA is not chromatinized and replication-coupled chromatin assembly is unlikely to precede NHEJ. Despite that, the expanding experimental evidence suggests the direct role of chromatin modifications in NHEJ in *S. cerevisiae*.

Histone Phosphorylation as a Marker of DNA Repair

One of the very first events, in response to the formation of DSB, is a phosphorylation of the C-terminal tail of histone H2A. The extent of this modification is very large, covering tens of kilobases of DNA in yeast [179,180] and megabase regions in higher eukaryotes [181,182]. This specific histone modification is carried out by members of PIKK family [65]. In mammals, all three members of this family (ATM, AT-related (ATR) and DNA-PKcs) have been implicated directly in DNA repair and can phosphorylate H2AX. ATM and DNA-PKcs were shown to function redundantly to phosphorylate H2AX after DNA damage in non-replicating cells [183,184], whereas ATR is responsible for H2AX phosphorylation following the replication blocks [185]. In the absence of a clear DNA-PKcs homologue in *S. cerevisiae*, the two ATM and ATR homologues (Tel1 and Mec1, respectively) (see text above), are both required for H2A phosphorylation [186,187]. To verify the role of H2A phosphorylation in *S. cerevisiae*, Downs *et al.* [186] tested several yeast mutant strains for their sensitivity to DSB-inducing agents. Yeast mutants in which the recognition motif for PIKK kinases (Ser-Gln-Glu) was deleted or mutated, all displayed sensitivity to phleomycin and MMS, drugs that introduce DSB *in vivo*. These H2A mutants also showed genetic interactions with components of both HR and NHEJ, demonstrating that H2A phosphorylation is a common requirement for both pathways. In a similar study, the H2A phosphorylation was required for efficient repair of topoisomerase I-mediated DSB during S phase [187]. Despite a large volume of accumulated data, an understanding of mechanistic details by which H2A phosphorylation contributes to DSB repair remains incomplete. The extremely rapid H2AX phosphorylation in response to DSB, followed by the detection of many repair factors in the radiation-induced nuclear foci, has led to assumption that this specific histone modification was required for the initial recruitment of repair factors to the site of DSB [181]. However, recent studies had shown that this was not the case and the recruitment of many known repair factors, including 53BP1, BRCA1 and subunits of MRE11/RAD50/NBS1 complex was shown to be independent of H2AX phosphorylation [188]. Also, in the budding yeast, it has been shown that the Tel1/Mec1-mediated activation of the intra S phase checkpoint is H2AX phosphorylation independent, indicating in both cases that H2AX phosphorylation is not required for initial recognition of DSB. The nature of this initial break-sensing mechanism preceding H2AX phosphorylation still remains a mystery. In this context, it is intriguing that in mammalian cells, where ATM is the major kinase responsible for histone H2AX phosphorylation, the global ATM activation can be achieved even in the absence of DSB formation by treating cells with agents inducing large scale chromatin/chromosomes structure changes (*e. g.* trichostatine A, hypotonic conditions, chloroquinine). This, together with the fact that very few DSB are required for the activation of many thousands of ATM molecules, has led to a proposal of DNA damage surveillance mechanism based on indirect activation of ATM by large-scale topological chromatin changes resulting from small number of DSB [189]. In this model, a local disturbance of chromatin structure, instead of direct interaction with DSB, would constitute the signal required for kinase activation. This could also explain why many thousands of H2AX molecules are being phosphorylated around DSB. If the initial activation event, triggered by a change in the chromatin structure, happens to be relatively distant to the site of DSB, the chain/cascade like phosphorylation may continue until the actual site of the break is encountered and the repair is initiated. This scenario could also provide an explanation why so many thousands of repair factors molecules are being recruited in response to a single DSB, where otherwise very few would be required if repair machinery was targeted to the break itself. It could also explain the relative heterogeneity in size of early radiation-induced foci (at least in vertebrates) that is often observed, although the sequestration of several DSB into large foci at the later repair intervals has recently been reported in both yeast and mammalian cells [190,191].

Role of Histone Acetylation and ATP-Dependent Chromatin Remodelling in Non-Homologous End-Joining

Chromatin-remodelling has been historically most studied in the context of the transcriptional control and accessibility of chromatin to transcriptional factors (for reviews, see *e. g.* [192,193]). Genetic and biochemical analysis of these regulatory mechanisms have revealed the existence of two main groups of chromatin modification activities. The first group is made up of ATP-dependent chromatin modification activities where "chromatin remodelling" refers to a wide variety of ATP hydrolysis-dependent changes within chromatin, including disruption of histone-DNA contacts within nucleosomes, movement of histone octamers both in *cis* and *trans* and other structural changes, generally enhancing the accessibility of nucleosomal DNA [194].

The second group includes the activities that catalyse covalent post-translational modifications (acetylation, phosphorylation, methylation, ubiquitination and ADP-ribosylation) of the core histone proteins on the N-terminal "tail" domain extending from the surface of the nucleosome. Apart from H2AX phosphorylation of serine 129 (importance of which is outlined in the previous section), lysine acetylation/deacetylation has recently emerged as an important modification involved in DSB repair in *S. cerevisiae* and mammalian cells and histone acetylase-deacetylase complexes are now directly implicated as key components of the DNA repair machinery. Acetylation neutralizes the positive charge of histones thus impinging directly on histone-DNA electrostatic interactions, relaxing the chromatin structure and promoting specific interactions with non-histone proteins. The importance of histone acetylation for NHEJ in *S. cerevisiae* had been recognized by finding that lysine substitution within the N-terminal tail of histone H4 caused sensitivity to DSB-inducing agents and defects in NHEJ [195]. Interestingly, these H4 alelles were not sensitive to UV-irradiation, suggesting the possibility that H4 lysine acetylation might be DSB specific. The same report also established that Esa1, the catalytic subunit of the yeast NuA4 histone acetyltransferase (HAT) complex, whose substrates also include the same H4 lysines, is also required for DSB by NHEJ. Interestingly, Esa1 is believed to be a yeast homologue of the human TIP60 protein, and human cells over-expressing HAT-deficient complexes with mutated TIP60 protein were shown earlier to be DSB repair deficient and unable to undergo apoptosis [196]. Similarly, the inactivation of another subunit of NuA4 complex, Yng2, also eliminates HAT activity of the complex [197] and renders cells sensitive to agents that induce replication fork collapse and the deletion of *YNG2* makes NHEJ essential for cell survival [198]. Moreover, inactivation of *GCN5* (encodes histone H3 HAT) leads to phenotype similar to that of the *yng2* mutant, suggesting that lysine acetylation in DSB repair is not unique for histone H4 and the emerging complex acetylation pattern may play a role in several pathways of DSB repair [198]. Although these reports established the importance of global histone acetylation for NHEJ, they have not addressed the question of changes in histone acetylation in the chromatin surrounding DSB. A recent report, perhaps surprisingly, showed that histone H4 becomes deacetylated in the closed proximity of the chromosomal break [174]. The specific lysine 16 of histone H4 deacetylation is dependent on Sin3 and Rpd3 components of the predominant histone deacetylase (HDAC) complex in *S. cerevisiae* and is required for efficient NHEJ. To complicate the situation further, the NuA4 HAT complex appears to be recruited to the phosphorylated histone H2A at the DNA damage sites, where a single NuA4 subunit, Arp4, is responsible for the interaction with phospho-H2A peptides [180].

What possible model could explain the conflicting simultaneous requirements for histone acetylation/deacetylation for efficient DSB repair? One possibility is that the concerted action of HAT and HDAC complexes, presumably at the different stages of the DSB repair process, may be required for efficient end-joining. Initial relaxing of chromatin structure in the vicinity of the break by HAT may be followed by the recruitment of HDAC to promote creation of a more "rigid" temporary environment to stabilize the broken DNA ends, possibly followed by another round of action of HAT after the completion of repair process in order to re-establish the correct histone code to allow normal gene activity. An attractive proposal would be to recruit both activities to the site of DNA break as a part of a single, tightly regulated complex. The existence of such close physical interaction has been reported in living human cells [199], suggesting that such a possibility is not without precedence.

What is the connection between ATP-dependent chromatin remodelling complexes and histone acetylation/deacetylation? Despite the fact that it is not clear how histone acetylation facilitates DSB repair, it is intriguing to notice that a human homologue of the NuA4 complex (called TIP60 or hNUA4) also contains SWI/SNF-like ATPase subunit and a subunit that harbours two bromodomains [200]. Given the known connection between lysine acetylation and bromodomains in an assembly of different chromatin remodelling complexes required for transcriptional regulation, it is reasonable to expect that a similar interplay between histone acetylation and retention of ATP-remodelling complexes might take place during DSB repair.

Two recent studies had shown that the yeast Swr1 ATPase is the catalytic subunit of a chromatin remodelling complex [201,202], which shows strong genetic interaction with subunits of the NuA4 complex [201,203]. One of the subunits of the Swr1 complex, Bdf1, contains a double bromodomain and both *swr1* and *bdf1* mutant strains are hypersenstitive to DSB-inducing agents [202-205]. Although the Swr1 complex has been linked with DSB repair by HR rather than NHEJ (due to the presence of associated RuvB-like helicases – a hexameric DNA helicases required for branch migration and HR-dependent repair in prokaryotes [206]), it is intriguing to note that it also controls the incorporation of another rare variant of histone H2A, namely H2A.Z, into chromatin [201,202]. Although the relevance of H2A.Z for the DNA repair is still uncertain, it is noteworthy that yeast strains lacking H2A.Z have been reported to be hypersensitive to a variety of DNA damaging agents [202].

The notion of interplay between histone acetylation/deacetylation and ATP-dependent chromatin remodelling is further supported by the recent discovery of complexes linked *via* a common subunit that are recruited to the site of DSB. The interaction between Arp4 subunit of NuA4 complex with phosphorylated form of histone H2A [180] has already been mentioned. The Arp4 is also a subunit of two additional ATP-dependent chromatin remodelling complexes, Ino80 and Swr1 [201,202,207]. These complexes also interact with phosphorylated histone H2A and are recruited to DSB [180,208,209].

In conclusion, the functional interplay between HAT/HDAC and ATP-remodelling complexes, originally established in the field of transcriptional regulation, is likely to function in DSB repair as well.

NON-HOMOLOGOUS END-JOINING IN PROKARYOTES

Discovery of Ku Homologues in Prokaryotes and Phage

As has been discussed above, NHEJ is found throughout the eukarya but until recently it was believed that this repair pathway was absent from lower organisms such as prokaryotes and archaea. The first clues to the presence of NHEJ in bacteria came from *in silico* studies that identified the presence of Ku homologues in many bacterial genomes and a single archaeal genome [34,35]. Unlike the heterodimeric Ku70/80 of eukaryotes, the bacterial Ku's appeared to be encoded by a single gene suggesting they are possibly homodimeric in structure. Recent biochemical studies have confirmed that the bacterial Ku's exist in solution as homodimers and these structures bind preferentially, in a sequence-independent manner, to the ends of dsDNA [37]. The bacterial Ku proteins are composed of approximately 250-350 amino acids. In contrast, the eukaryotic Ku proteins are much larger (70-80 kDa). The bacterial Ku homologues represent a conserved domain at the centre of the larger Ku proteins but lack the conserved N- and C-terminal extensions found in the eukaryotic proteins (Figure 2). This common core structure is conserved between amino acids 250-550 approximately in the eukaryotic Ku proteins. Biochemical studies have shown that this region of Ku70 and Ku80 is one of the major determinants for both heterodimerization and DNA binding [35]. In the crystal structure of the human KU70/80 heterodimer, this core region dimerizes to form a ring-like structure through which the end of the DNA is threaded [40,43].

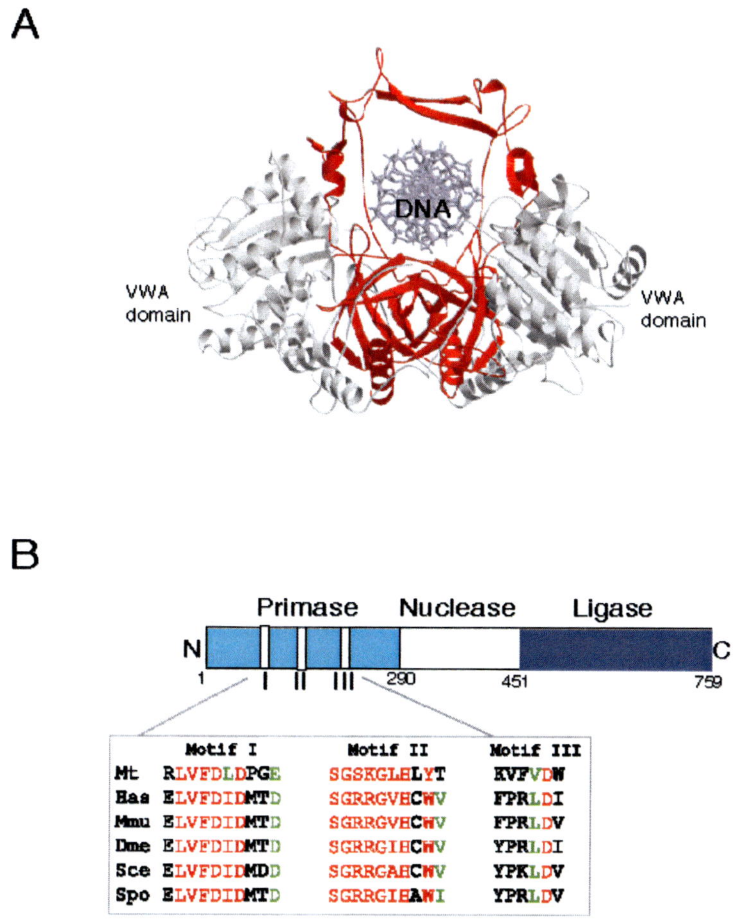

Figure 2. Structure of the prokaryotic non-homologous end-joining repair proteins
(A) Structure of the bacterial Ku homodimer. A structure-based sequence alignment was used to model the structure of *Mycobacterium tuberculosis* Ku (Mt-Ku). The smaller bacterial Ku homologues do not possess the extreme N-terminal von Willebrand domains (VWA) or the C-terminal SAP (Ku70) and DNA-PKcs binding (Ku80) domains present in the larger eukaryotic Ku proteins. These domains are coloured grey in the structure of the human KU complex (PDB ID: 1jey) to leave a truncated KU complex (red) consisting of truncated KU70: residues 262-521 and KU80: residues 254-515. This represents a close approximation of the structure of the bacterial Ku homodimer.
(B) Domain organisation of the *Mycobacterium tuberculosis* NHEJ DNA ligase D (Mt-LigD). The location of the predicted polymerase, nuclease and ligase domains are shown. A sequence alignment of three catalytic motifs (I-III) conserved across all of the eukaryotic-like DNA primases (small catalytic subunit) and the Mt-LigD pol domain is also shown. The organisms and the protein accession numbers are: Sce, *Saccharomyces cerevisiae* (P10363); Spo, *Schizosaccharomyces pombe* (O14215); Hsa, *Homo sapiens* (P49642); Mmu, *Mus musculus* (P20664); Dme, *Drosophila melanogaster* (Q24317) and Mt, *Mycobacterium tuberculosis* (Rv0938).

In silico studies have also identified that the early protein Gam of bacteriophage Mu shares significant sequence homology with both subunits of eukaryotic Ku [210]. Gam orthologues have also been discovered in a number of bacterial genomes but the locations of these genes suggest they are remnants of prophage insertions. The *gam* genes also share homology with the genes encoding prokaryotes Ku-like proteins. Gam is a homodimer that recognize DNA ends in a manner similar to eukaryotic and prokaryotic Ku and protects linear Mu phage DNA from nuclease digestion [210]. These properties of Gam suggest that its main role following infection is to bind to the ends of linear viral DNA, preventing its degradation by bacterial exonucleases and thus enhancing transposition of the phage genome into the host's chromosome. Interestingly, *E. coli* expressing Mu Gam take up linearized plasmid 100-fold more efficiently than uninfected cells [211], whilst in yeast cells, linear plasmids can be efficiently introduced and maintained only when Yku70/80 is present [26]. These observations reveal functional parallels between bacteriophage Gam and eukaryotic Ku and suggest that their functions may have been evolutionarily conserved.

The resemblance between the bacterial, phage and eukaryotic Ku's suggests they evolved from a common ancestral *ku* gene (Figure 2). This early Ku progenitor probably functioned as a homodimeric complex that bound to DNA ends and protected the DNA ends from nucleolytic degradation. Subsequently it evolved by fusion with other domains to facilitate the recruitment of other factors, including DNA ligases, forming a primordial NHEJ repair complex. It is notable that only one characterized archaeal species, *Archaeglobus fulgidus,* contains the Ku system, suggesting that this was acquired by lateral transfer from the prokarya. Although the eukaryotic Ku protein homologues have diverged substantially in sequence, they have maintained the same overall size, subunit structure and DNA binding properties [41]. Interestingly, human KU70 can complement yeast *yku70* null mutant [41]. Although the Ku70 and Ku80 subunits are biochemically distinct, they appear to have arisen from a common ancestral gene [41]. This is likely as the sequences of the two subunits of Ku protein are more divergent from each other in higher eukaryotes than in yeast. Moreover, a number of bacterial genomes, including *Mesorhizobium loti* and *Sinorhizobium loti*, encode two distinct bacterial *ku* genes within the same operon [35]. It has been proposed that this may represent the outcome of an early gene duplication event that has given rise to the first examples of the heterodimeric Ku complexes.

It now appears very likely that the eukaryotic Ku proteins directly evolved from an ancestral prokaryotic *ku* gene that have subsequently acquired additional domains, presumably to enhance their role in DNA repair [34,35]. The N-terminal region of Ku70 and Ku80 (amino acids 1-240 approx.), lacking in bacterial Ku or phage Gam, comprises a Von Willebrant factor A (VWA) domain [40] that has been implicated in recruiting additional proteins to sites of DNA damage, forming larger protein complexes required for concerted DNA repair and signaling. A SAP motif is present in the C-terminal region of eukaryotic Ku70 proteins (amino acids 561-609). The SAP motif is composed of three α-helices that form a helical DNA binding motif [40], although this domain has only weak affinity for DNA. It has been proposed that it may act like a break, preventing Ku from translocating away from the ends of the break [40]. The C-terminal region of the *Streptomyces coelicolor* Ku protein (SCF55.25c) shares significant homology with the SAP motif of Ku70 [34,35]. This SAP-like motif of *S. coelicolor* Ku is located in a 40 amino acid extension (amino acids 290-330) and it is likely that this domain is involved in DNA binding. This α-helical motif is not detected in the other bacterial Ku or Gam homologues.

Prokaryotic Ku Recruits a DNA Repair Ligase to DNA and Promotes End-Joining

Circumstantial evidence for the association of the bacterial Ku proteins with other DNA repair proteins is evident from the operon structure of the bacterial Ku genes. Many of the prokaryotic Ku homologues are organized into operons containing putative ATP-dependent DNA ligases [34,36]. Operons frequently co-regulate functionally interacting proteins raising the possibility that these putative ligases might interact with the Ku-like proteins. These operon-associated ligases share a common core region with other ATP-dependent ligases that encompasses the signature catalytic motifs [34,36]. Weller *et al.* established that *Mycobacterium tuberculosis* ligase D (Mt-LigD), and by inference other members of this Ku-associated ligase, are functional DNA ligases capable of catalyzing DSB rejoining in an ATP-dependent manner [37]. Moreover, they demonstrated that DNA ligation activity of Mt-LigD was specifically and significantly stimulated by the addition of Mt-Ku but not by human KU. Human KU recruits DNA ligase IV/XRCC4 to the termini of DSB and similarly Mt-Ku directly recruits Mt-LigD, but not human DNA ligase IV/XRCC4, to DNA ends [37].

Weller *et al.* also generated strains of *Bacillus subtilis* bearing inactivating mutations in *ykoV* (encodes Ku) and *ykoU* (encodes ligase D) [37]. These strains were not sensitive to UV or MMS suggesting that nucleotide excision repair functions normally and alkylation damage induced by MMS is repaired efficiently. In contrast, the mutant strains were sensitive to IR during stationary phase. As the major cytotoxic lesions induced by IR are DSB, these results suggested that Ku and ligase D play a role in DSB repair in prokaryotes.

Prokaryotic Ku and Ligase D Constitute a Two-Component Non-Homologous End-Joining Repair Machine

The majority of DSB produced in cells have termini that are non-complementary. These ends require processing by polymerases and nucleases to generate ligatable termini. As discussed above, many of the factors involved in this processing of the ends of DSB have now been identified in eukaryotes, including the polymerases, human POL μ, λ and yeast Pol4, and the nucleases ARTEMIS and yeast Rad27. In contrast, only two NHEJ repair factors, Ku and ligase D have been identified in prokaryotes, which raises the question of how bacteria compensate for the lack of additional repair factors. Insights into this mystery came from *in silico* sequence analysis, which revealed that many of the Ku-associated ligases possess, in addition to a ligase domain, additional putative catalytic domains [34,36]. One of these domains (pol) exhibits significant sequence homology with eukaryotic polymerases. This polymerase domain belongs to the Pol X family whose members include human POL μ and λ, TdT, and yeast Pol4, enzymes that have been implicated in the gap-filling reactions that occur during V(D)J recombination and NHEJ. Recently, Della *et al.* demonstrated that this pol domain of Mt-LigD possesses a remarkable and unique variety of nucleotidyl transferase activities including primase, terminal transferase, gap-filling polymerase activity [38]. In addition, Mt-LigD also possesses a 3' to 5' ssDNA exonuclease activity. Thus, Della *et al.* confirmed these putative functions, establishing Mt-LigD as a versatile multi-functional DNA repair enzyme that possesses a wide variety of processing activities. Mt-LigD possesses all of the enzymatic activities required to process and join incompatible breaks, irrespective of the ends, effectively allowing NHEJ-mediated DSB repair to be reconstituted *in vitro* [38]. Ku is essential for recruiting the ligase D to the DSB, thus together these proteins constitute a two-component NHEJ repair complex (Figure 3). Although Ku is not directly involved in end-processing, there is some evidence that Ku modulates these processing activities [38].

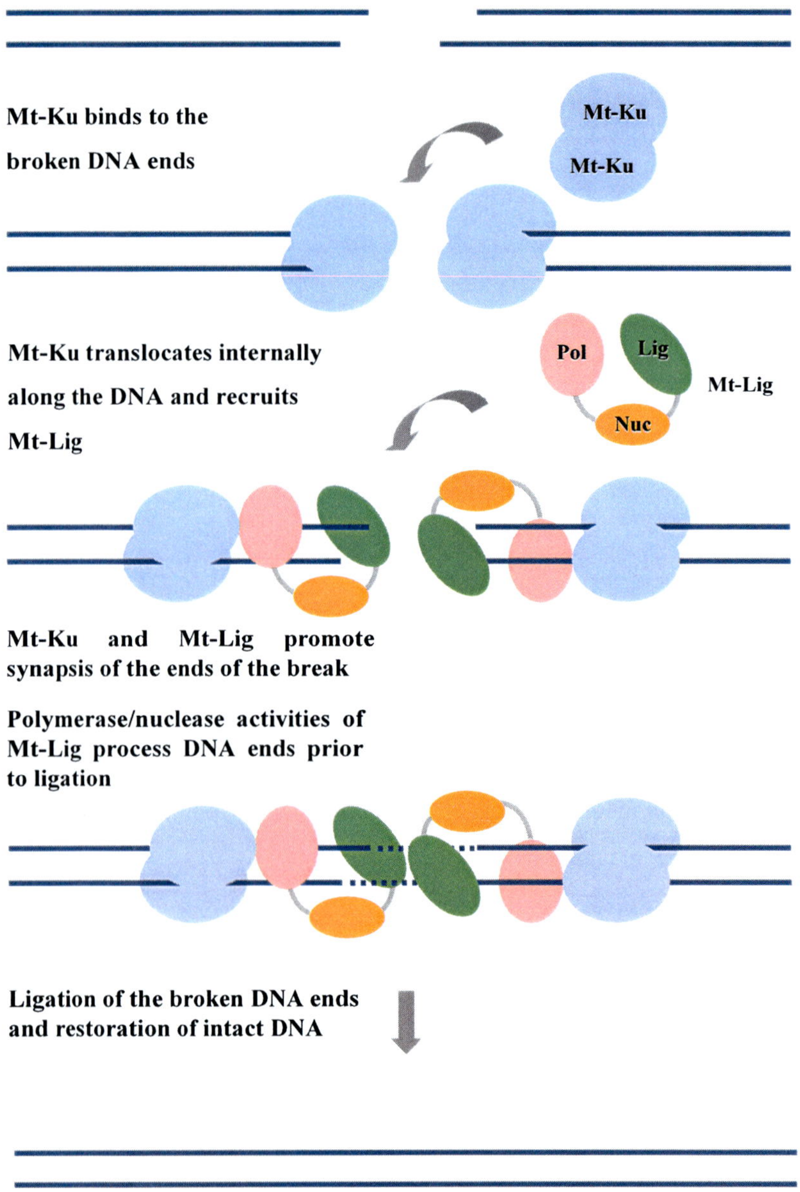

Figure 3. Model for prokaryotic non-homologous end-joining *Mycobacterium tuberculosis* Ku (Mt-Ku) homodimer is the first component that locates DSB. Following its binding to the broken DNA ends, Mt-Ku recruits the multifunctional DNA repair ligase D (Mt-LigD) to the DSB. Subsequently, the broken ends are aligned by this NHEJ complex prior to ligation of the break. When DNA ends are non-complementary and/or are damaged, the DNA end-processing, gap-filling and nucleolytic activities, of Mt-LigD (Pol and Nuc, respectively) generate ligatable DNA termini prior to ligation.

In addition to acting as a DNA polymerase, Mt-LigD also catalyses DNA-dependent RNA polymerisation suggesting that repair of DSB in prokaryotic cells may result in the incorporation of short stretches of RNA into genomic DNA. Indeed, human POL μ, which has been implicated in mammalian NHEJ, also possesses RNA polymerase activity [212]. These patches of RNA at the junctions of repaired DSB may be tolerated in a number of ways. The replicative DNA polymerases may be capable of synthesizing through short stretches of hybrid RNA/DNA duplex. Alternatively, the incorporated RNA may be excised from DNA prior to subsequent replication of the genomic DNA by short-patch repair processes such as the recently described type 2 RNase H pathway [213]. Indeed it has been reported that plasmid DNA containing RNA introduced by the bacterial NHEJ system could be replicated in *E. coli* [38]. Why would a DSB repair enzyme possess an RNA polymerase activity to support NHEJ repair? The answer may lie in the life-cycle of these bacteria. Many of the prokaryotic organisms that possess the Ku and ligase D operon system exist for large periods in a stationary or sporulated phase, when intracellular pools of dNTPs are likely to be low and DSB repair cannot be achieved *via* HR. They may have evolved this strategy of utilizing NTPs to overcome this problem in the short term.

Prokaryotic Non-Homologous End-Joining Complex Functions as a *Bona Fide* Non-Homologous End-Joining System *in vivo*

Doherty and colleagues have definitively established that the bacterial Ku and ligase D proteins form a complex with high capacity for processing and religation of DSB ends *in vitro* [37,38]. However, while radiosensitization of *B. subtilis ykoU*⁻ and *ykoV*⁻ mutants was also observed [37], it did not prove that the corresponding bacterial proteins mediate true NHEJ repair of DSB *in vivo*. This is critical, since repair of single chromatinized DSB may be far more challenging. Della *et al.* demonstrated, using a "suicide deletion" assay for monitoring of NHEJ at a site-specific chromosomal DSB, that ectopically expressed Mt-Ku and Mt-LigD could complement *YKU70* and *LIG4* mutations in *S. cerevisiae* [38], which abolish NHEJ mediated DSB repair, restoring NHEJ to ~50% of wild type levels. Importantly, they showed that this reconstitution required the presence of both *M. tuberculosis ku* and *ligD* genes. *S. cerevisiae* NHEJ also depends on the Mre11/Rad50/Xrs2 complex [53,55,57,66-70] and it has been proposed that this complex acts to maintain the ends of DSB in proximity for repair [70]. Surprisingly, expression of the *M. tuberculosis* NHEJ proteins in the *rad50* mutant cells led to a substantial recovery of NHEJ [38].

These remarkable complementation experiments in yeast have numerous implications for both bacterial and yeast DSB repair. Firstly, it demonstrates that prokaryotic Ku and ligase D proteins represent a *bona fide* NHEJ system. It also confirms that prokaryotic NHEJ complex is a two-component system, consisting of a DSB targeting subunit (Ku) and a catalytic subunit (ligase D) (Figure 3). Furthermore, these findings suggest that other bacterial proteins are not required for NHEJ and that the *M. tuberculosis* Ku-ligase D complex can catalyze significant NHEJ in the absence of the Mre11/Rad50/Xrs2 complex. Despite having opposing end-processing activities in the joining complex, the *M. tuberculosis* NHEJ apparatus is highly precise when presented with compatible overhangs. This implies that the processing activities are tightly co-ordinated. It verifies that bacterial NHEJ proteins reconstitute a repair system that rarely forms mutations and suggests that NHEJ repair in eukaryotes is more precise than has been reported. During eukaryotic NHEJ-mediated DSB repair, short stretches of sequence homology, know as microhomologies (see text above), are often employed to align the termini of breaks prior to the processing of the end [96]. Similarly, prokaryotic NHEJ repair of DSB, both *in vitro* and *in vivo*, also involves the preferential utilization of recessed microhomologies internal to the overhangs of incompatible ends that facilitate resection and ligation of the DSB.

A number of histone modifications have been implicated in NHEJ processes in yeast (see text above) [174,186,195]. It is therefore somewhat surprising that the prokaryotic NHEJ proteins catalyzed relatively efficient repair of chromatinized DSB in yeast as bacteria lack the equivalent histone proteins. Clearly, efficient NHEJ does not absolutely require a site-specific and modification-dependent interaction of the NHEJ complex with histones, as there is no reason for the *M. tuberculosis* proteins to carry this function. Thus, either bacterial NHEJ occurs entirely independently of histones, or a role of histone modification in NHEJ is indirect and once the necessary modifications of chromatin in the vicinity of the DSB have taken place, the breaks are accessible for bacterial NHEJ machinery. This would suggest that histone modifications may not be required for the completion of the end-joining process.

A Role of Non-Homologous End-Joining in Stationary Phase Mutagenesis

Due to its importance for genome diversification in, and evolution of, microorganisms, as well as aging and carcinogenesis in mammals, a process of stationary phase (also referred to as adaptive, directed, selection-induced, or DNA replication-independent) mutagenesis has intensively been investigated. Although these studies were initiated in the bacteria *E. coli* [214,215] and *Salmonella typhimurium* [216], thereafter they have been extended to simple eukaryotic organism, the budding yeast *S. cerevisiae* [217-221].

In principle, two ways of mutation generation exist in a cell, DNA polymerase errors that have been fixed as mutations and DNA damage that has been processed or tolerated to give raise mutations. Although both of them are active in stationary phase cells, the former is restricted to mutations introduced by repair DNA synthesis, which again is likely to be evoked by DNA damage. Hence, DNA damage seems to account for virtually all mutations induced in stationary phase cells [62].

When spontaneous mutation spectra from exponential and stationary cells were compared in *S. cerevisiae*, it has been found out that nucleotide gains and losses were reasonably balanced in revertants obtained from dividing cells, while in those from non-dividing cells mainly simple deletions in mononucleotide repeats were observed. This was the first evidence for difference between stationary phase and DNA replication-dependent mutagenesis in eukaryotic cells. Since simple deletions in mononucleotide repeats are expected to be generated by DNA polymerase slippage errors that escaped correction by mismatch repair or proofreading, error-prone repair DNA synthesis induced by the processing of spontaneous DNA damage was suggested to be the responsible mechanism. Alternatively, DSB repair by NHEJ was also considered to be the cause [222]. The latter eventually turned to be the case as it has been shown that disruption of either the *YKU70* or *LIG4* gene resulted in a 50%

reduction of mutation frequency in stationary phase haploid cells and that typical NHEJ-dependent mutations are small deletions within mononucleotide repeats. These findings provided evidence that NHEJ has a mutagenic potential in non-dividing cells and that non-dividing cells pay for a large-scale chromosomal stability with an increased frequency of small-scale mutations [62].

As stated above, Pol4 has been reported to be required for joining of aligned intermediates containing 5' or 3' terminal mismatches [99] and to interact with Lig4/Lif1, which stimulates the DNA synthesis and ligation activities of Pol4 and Lig4/Lif1, respectively [107]. As the sole polymerase involved in the *S. cerevisiae* NHEJ, Pol4 has been examined for its possible role in the process of spontaneous stationary phase mutagenesis. In contrast to Yku70- and Lig4-deficient cells, the spontaneous stationary phase mutagenesis in Pol4-deficient cells was not decreased. Instead, Pol4 appears to antagonize the process, as Pol4-deficient cells exhibit significantly increased mutation frequency. As suggested by Heidenreich and Eisler, the antagonistic effect of Pol4 towards spontaneous stationary phase mutagenesis could be somewhat caused by increased NHEJ accuracy, but could also include an unknown component, which may be related to the so far unidentified mechanism producing the remaining 50% of mutations [110].

To address the question whether effect of NHEJ on process of stationary phase mutagenesis could be amplified and accordingly reproduced by an exogenous introduction of DSB, IR exposure was used in next experiments. In a similar manner to unexposed cells (see text above), IR-exposed Lig4-deficient cells displayed significantly reduced stationary phase mutation frequency, indicating that a functional NHEJ is necessary for the formation of IR-induced mutations in stationary phase cells. In other words, this parallel provides evidence that NHEJ, being responsible for mutagenic repair of endogenously occurring DSB [62], is also active in mutagenic repair of DSB induced by exogenous sources. Taken together, there is no doubt that DSB repair by NHEJ is remarkably important mechanism that introduces mutations in stationary phase eukaryotic cells.

Assuming that the NHEJ pathway is conserved between prokaryotic and eukaryotic organisms, its participation in the process of stationary phase mutagenesis in bacteria may also be expected. Notably, many of the bacteria disposing of NHEJ spend at least part of their life cycle in stationary phase (*M. tuberculosis*, *M. loti* and *S. loti*), where HR is not possible. Therefore, NHEJ could offer a particular advantage to these bacterial species to maintain integrity of their genomes when they are not replicating. In addition, bacterial species such as *B. subtilis* and *S. coelicolor* may also take advantage of NHEJ to preserve their genomes when existing as spores [15,37,223,224].

Conclusion

The NHEJ pathway appears to be highly conserved throughout evolution and to operate from bacteria to man. The exact origins of NHEJ remains to be established, however the presence of Ku-like proteins in bacteriophage suggests that the NHEJ repair pathway may have its origins in processes designed to protect the ends of linerized genomes from nucleolytic degradation and also possibly to enhance DNA integration. Therefore, it is not surprising that NHEJ machinery also seems to be utilized in unicellular organisms for processes other than DNA repair, that include class switching, stationary phase mutagenesis, integration processes allowing horizontal gene transfer, *etc.* Subsequently, the primordial Ku evolved into a recruiting complex that promotes the binding of additional repair proteins, including ligases, polymerases and nucleases, to the termini of DSB. NHEJ has been reconstituted *in vitro* using bacterial [37,38], yeast [70] and mammalian [16] proteins and *in vivo* studies of these processes in unicellular organisms have provided compelling evidence that only a subset of repair factors are essential for the repair of DSB *in vivo* [38]. Although these studies have shown that the basic functions of the core NHEJ components are highly conserved and that the fundamental steps of NHEJ reaction are in principle the same, they also highlight the substantial variation between different organisms in the details of the NHEJ process.

The utilization of NHEJ appears to be one of the most important differences between unicellular and multicellular organisms. Since unicellular organisms have very little non-coding DNA, it is essential that DNA damage is repaired primarily through a pathway that is highly accurate. Moreover, these organisms are rapidly replicating when present in an environment with unrestricted nutrients, so they might possess multiple copies of the genome as is the case for many bacteria. These two factors may be the major reasons why unicellular organisms repair DSB primarily by HR. However, when nutrients are limiting in the environment, DNA replication is inhibited and hence NHEJ is required to take over, as a back-up pathway, for repairing of DSB. Notably, in *S. cerevisiae* additional regulatory NHEJ factors are present, which are dependent on nutrition conditions and which directly influence efficiency of NHEJ (see text above) [153].

A great deal of attention has recently been paid to the role of histones and their modifications in the NHEJ process, combining *in vitro* and *in vivo* experimental strategies. As mentioned above, several core histone modifications have already been shown to occur in NHEJ in *S. cerevisiae* [174,186,195], with some of them being conserved in mammals. In mammalian cells, linker histone H1 modification has also recently been reported to play a possible role in NHEJ. Histone H1 inhibits NHEJ reaction *in vitro* and its phosphorylation by DNA-PKcs is required for its release from DNA and partial restoration of NHEJ reaction [225]. In *S. cerevisiae*, histone H1-like protein, Hho1, is inhibitory to DSB repair by HR [226], and consequently HR inhibition occurs particularly in regions of the yeast genome in which Hho1 is overrepresented [227]. In these regions, DSB repair may be directed into NHEJ, which, in a similar manner to mammalian NHEJ, could release Hho1-mediated inhibition by its phosphorylation. In the absence of a clear homologue of DNA-PKcs in *S. cerevisiae*, Hho1 phosphorylation could be mediated by other PIKK kinases, such as Tel1 or Mec1 (see text above). Taken together, the chromatin modifications/remodeling machinery at DSB seems extraordinary, although, to date, it is still not clear which step(s) of the process are facilitated by these enzymatic activities. The fact that bacterial NHEJ complex can rejoin DSB in the chromosomal DNA in *S. cerevisiae* [38] indicates that chromatin modification(s) may be involved more in a "preparatory phase", making DSB in chromatin topologically and/or biochemically available for the core components of NHEJ machinery.

In retrospect, it is not surprising that NHEJ has recently been receiving a lot of attention from the wider scientific community. Its high degree of evolutionary conservation, coupled to its flexibility as demonstrated by its involvement in various DSB-mediated processes other than DNA repair itself, make it an increasingly interesting and complex scientific problem. Although the core components of the system have been well characterized, further important developments are likely to be expected with respect to its interactions with other DNA repair pathways, modulating and regulatory activities, interfaces with processes that facilitate DNA repair in chromatin and evolutionary origins.

ACKNOWLEDGMENTS

We apologize to all colleagues whose work has not been cited due to space limitations. Work in the MC laboratory is supported by the VEGA Grant Agency of the Slovak Republic (grant no. 2/3091/24) and project 2003 SP 51 028 08 00/028 08 01 from the national program Use of Cancer Genomics to Improve the Human Population Health. BK is supported by BBSRC and the Royal Society. AJD is a Royal Society University Research Fellow, supported by Association for International Cancer Research, BBSRC, Cancer Research UK and the Royal Society.

REFERENCES

[1] Pastink, A. and Lohman, P. H. M. (1999). Repair and consequences of double-strand breaks in DNA. *Mutat. Res. - Fundam. Mol. Mech. Mut.*, 428, 141-156.

[2] Pastink, A., Eeken, J. C. J. and Lohman, P. H. M. (2001). Genomic integrity and the repair of double-strand DNA breaks. *Mutat. Res. - Fundam. Mol. Mech. Mut.*, 480, 37-50.

[3] van den Bosch, M., Lohman, P. H. M. and Pastink, A. (2002). DNA double-strand break repair by homologous recombination. *Biol. Chem.*, 383, 873-892.

[4] Valerie, K. and Povirk, L. F. (2003). Regulation and mechanisms of mammalian double-strand break repair. *Oncogene*, 22, 5792-5812.

[5] West, S. C. (2003). Molecular views of recombination proteins and their control. *Nat. Rev. Mol. Cell. Biol.*, 4, 435-445.

[6] Dudáš, A. and Chovanec, M. (2004). DNA double-strand break repair by homologous recombination. *Mut. Res. - Rev. Mutat. Res.*, 566, 131-167.

[7] Dudášová, Z. and Chovanec, M. (2003). Artemis, a novel guardian of the genome. *Neoplasma*, 50, 311-318.

[8] Lees-Miller, S. P. and Meek, K. (2003). Repair of DNA double-strand breaks by non-homologous end joining. *Biochimie*, 85, 1161-1173.

[9] Lieber, M. R., Ma, Y. M., Pannicke, U. and Schwarz, K. (2003). Mechanism and regulation of human non-homologous DNA end-joining. *Nat. Rev. Mol. Cell. Biol.*, 4, 712-720.

[10] Collis, S. J., Deweese, T. L., Jeggo, P. A. and Parker, A. R. (2005). The life and death of DNA-PK. *Oncogene,* 24, 949-961.

[11] Dudášová, Z., Dudáš, A. and Chovanec, M. (2004). Non-homologous end-joining factors of *Saccharomyces cerevisiae*. *FEMS Microbiol. Rev.*, 28, 581-601.

[12] Lieber, M. R., Ma, Y., Pannicke, U. and Schwarz, K. (2004). The mechanism of vertebrate nonhomologous DNA end joining and its role in V(D)J recombination. *DNA Repair (Amst)*, 3, 817-826.

[13] Weterings, E. and van Gent, D. C. (2004). The mechanism of non-homologous end-joining: a synopsis of synapsis. *DNA Repair (Amst)*, 3, 1425-1435.

[14] Jeggo, P. A. (2004). *The mechanisms of DNA non-homologous end-joining: lessons learned from biophysical, biochemical and cellular studies*. In: K. W. Caldecott (Ed.), *Eukaryotic DNA damage surveillance and repair* (pp.146-158) Georgetown, Texas, USA: Landes Bioscience/Eurekah.com and Kluwer Academic/Plenum Publishers.

[15] Hefferin, M. L. and Tomkinson, A. E. (2005). Mechanism of DNA double-strand break repair by non-homologous end joining. *DNA Repair (Amst)*, 4, 629-648.

[16] Ma, Y., Lu, H., Tippin, B., Goodman, M. F., Shimazaki, N., Koiwai, O., Hsieh, C. L., Schwarz, K. and Lieber, M. R. (2004). A biochemically defined system for mammalian nonhomologous DNA end joining. *Mol. Cell*, 16, 701-713.

[17] Weterings, E., Verkaik, N. S., Bruggenwirth, H. T., Hoeijmakers, J. H. J. and van Gent, D. C. (2003). The role of DNA dependent protein kinase in synapsis of DNA ends. *Nucl. Acids Res.*, 31, 7238-7246.

[18] Meek, K., Gupta, S., Ramsden, D. A. and Lees-Miller, S. P. (2004). The DNA-dependent protein kinase: the director at the end. *Immunol. Rev.*, 200, 132-141.

[19] Riballo, E., Kuhne, M., Rief, N., Doherty, A., Smith, G. C., Recio, M. J., Reis, C., Dahm, K., Fricke, A., Krempler, A., Parker, A. R., Jackson, S. P., Gennery, A., Jeggo, P. A. and Lobrich, M. (2004). A pathway of double-strand break rejoining dependent upon ATM, Artemis, and proteins locating to γ-H2AX foci. *Mol. Cell*, 16, 715-724.

[20] Ma, Y., Pannicke, U., Schwarz, K. and Lieber, M. R. (2002). Hairpin opening and overhang processing by an Artemis/DNA-dependent protein kinase complex in nonhomologous end joining and V(D)J recombination. *Cell*, 108, 781-794.

[21] Mahajan, K. N., Nick McElhinny, S. A., Mitchell, B. S. and Ramsden, D. A. (2002). Association of DNA polymerase μ (pol μ) with Ku and ligase IV: Role for pol μ in end-joining double-strand break repair. *Mol. Cell. Biol.*, 22, 5194-5202.

[22] Bebenek, K., Garcia-Diaz, M., Blanco, L. and Kunkel, T. A. (2003). The frameshift infidelity of human DNA polymerase λ. Implications for function. *J. Biol. Chem.*, 278, 34685-34690.

[23] Lee, J. W., Blanco, L., Zhou, T., Garcia-Diaz, M., Bebenek, K., Kunkel, T. A., Wang, Z. and Povirk, L. F. (2004). Implication of DNA polymerase λ in alignment-based gap filling for nonhomologous DNA end joining in human nuclear extracts. *J. Biol. Chem.*, 279, 805-811.

[24] Feldmann, H. and Winnacker, E. L. (1993). A putative homologue of the human autoantigen Ku from *Saccharomyces cerevisiae*. *J. Biol. Chem.*, 268, 12895-12900.

[25] Boulton, S. J. and Jackson, S. P. (1996). *Saccharomyces cerevisiae* Ku70 potentiates illegitimate DNA double-strand break repair and serves as a barrier to error-prone DNA repair pathways. *EMBO J.*, 15, 5093-5103.

[26] Boulton, S. J. and Jackson, S. P. (1996). Identification of a *Saccharomyces cerevisiae* Ku80 homologue: roles in DNA double strand break rejoining and in telomeric maintenance. *Nucl. Acids Res.*, 24, 4639-4648.

[27] Feldmann, H., Driller, L., Meier, B., Mages, G., Kellermann, J. and Winnacker, E. L. (1996). HDF2, the second subunit of the Ku homologue from *Saccharomyces cerevisiae*. *J. Biol. Chem.*, 271, 27765-27769.

[28] Mages, G. J., Feldmann, H. M. and Winnacker, E. L. (1996). Involvement of the *Saccharomyces cerevisiae HDF1* gene in DNA double-strand break repair and recombination. *J. Biol. Chem.*, 271, 7910-7915.

[29] Schär, P., Herrmann, G., Daly, G. and Lindahl, T. (1997). A newly identified DNA ligase of *Saccharomyces cerevisiae* involved in *RAD52*-independent repair of DNA double-strand breaks. *Genes Dev.,* 11, 1912-1924.

[30] Teo, S. H. and Jackson, S. P. (1997). Identification of *Saccharomyces cerevisiae* DNA ligase IV: involvement in DNA double-strand break repair. *EMBO J.*, 16, 4788-4795.

[31] Wilson, T. E., Grawunder, U. and Lieber, M. R. (1997). Yeast DNA ligase IV mediates non-homologous DNA end joining. *Nature,* 388, 495-498.

[32] Herrmann, G., Lindahl, T. and Schär, P. (1998). *Saccharomyces cerevisiae* LIF1: a function involved in DNA double-strand break repair related to mammalian XRCC4. *EMBO J.*, 17, 4188-4198.

[33] Ramos, W., Liu, G., Giroux, C. N. and Tomkinson, A. E. (1998). Biochemical and genetic characterization of the DNA ligase encoded by *Saccharomyces cerevisiae* open reading frame YOR005c, a homolog of mammalian DNA ligase IV. *Nucl. Acids Res.,* 26, 5676-5683.

[34] Aravind, L. and Koonin, E. V. (2001). Prokaryotic homologs of the eukaryotic DNA-end-binding protein Ku, novel domains in the Ku protein and prediction of a prokaryotic double-strand break repair system. *Genome Res.*, 11, 1365-1374.

[35] Doherty, A. J., Jackson, S. P. and Weller, G. R. (2001). Identification of bacterial homologues of the Ku DNA repair proteins. *FEBS Lett.*, 500, 186-188.

[36] Weller, G. R. and Doherty, A. J. (2001). A family of DNA repair ligases in bacteria? *FEBS Lett.*, 505, 340-342.

[37] Weller, G. R., Kysela, B., Roy, R., Tonkin, L. M., Scanlan, E., Della, M., Devine, S. K., Day, J. P., Wilkinson, A., d'Adda di Fagagna, F. D., Devine, K. M., Bowater, R. P., Jeggo, P. A., Jackson, S. P. and Doherty, A. J. (2002). Identification of a DNA nonhomologous end-joining complex in bacteria. *Science*, 297, 1686-1689.

[38] Della, M., Palmbos, P. L., Tseng, H. M., Tonkin, L. M., Daley, J. M., Topper, L. M., Pitcher, R. S., Tomkinson, A. E., Wilson, T. E. and Doherty, A. J. (2004). Mycobacterial Ku and ligase proteins constitute a two-component NHEJ repair machine. *Science*, 306, 683-685.

[39] Mimori, T., Akizuki, M., Yamagata, H., Inada, S., Yoshida, S. and Homma, M. (1981). Characterization of a high molecular weight acidic nuclear protein recognized by autoantibodies in sera from patients with polymyositis-scleroderma overlap. *J. Clin. Invest.*, 68, 611-620.

[40] Walker, J. R., Corpina, R. A. and Goldberg, J. (2001). Structure of the Ku heterodimer bound to DNA and its implications for double-strand break repair. *Nature,* 412, 607-614.

[41] Featherstone, C. and Jackson, S. P. (1999). Ku, a DNA repair protein with multiple cellular functions? *Mutat. Res. - DNA Repair*, 434, 3-15.

[42] Tuteja, R. and Tuteja, N. (2000). Ku autoantigen: A multifunctional DNA-binding protein. *Crit. Rev. Biochem. Mol. Biol.,* 35, 1-33.

[43] Doherty, A. J. and Jackson, S. P. (2001) DNA repair: How Ku makes ends meet. *Curr. Biol.*, 11, R920-R924.

[44] Koike, M. (2002). Dimerization, translocation and localization of Ku70 and Ku80 proteins. *J. Radiat. Res.*, 43, 223-236.

[45] Downs, J. A. and Jackson, S. P. (2004). A means to a DNA end: The many roles of Ku. *Nat. Rev. Mol. Cell. Biol.*, 5, 367-378.

[46] Ramsden, D. A. and Gellert, M. (1998). Ku protein stimulates DNA end joining by mammalian DNA ligases: a direct role for Ku in repair of DNA double-strand breaks. *EMBO J.*, 17, 609-614.

[47] Gottlieb, T. M. and Jackson, S. P. (1993). The DNA-dependent protein kinase: requirement for DNA ends and association with Ku antigen. *Cell*, 72, 131-142.

[48] Hammarsten, O. and Chu, G. (1998). DNA-dependent protein kinase: DNA binding and activation in the absence of Ku. *Proc. Natl. Acad. Sci. U.S.A.*, 95, 525-530.

[49] Yaneva, M., Kowalewski, T. and Lieber, M. R. (1997). Interaction of DNA-dependent protein kinase with DNA and with Ku: biochemical and atomic-force microscopy studies. *EMBO J.*, 16, 5098-5112.

[50] West, R. B., Yaneva, M. and Lieber, M. R. (1998). Productive and nonproductive complexes of Ku and DNA-dependent protein kinase at DNA termini. *Mol. Cell. Biol.*, 18, 5908-5920.

[51] Chen, L., Trujillo, K., Sung, P. and Tomkinson, A. E. (2000). Interactions of the DNA ligase IV-XRCC4 complex with DNA ends and the DNA-dependent protein kinase. *J. Biol. Chem.*, 275, 26196-26205.

[52] Nick McElhinny, S. A., Snowden, C. M., McCarville, J. and Ramsden, D. A. (2000). Ku recruits the XRCC4-ligase IV complex to DNA ends. *Mol. Cell. Biol.*, 20, 2996-3003.

[53] Milne, G. T., Jin, S., Shannon, K. B. and Weaver, D. T. (1996). Mutations in two Ku homologs define a DNA end-joining repair pathway in *Saccharomyces cerevisiae*. *Mol. Cell. Biol.*, 16, 4189-4198.

[54] Siede, W., Friedl, A. A., Dianova, I., Eckardt-Schupp, F. and Friedberg, E. C. (1996). The *Saccharomyces cerevisiae* Ku autoantigen homologue affects radiosensitivity only in the absence of homologous recombination. *Genetics*, 142, 91-102.

[55] Moreau, S., Ferguson, J. R. and Symington, L. S. (1999). The nuclease activity of Mre11 is required for meiosis but not for mating type switching, end joining, or telomere maintenance. *Mol. Cell. Biol.*, 19, 556-566.

[56] Laroche, T., Martin, S. G., Gotta, M., Gorham, H. C., Pryde, F. E., Louis, E. J. and Gasser, S. M. (1998). Mutation of yeast Ku genes disrupts the subnuclear organization of telomeres. *Curr. Biol.*, 8, 653-656.

[57] Boulton, S. J. and Jackson, S. P. (1998). Components of the Ku-dependent non-homologous end-joining pathway are involved in telomeric length maintenance and telomeric silencing. *EMBO J.*, 17, 1819-1828.

[58] Gravel, S., Larrivee, M., Labrecque, P. and Wellinger, R. J. (1998). Yeast Ku as a regulator of chromosomal DNA end structure. *Science*, 280, 741-744.

[59] Nugent, C. I., Bosco, G., Ross, L. O., Evans, S. K., Salinger, A. P., Moore, J. K., Haber, J. E. and Lundblad, V. (1998). Telomere maintenance is dependent on activities required for end repair of double-strand breaks. *Curr. Biol.*, 8, 657-660.

[60] Barnes, G. and Rio, D. (1997). DNA double-strand-break sensitivity, DNA replication, and cell cycle arrest phenotypes of Ku-deficient *Saccharomyces cerevisiae*. *Proc. Natl. Acad. Sci. U. S. A.*, 94, 867-872.

[61] Porter, S. E., Greenwell, P. W., Ritchie, K. B. and Petes, T. D. (1996). The DNA-binding protein Hdf1p (a putative Ku homologue) is required for maintaining normal telomere length in *Saccharomyces cerevisiae*. *Nucl. Acids Res.*, 24, 582-585.
[62] Heidenreich, E., Novotny, R., Kneidinger, B., Holzmann, V. and Wintersberger, U. (2003). Non-homologous end joining as an important mutagenic process in cell cycle-arrested cells. *EMBO J.*, 22, 2274-2283.
[63] Greenwell, P. W., Kronmal, S. L., Porter, S. E., Gassenhuber, J., Obermaier, B. and Petes, T. D. (1995). *TEL1*, a gene involved in controlling telomere length in *S. cerevisiae*, is homologous to the human ataxia telangiectasia gene. *Cell*, 82, 823-829.
[64] Jackson, S. P. (1996). DNA damage detection by DNA dependent protein kinase and related enzymes. *Cancer Surv.*, 28261-28279.
[65] Smith, G. C. M. and Jackson, S. P. (1999). The DNA-dependent protein kinase. *Genes Dev.*, 13, 916-934.
[66] Schiestl, R. H., Zhu, J. and Petes, T. D. (1994). Effect of mutations in genes affecting homologous recombination on restriction enzyme-mediated and illegitimate recombination in *Saccharomyces cerevisiae*. *Mol. Cell. Biol.*, 14, 4493-4500.
[67] Tsukamoto, Y., Kato, J. and Ikeda, H. (1996). Effects of mutations of *RAD50*, *RAD51*, *RAD52*, and related genes on illegitimate recombination in *Saccharomyces cerevisiae*. *Genetics*, 142, 383-391.
[68] Tsukamoto, Y., Kato, J. and Ikeda, H. (1997). Budding yeast Rad50, Mre11, Xrs2, and Hdf1, but not Rad52, are involved in the formation of deletions on a dicentric plasmid. *Mol. Gen. Genet.*, 255, 543-547.
[69] Moore, J. K. and Haber, J. E. (1996). Cell cycle and genetic requirements of two pathways of nonhomologous end-joining repair of double-strand breaks in *Saccharomyces cerevisiae*. *Mol. Cell. Biol.*, 16, 2164-2173.
[70] Chen, L., Trujillo, K., Ramos, W., Sung, P. and Tomkinson, A. E. (2001). Promotion of Dnl4-catalyzed DNA end-joining by the Rad50/Mre11/Xrs2 and Hdfl/Hdf2 complexes. *Mol. Cell*, 8, 1105-1115.
[71] Ivanov, E. L., Sugawara, N., White, C. I., Fabre, F. and Haber, J. E. (1994). Mutations in *XRS2* and *RAD50* delay but do not prevent mating-type switching in *Saccharomyces cerevisiae*. *Mol. Cell. Biol.*, 14, 3414-3425.
[72] Kozhin, S. A., Chepurnaya, O. V. and Korolev, V. G. (1995). The *RAD58* (*XRS4*) gene: map position on the right arm of chromosome XIII. *Yeast*, 11, 1211-1213.
[73] Kupiec, M. and Simchen, G. (1984). Cloning and mapping of the *RAD50* gene of *Saccharomyces cerevisiae*. *Mol. Gen. Genet.*, 193, 525-531.
[74] Alani, E., Subbiah, S. and Kleckner, N. (1989). The yeast *RAD50* gene encodes a predicted 153-kD protein containing a purine nucleotide-binding domain and two large heptad-repeat regions. *Genetics*, 122, 47-57.
[75] Petrini, J. H., Walsh, M. E., DiMare, C., Chen, X. N., Korenberg, J. R. and Weaver, D. T. (1995). Isolation and characterization of the human MRE11 homologue. *Genomics*, 29, 80-86.
[76] Dolganov, G. M., Maser, R. S., Novikov, A., Tosto, L., Chong, S., Bressan, D. A. and Petrini, J. H. J. (1996). Human Rad50 is physically associated with human Mre11: Identification of a conserved multiprotein complex implicated in recombinational DNA repair. *Mol. Cell. Biol.*, 16, 4832-4841.

[77] Carney, J. P., Maser, R. S., Olivares, H., Davis, E. M., Lebeau, M., Yates, J. R., Hays, L., Morgan, W. F. and Petrini, J. H. J. (1998). The hMre11/hRad50 protein complex and Nijmegen breakage syndrome: Linkage of double-strand break repair to the cellular DNA damage response. *Cell*, 93, 477-486.

[78] Matsuura, S., Tauchi, H., Nakamura, A., Kondo, N., Sakamoto, S., Endo, S., Smeets, D., Solder, B., Belohradsky, B. H., Der Kaloustian, V. M., Oshimura, M., Isomura, M., Nakamura, Y. and Komatsu, K. (1998). Positional cloning of the gene for Nijmegen breakage syndrome. *Nat. Genet.*, 19, 179-181.

[79] Varon, R., Vissinga, C., Platzer, M., Cerosaletti, K. M., Chrzanowska, K. H., Saar, K., Beckmann, G., Seemanova, E., Cooper, P. R., Nowak, N. J., Stumm, M., Weemaes, C. M. R., Gatti, R. A., Wilson, R. K., Digweed, M., Rosenthal, A., Sperling, K., Concannon, P. and Reis, A. (1998). Nibrin, a novel DNA double-strand break repair protein, is mutated in Nijmegen breakage syndrome. *Cell*, 93, 467-476.

[80] Furuse, M., Nagase, Y., Tsubouchi, H., Murakamimurofushi, K., Shibata, T. and Ohta, K. (1998). Distinct roles of two separable *in vitro* activities of yeast Mre11 in mitotic and meiotic recombination. *EMBO J.*, 17, 6412-6425.

[81] Tsubouchi, H. and Ogawa, H. (1998). A novel *mre11* mutation impairs processing of double-strand breaks of DNA during both mitosis and meiosis. *Mol. Cell. Biol.*, 18, 260-268.

[82] Usui, T., Ohta, T., Oshiumi, H., Tomizawa, J., Ogawa, H. and Ogawa, T. (1998). Complex formation and functional versatility of Mre11 of budding yeast in recombination. *Cell*, 95, 705-716.

[83] Trujillo, K. M. and Sung, P. (2001). DNA structure-specific nuclease activities in the *Saccharomyces cerevisiae* Rad50 center dot Mre11 complex. *J. Biol. Chem.*, 276, 35458-35464.

[84] Raymond, W. E. and Kleckner, N. (1993). RAD50 protein of *S. cerevisiae* exhibits ATP-dependent DNA binding. *Nucl. Acids Res.*, 21, 3851-3856.

[85] Moncalian, G., Lengsfeld, B., Bhaskara, V., Hopfner, K. P., Karcher, A., Alden, E., Tainer, J. A. and Paull, T. T. (2004). The Rad50 signature motif: Essential to ATP binding and biological function. *J. Mol. Biol.*, 335, 937-951.

[86] Trujillo, K. M., Roh, D. H., Chen, L., van Komen, S., Tomkinson, A. and Sung, P. (2003). Yeast Xrs2 binds DNA and helps target Rad50 and Mre11 to DNA ends. *J. Biol. Chem.*, 278, 48957-48964.

[87] Hopfner, K. P., Karcher, A., Shin, D. S., Craig, L., Arthur, L. M., Carney, J. P. and Tainer, J. A. (2000). Structural biology of Rad50 ATPase: ATP-driven conformational control in DNA double-strand break repair and the ABC-ATPase superfamily. *Cell*, 101, 789-800.

[88] Hopfner, K. P., Karcher, A., Craig, L., Woo, T. T., Carney, J. P. and Tainer, J. A. (2001). Structural biochemistry and interaction architecture of the DNA double-strand break repair Mre11 nuclease and Rad50-ATPase. *Cell*, 105, 473-485.

[89] Hopfner, K. P., Craig, L., Moncalian, G., Zinkel, R. A., Usui, T., Owen, B. A. L., Karcher, A., Henderson, B., Bodmer, J. L., Mcmurray, C. T., Carney, J. P., Petrini, J. H. J. and Tainer, J. A. (2002). The Rad50 zinc-hook is a structure joining Mre11 complexes in DNA recombination and repair. *Nature*, 418, 562-566.

[90] Lewis, L. K., Karthikeyan, G., Westmoreland, J. W. and Resnick, M. A. (2002). Differential suppression of DNA repair deficiencies of yeast *rad50, mre11* and *xrs2* mutants by *EXO1* and *TLC1* (the RNA component of telomerase). *Genetics*, 160, 49-62.

[91] Alani, E., Padmore, R. and Kleckner, N. (1990). Analysis of wild-type and *rad50* mutants of yeast suggests an intimate relationship between meiotic chromosome synapsis and recombination. *Cell*, 61, 419-436.

[92] Ivanov, E. L., Fedorova, I. V. and Kovaltsova, S. V. (1992). Isolation and characterization of new mutants with enhanced spontaneous mutability in the yeast *Saccharomyces cerevisiae. Genetika*, 28, 47-55.

[93] Kironmai, K. M. and Muniyappa, K. (1997). Alteration of telomeric sequences and senescence caused by mutations in *RAD50* of *Saccharomyces cerevisiae. Genes Cells*, 2, 443-455.

[94] Bressan, D. A., Baxter, B. K. and Petrini, J. H. J. (1999). The Mre11-Rad50-Xrs2 protein complex facilitates homologous recombination-based double-strand break repair in *Saccharomyces cerevisiae. Mol. Cell. Biol.*, 19, 7681-7687.

[95] Chen, C. and Kolodner, R. D. (1999). Gross chromosomal rearrangements in *Saccharomyces cerevisiae* replication and recombination defective mutants. *Nat. Genet.*, 23, 81-85.

[96] Roth, D. B. and Wilson, J. H. (1986). Nonhomologous recombination in mammalian cells: role for short sequence homologies in the joining reaction. *Mol. Cell. Biol.*, 6, 4295-4304.

[97] Kramer, K. M., Brock, J. A., Bloom, K., Moore, J. K. and Haber, J. E. (1994). Two different types of double-strand breaks in *Saccharomyces cerevisiae* are repaired by similar *RAD52*-independent, nonhomologous recombination events. *Mol. Cell. Biol.*, 14, 1293-1301.

[98] Wu, X. T., Wilson, T. E. and Lieber, M. R. (1999). A role for FEN-1 in nonhomologous DNA end joining: The order of strand annealing and nucleolytic processing events. *Proc. Nat. Acad. Sci. U. S. A.*, 96, 1303-1308.

[99] Wilson, T. E. and Lieber, M. R. (1999). Efficient processing of DNA ends during yeast nonhomologous end joining. Evidence for a DNA polymerase β (*POL4*)-dependent pathway. *J. Biol. Chem.*, 274, 23599-23609.

[100] Prasad, R., Widen, S. G., Singhal, R. K., Watkins, J., Prakash, L. and Wilson, S. H. (1993). Yeast open reading frame YCR14C encodes a DNA β-polymerase-like enzyme. *Nucl. Acids Res.*, 21, 5301-5307.

[101] Leem, S. H., Ropp, P. A. and Sugino, A. (1994). The yeast *Saccharomyces cerevisiae* DNA polymerase IV: Possible involvement in double-strand break DNA repair. *Nucl. Acids Res.*, 22, 3011-3017.

[102] Shimizu, K., Santocanale, C., Ropp, P. A., Longhese, M. P., Plevani, P., Lucchini, G. and Sugino, A. (1993). Purification and characterization of a new DNA polymerase from budding yeast *Saccharomyces cerevisiae*. A probable homolog of mammalian DNA polymerase β. *J. Biol. Chem.*, 268, 27148-27153.

[103] Budd, M. E. and Campbell, J. L. (1995). DNA polymerases required for repair of UV-induced damage in *Saccharomyces cerevisiae. Mol. Cell. Biol.*, 15, 2173-2179.

[104] Burgers, P. M. J., Koonin, E. V., Bruford, E., Blanco, L., Burtis, K. C., Christman, M. F., Copeland, W. C., Friedberg, E. C., Hanaoka, F., Hinkle, D. C., Lawrence, C. W., Nakanishi, M., Ohmori, H., Prakash, L., Prakash, S., Reynaud, C. A., Sugino, A., Todo,

T., Wang, Z. G., Weill, J. C. and Woodgate, R. (2001). Eukaryotic DNA polymerases: Proposal for a revised nomenclature. *J. Biol. Chem.*, 276, 43487-43490.

[105] Aoufouchi, S., Flatter, E., Dahan, A., Faili, A., Bertocci, B., Storck, S., Delbos, F., Cocea, L., Gupta, N., Weill, J. C. and Reynaud, C. A. (2000). Two novel human and mouse DNA polymerases of the polX family. *Nucl. Acids Res.*, 28, 3684-3693.

[106] Garcia-Diaz, M., Dominguez, O., Lopez-Fernandez, L. A., de Lera, L. T., Saniger, M. L., Ruiz, J. F., Parraga, M., Garcia-Ortiz, M. J., Kirchhoff, T., del Mazo, J., Bernad, A. and Blanco, L. (2000). DNA polymerase lambda (pol λ), a novel eukaryotic DNA polymerase with a potential role in meiosis. *J. Mol. Biol.*, 301, 851-867.

[107] Tseng, H. M. and Tomkinson, A. E. (2002). A physical and functional interaction between yeast Pol4 and Dnl4-Lif1 links DNA synthesis and ligation in nonhomologous end joining. *J. Biol. Chem.*, 277, 45630-45637.

[108] Ruiz, J. F., Dominguez, O., de Lera, T. L., Garcia-Diaz, M., Bernad, A. and Blanco, L. (2001) DNA polymerase μ, a candidate hypermutase? *Phil. Trans. Roy. Soc. London B*, 356, 99-109.

[109] McInnis, M., O'Neill, G., Fossum, K. and Reagan, M. S. (2002) Epistatic analysis of the roles of the *RAD27* and *POL4* gene products in DNA base excision repair in *S. cerevisiae*. *DNA Repair (Amst)*, 1, 311-315.

[110] Heidenreich, E. and Eisler, H. (2004). Non-homologous end joining dependency of gamma-irradiation-induced adaptive frameshift mutation formation in cell cycle-arrested yeast cells. *Mutat. Res.*, 556, 201-208.

[111] Reagan, M. S., Pittenger, C., Siede, W. and Friedberg, E. C. (1995). Characterization of a mutant strain of *Saccharomyces cerevisiae* with a deletion of the *RAD27* gene, a structural homolog of the *RAD2* nucleotide excision repair gene. *J. Bacteriol.*, 177, 364-371.

[112] Sommers, C. H., Miller, E. J., Dujon, B., Prakash, S. and Prakash, L. (1995). Conditional lethality of null mutations in *RTH1* that encodes the yeast counterpart of a mammalian 5'- to 3'-exonuclease required for lagging strand DNA synthesis in reconstituted systems. *J. Biol. Chem.*, 270, 4193-4196.

[113] Vallen, E. A., and Cross, F. R. (1995). Mutations in *RAD27* define a potential link between G1 cyclins and DNA replication. *Mol. Cell. Biol.*, 15, 4291-4302.

[114] Tishkoff, D. X., Filosi, N., Gaida, G. M. and Kolodner, R. D. (1997). A novel mutation avoidance mechanism dependent on *S. cerevisiae RAD27* is distinct from DNA mismatch repair. *Cell*, 88, 253-263.

[115] Symington, L. S. (1998). Homologous recombination is required for the viability of *rad27* mutants. *Nucl. Acids Res.*, 26, 5589-5595.

[116] 1Wu, X. H. and Wang, Z. G. (1999). Relationships between yeast Rad27 and Apn1 in response to apurinic/apyrimidinic (AP) sites in DNA. *Nucl. Acids Res.*, 27, 956-962.

[117] Hansen, R. J., Friedberg, E. C. and Reagan, M. S. (2000). Sensitivity of a *S. cerevisiae RAD27* deletion mutant to DNA-damaging agents and *in vivo* complementation by the human *FEN-1* gene. *Mutat. Res. - DNA Repair*, 461, 243-248.

[118] Loeillet, S., Palancade, B., Cartron, M., Thierry, A., Guy-Franck, R., Dujon, B., Doye, V. and Nicolas, A. (2005). Genetic network interactions among replication, repair and nuclear pore deficiencies in yeast. *DNA Repair (Amst)*, 4, 459-468.

[119] Barber, L. J., Ward, T. A., Hartley, J. A. and McHugh, P. J. (2005). DNA interstrand cross-link repair in the *Saccharomyces cerevisiae* cell cycle: overlapping roles for

PSO2 (*SNM1*) with MutS factors and *EXO1* during S phase. *Mol. Cell. Biol.*, 25, 2297-2309.

[120] Debrauwere, H., Loeillet, S., Lin, W., Lopes, J. and Nicolas, A. (2001). Links between replication and recombination in *Saccharomyces cerevisiae*: A hypersensitive requirement for homologous recombination in the absence of Rad27 activity. *Proc. Nat. Acad. Sci. U. S. A.*, 98, 8263-8269.

[121] Holmes, A. M. and Haber, J. E. (1999). Double-strand break repair in yeast requires both leading and lagging strand DNA polymerases. *Cell*, 96, 415-424.

[122] Freudenreich, C. H., Kantrow, S. M. and Zakian, V. A. (1998). Expansion and length-dependent fragility of CTG repeats in yeast. *Science*, 279, 853-856.

[123] Kokoska, R. J., Stefanovic, L., Tran, H. T., Resnick, M. A., Gordenin, D. A. and Petes, T. D. (1998). Destabilization of yeast micro- and minisatellite DNA sequences by mutations affecting a nuclease involved in Okazaki fragment processing (*rad27*) and DNA polymerase δ (*pol3-t*). *Mol. Cell. Biol.*, 18, 2779-2788.

[124] Maurer, D. J., O'Callaghan, B. L. and Livingston, D. M. (1998). Mapping the polarity of changes that occur in interrupted CAG repeat tracts in yeast. *Mol. Cell. Biol.*, 18, 4597-4604.

[125] Schweitzer, J. K. and Livingston, D. M. (1998). Expansions of CAG repeat tracts are frequent in a yeast mutant defective in Okazaki fragment maturation. *Hum. Mol. Genet.*, 7, 69-74.

[126] Richard, G. F., Dujon, B. and Haber, J. E. (1999). Double-strand break repair can lead to high frequencies of deletions within short CAG/CTG trinucleotide repeats. *Mol. Gen. Genet.*, 261, 871-882.

[127] Negritto, M. C., Qiu, J. Z., Ratay, D. O., Shen, B. H. and Bailis, A. M. (2001). Novel function of Rad27 (FEN-1) in restricting short-sequence recombination. *Mol. Cell. Biol.*, 21, 2349-2358.

[128] Xie, Y., Liu, Y., Argueso, J. L., Henricksen, L. A., Kao, H. I., Bambara, R. A. and Alani, E. (2001). Identification of *rad27* mutations that confer differential defects in mutation avoidance, repeat tract instability, and flap cleavage. *Mol. Cell. Biol.*, 21, 4889-4899.

[129] Callahan, J. L., Andrews, K. J., Zakian, V. A. and Freudenreich, C. H. (2003). Mutations in yeast replication proteins that increase CAG/CTG expansions also increase repeat fragility. *Mol. Cell. Biol.*, 23, 7849-7860.

[130] Liu, Y., Zhang, H., Veeraraghavan, J., Bambara, R. A. and Freudenreich, C. H. (2004). *Saccharomyces cerevisiae* flap endonuclease 1 uses flap equilibration to maintain triplet repeat stability. *Mol. Cell. Biol.*, 24, 4049-4064.

[131] Parenteau, J. and Wellinger, R. J. (1999). Accumulation of single-stranded DNA and destabilization of telomeric repeats in yeast mutant strains carrying a deletion of *RAD27*. *Mol. Cell. Biol.*, 19, 4143-4152.

[132] Sundararajan, A., Lee, B. S. and Garfinkel, D. J. (2003). The Rad27 (Fen-1) nuclease inhibits Ty1 mobility in *Saccharomyces cerevisiae*. *Genetics*, 163, 55-67.

[133] Hoopes, L. L. M., Budd, M., Choe, W., Tao, W. T. and Campbell, J. L. (2002). Mutations in DNA replication genes reduce yeast life span. *Mol. Cell. Biol.*, 22, 4136-4146.

[134] Bambara, R. A., Murante, R. S. and Henricksen, L. A. (1997). Enzymes and reactions at the eukaryotic DNA replication fork. *J. Biol. Chem.*, 272, 4647-4650.

[135] Waga, S. and Stillman, B. (1998). The DNA replication fork in eukaryotic cells. *Annu. Rev. Biochem.*, 67721-67751.

[136] Harrington, J. J. and Lieber, M. R. (1994). Functional domains within FEN-1 and RAD2 define a family of structure-specific endonucleases: Implications for nucleotide excision repair. *Genes Dev.*, 8, 1344-1355.

[137] Tseng, H. M. and Tomkinson, A. E. (2004). Processing and joining of DNA ends coordinated by interactions among Dnl4/Lif1, Pol4, and FEN-1. *J. Biol. Chem.*, 279, 47580-47588.

[138] Teo, S. H. and Jackson, S. P. (2000). Lif1p targets the DNA ligase Lig4p to sites of DNA double-strand breaks. *Curr. Biol.*, 10, 165-168.

[139] Sibanda, B. L., Critchlow, S. E., Begun, J., Pei, X. Y., Jackson, S. P., Blundell, T. L. and Pellegrini, L. (2001). Crystal structure of an Xrcc4-DNA ligase IV complex. *Nat. Struct. Biol.*, 8, 1015-1019.

[140] Junop, M. S., Modesti, M., Guarne, A., Ghirlando, R., Gellert, M. and Yang, W. (2000). Crystal structure of the Xrcc4 DNA repair protein and implications for end joining. *EMBO J.*, 19, 5962-5970.

[141] Karathanasis, E. and Wilson, T. E. (2002). Enhancement of *Saccharomyces cerevisiae* end joining efficiency by cell growth stage but not by impairment of recombination. *Genetics*, 161, 1015-1027.

[142] Åström, S. U., Okamura, S. M. and Rine, J. (1999). Yeast cell-type regulation of DNA repair. *Nature*, 397, 310.

[143] Gasser, S. M. and Cockell, M. M. (2001). The molecular biology of the SIR proteins. *Gene*, 279, 1-16.

[144] Tsukamoto, Y., Kato, J. and Ikeda, H. (1997). Silencing factors participate in DNA repair and recombination in *Saccharomyces cerevisiae*. *Nature*, 388, 900-903.

[145] Lee, S. E., Pâques, F., Sylvan, J. and Haber, J. E. (1999). Role of yeast *SIR* genes and mating type in directing DNA double-strand breaks to homologous and non-homologous repair paths. *Curr. Biol.*, 9, 767-770.

[146] Kegel, A., Sjostrand, J. O. O. and Åström, S. U. (2001). Nej1p, a cell type-specific regulator of nonhomologous end joining in yeast. *Curr. Biol.*, 11, 1611-1617.

[147] Ooi, S. L., Shoemaker, D. D. and Boeke, J. D. (2001). A DNA microarray-based genetic screen for nonhomologous end-joining mutants in *Saccharomyces cerevisiae*. *Science*, 294, 2552-2556.

[148] Valencia, M., Bentele, M., Vaze, M. B., Herrmann, G., Kraus, E., Lee, S. E., Schär, P. and Haber, J. E. (2001). *NEJ1* controls non-homologous end joining in *Saccharomyces cerevisiae*. *Nature*, 414, 666-669.

[149] Frank-Vaillant, M. and Marcand, S. (2001). NHEJ regulation by mating type is exercised through a novel protein, Lif2p, essential to the Ligase IV pathway. *Genes Dev.*, 15, 3005-3012.

[150] Jazayeri, A. and Jackson, S.P. (2002). Screening the yeast genome for new DNA-repair genes. *Genome Biol.*, 3, REVIEWS1009.

[151] Ito, T., Chiba, T., Ozawa, R., Yoshida, M., Hattori, M. and Sakaki, Y. (2001). A comprehensive two-hybrid analysis to explore the yeast protein interactome. *Proc. Natl. Acad. Sci. U. S. A.*, 98, 4569-4574.

[152] Yu, J. H., Marshall, K., Yamaguchi, M., Haber, J. E. and Weil, C. F. (2004). Microhomology-dependent end joining and repair of transposon-induced DNA hairpins by host factors in *Saccharomyces cerevisiae. Mol. Cell. Biol.*, 24, 1351-1364.
[153] Wilson, T. E. (2002). A genomics-based screen for yeast mutants with an altered recombination/end-joining repair ratio. *Genetics*, 162, 677-688.
[154] Liti, G. and Louis, E. J. (2003). *NEJ1* prevents NHEJ-dependent telomere fusions in yeast without telomerase. *Mol. Cell*, 11, 1373-1378.
[155] Williams, B. and Lustig, A. J. (2003). The paradoxical relationship between NHEJ and telomeric fusion. *Mol. Cell*, 11, 1125-1126.
[156] Pâques, F. and Haber, J. E. (1999). Multiple pathways of recombination induced by double-strand breaks in *Saccharomyces cerevisiae. Microbiol Mol. Biol. Rev.*, 63, 349-404.
[157] Jackson, S. P. (2001). Detecting, signalling and repairing DNA double-strand breaks. *Biochem. Soc. Trans.*, 29, 655-661.
[158] Symington, L. S. (2002). Role of *RAD52* epistasis group genes in homologous recombination and double-strand break repair. *Microbiol. Mol. Biol. Rev.*, 66, 630-670.
[159] Krejči, L., Chen, L., van Komen, S., Sung, P. and Tomkinson, A. (2003). Mending the break: two DNA double-strand break repair machines in eukaryotes. *Prog. Nucl. Acid. Res. Mol. Biol.*, 74, 159-201.
[160] Nickoloff, J. A. and Hoekstra, M. F. (1998). *Double-strand break and recombinational repair in Saccharomyces cerevisiae*. In: J. A. Nickoloff, and M. F. Hoekstra (Eds.), DNA damage and repair, Vol. 1. DNA repair in prokaryotes and lower eukaryotes (pp. 335-362). New York, USA: Human press, Inc. Totowa.
[161] Schwarz, K., Ma, Y., Pannicke, U. and Lieber, M. R. (2003). Human severe combined immune deficiency and DNA repair. *Bioessays*, 25, 1061-1070.
[162] Fairhead, C., Llorente, B., Denis, F., Soler, M. and Dujon, B. (1996). New vectors for combinatorial deletions in yeast chromosomes and for gap-repair cloning using 'split-marker' recombination. *Yeast*, 12, 1439-1457.
[163] Lewis, L. K., Kirchner, J. M. and Resnick, M. A. (1998). Requirement for end-joining and checkpoint functions, but not *RAD52*-mediated recombination, after *Eco*RI endonuclease cleavage of *Saccharomyces cerevisiae* DNA. *Mol. Cell. Biol.*, 18, 1891-1902.
[164] Aylon, Y. and Kupiec, M. (2004). DSB repair: the yeast paradigm. *DNA Repair (Amst)*, 3, 797-815.
[165] Takata, M., Sasaki, M. S., Sonoda, E., Morrison, C., Hashimoto, M., Utsumi, H., Yamaguchiiwai, Y., Shinohara, A. and Takeda, S. (1998). Homologous recombination and non-homologous end-joining pathways of DNA double-strand break repair have overlapping roles in the maintenance of chromosomal integrity in vertebrate cells. *EMBO J.*, 17, 5497-5508.
[166] Clikeman, J. A., Khalsa, G. J., Barton, S. L. and Nickoloff, J. A. (2001). Homologous recombinational repair of double-strand breaks in yeast is enhanced by *MAT* heterozygosity through yKU-dependent and -independent mechanisms. *Genetics*, 157, 579-589.
[167] Hegde, V. and Klein, H. (2000). Requirement for the *SRS2* DNA helicase gene in non-homologous end joining in yeast. *Nucl. Acids Res.*, 28, 2779-2783.

[168] Aylon, Y., Liefshitz, B., Bitan-Banin, G. and Kupiec, M. (2003). Molecular dissection of mitotic recombination in the yeast *Saccharomyces cerevisiae*. *Mol. Cell. Biol.*, 23, 1403-1417.

[169] Ristic, D., Modesti, M., Kanaar, R. and Wyman, C. (2003). Rad52 and Ku bind to different DNA structures produced early in double-strand break repair. *Nucl. Acids Res.*, 31, 5229-5237.

[170] Frank-Vaillant, M. and Marcand, S. (2002). Transient stability of DNA ends allows nonhomologous end joining to precede homologous recombination. *Mol. Cell*, 10, 1189-1199.

[171] White, C. I. and Haber, J. E. (1990). Intermediates of recombination during mating type switching in *Saccharomyces cerevisiae*. *EMBO J.*, 9, 663-673.

[172] Sugawara, N. and Haber, J. E. (1992). Characterization of double-strand break-induced recombination: Homology requirements and single-stranded DNA formation. *Mol. Cell. Biol.*, 12, 563-575.

[173] Smith, J., Baldeyron, C., De Oliveira, I., Sala-Trepat, M. and Papadopoulo, D. (2001). The influence of DNA double-strand break structure on end-joining in human cells. *Nucl. Acids Res.*, 29, 4783-4792.

[174] Jazayeri, A., McAinsh, A. D. and Jackson, S. P. (2004). *Saccharomyces cerevisiae* Sin3p facilitates DNA double-strand break repair. *Proc. Nat. Acad. Sci. U. S. A.*, 101, 1644-1649.

[175] Thiagalingam, S., Cheng, K. H., Lee, H. J., Mineva, N., Thiagalingam, A. and Ponte, J. F. (2003). Histone deacetylases: unique players in shaping the epigenetic histone code. Ann. N. Y. *Acad. Sci.*, 983, 84-100.

[176] Martin, S. G., Laroche, T., Suka, N., Grunstein, M. and Gasser, S. M. (1999). Relocalization of telomeric Ku and SIR proteins in response to DNA strand breaks in yeast. *Cell*, 97, 621-633.

[177] McAinsh, A. D., Scott-Drew, S., Murray, J. A. H. and Jackson, S. P. (1999) DNA damage triggers disruption of telomeric silencing and Mec1p-dependent relocation of Sir3p. *Curr. Biol.*, 9, 963-966.

[178] Mills, K. D., Sinclair, D. A. and Guarente, L. (1999). MEC1-dependent redistribution of the Sir3 silencing protein from telomeres to DNA double-strand breaks. *Cell*, 97, 609-620.

[179] Shroff, R., Arbel-Eden, A., Pilch, D., Ira, G., Bonner, W. M., Petrini, J. H., Haber, J. E. and Lichten, M. (2004). Distribution and dynamics of chromatin modification induced by a defined DNA double-strand break. *Curr. Biol.*, 14, 1703-1711.

[180] Downs, J. A., Allard, S., Jobin-Robitaille, O., Javaheri, A., Auger, A., Bouchard, N., Kron, S. J., Jackson, S. P. and Cote, J. (2004). Binding of chromatin-modifying activities to phosphorylated histone H2A at DNA damage sites. *Mol. Cell*, 16, 979-990.

[181] Rogakou, E. P., Boon, C., Redon, C. and Bonner, W. M. (1999). Megabase chromatin domains involved in DNA double-strand breaks *in vivo*. *J. Cell. Biol.*, 146, 905-915.

[182] Rogakou, E. P., Pilch, D. R., Orr, A. H., Ivanova, V. S. and Bonner, W. M. (1998). DNA double-stranded breaks induce histone H2AX phosphorylation on serine 139. *J. Biol. Chem.*, 273, 5858-5868.

[183] Burma, S., Chen, B. P., Murphy, M., Kurimasa, A. and Chen, D. J. (2001). ATM phosphorylates histone H2AX in response to DNA double-strand breaks. *J. Biol. Chem.*, 276, 42462-42467.

[184] Stiff, T., O'Driscoll, M., Rief, N., Iwabuchi, K., Lobrich, M. and Jeggo, P. A. (2004) ATM and DNA-PK function redundantly to phosphorylate H2AX after exposure to ionizing radiation. *Cancer Res.*, 64, 2390-2396.

[185] Ward, I. M. and Chen, J. (2001). Histone H2AX is phosphorylated in an ATR-dependent manner in response to replicational stress. *J. Biol. Chem.*, 276, 47759-47762.

[186] Downs, J. A., Lowndes, N. F. and Jackson, S. P. (2000). A role for *Saccharomyces cerevisiae* histone H2A in DNA repair. *Nature*, 408, 1001-1004.

[187] Redon, C., Pilch, D. R., Rogakou, E. P., Orr, A. H., Lowndes, N. F. and Bonner, W. M. (2003). Yeast histone 2A serine 129 is essential for the efficient repair of checkpoint-blind DNA damage. *EMBO* Rep., 4, 678-684.

[188] Celeste, A., Fernandez-Capetillo, O., Kruhlak, M. J., Pilch, D. R., Staudt, D. W., Lee, A., Bonner, R. F., Bonner, W. M. and Nussenzweig, A. (2003). Histone H2AX phosphorylation is dispensable for the initial recognition of DNA breaks. *Nat. Cell. Biol.*, 5, 675-679.

[189] Bakkenist, C. J. and Kastan, M. B. (2003). DNA damage activates ATM through intermolecular autophosphorylation and dimer dissociation. *Nature*, 421, 499-506.

[190] Lisby, M., Mortensen, U. H. and Rothstein, R. (2003). Colocalization of multiple DNA double-strand breaks at a single Rad52 repair centre. *Nat. Cell. Biol.*, 5, 572-577.

[191] Aten, J. A., Stap, J., Krawczyk, P. M., van Oven, C. H., Hoebe, R. A., Essers, J. and Kanaar, R. (2004) Dynamics of DNA double-strand breaks revealed by clustering of damaged chromosome domains. *Science*, 303, 92-95.

[192] Kurdistani, S. K. and Grunstein, M. (2003). Histone acetylation and deacetylation in yeast. *Nat. Rev. Mol. Cell. Biol.*, 4, 276-284.

[193] Lusser, A. and Kadonaga, J. T. (2003). Chromatin remodeling by ATP-dependent molecular machines. *Bioessays*, 25, 1192-1200.

[194] Vignali, M., Hassan, A. H., Neely, K. E. and Workman, J. L. (2000). ATP-dependent chromatin-remodeling complexes. *Mol. Cell. Biol.*, 20, 1899-1910.

[195] Bird, A. W., Yu, D. Y., Pray-Grant, M. G., Qiu, Q. F., Harmon, K. E., Megee, P. C., Grant, P. A., Smith, M. M. and Christman, M. F. (2002). Acetylation of histone H4 by Esa1 is required for DNA double-strand break repair. *Nature*, 419, 411-415.

[196] Ikura, T., Ogryzko, V. V., Grigoriev, M., Groisman, R., Wang, J., Horikoshi, M., Scully, R., Qin, J. and Nakatani, Y. (2000). Involvement of the TIP60 histone acetylase complex in DNA repair and apoptosis. *Cell*, 102, 463-473.

[197] Nourani, A., Doyon, Y., Utley, R. T., Allard, S., Lane, W. S. and Cote, J. (2001). Role of an ING1 growth regulator in transcriptional activation and targeted histone acetylation by the NuA4 complex. *Mol. Cell. Biol.*, 21, 7629-7640.

[198] Choy, J. S. and Kron, S. J. (2002). NuA4 subunit Yng2 function in intra-S-phase DNA damage response. *Mol. Cell. Biol.*, 22, 8215-8225.

[199] Yamagoe, S., Kanno, T., Kanno, Y., Sasaki, S., Siegel, R. M., Lenardo, M. J., Humphrey, G., Wang, Y., Nakatani, Y., Howard, B. H. and Ozato, K. (2003). Interaction of histone acetylases and deacetylases *in vivo*. *Mol. Cell. Biol.*, 23, 1025-1033.

[200] Doyon, Y., Selleck, W., Lane, W.S., Tan, S. and Cote, J. (2004). Structural and functional conservation of the NuA4 histone acetyltransferase complex from yeast to humans. *Mol. Cell. Biol.*, 24, 1884-1896.

[201] Krogan, N. J., Keogh, M. C., Datta, N., Sawa, C., Ryan, O. W., Ding, H., Haw, R. A., Pootoolal, J., Tong, A., Canadien, V., Richards, D. P., Wu, X., Emili, A., Hughes, T. R., Buratowski, S. and Greenblatt, J. F. (2003). A Snf2 family ATPase complex required for recruitment of the histone H2A variant Htz1. *Mol. Cell*, 12, 1565-1576.
[202] Mizuguchi, G., Shen, X., Landry, J., Wu, W. H., Sen, S. and Wu, C. (2004). ATP-driven exchange of histone H2AZ variant catalyzed by SWR1 chromatin remodeling complex. *Science*, 303, 343-348.
[203] Matangkasombut, O. and Buratowski, S. (2003). Different sensitivities of bromodomain factors 1 and 2 to histone H4 acetylation. *Mol. Cell*, 11, 353-363.
[204] Chua, P. and Roeder, G. S. (1995) Bdf1, a yeast chromosomal protein required for sporulation. *Mol. Cell. Biol.*, 15, 3685-3696.
[205] Chang, M., Bellaoui, M., Boone, C. and Brown, G. W. (2002). A genome-wide screen for methyl methanesulfonate-sensitive mutants reveals genes required for S phase progression in the presence of DNA damage. *Proc. Natl. Acad. Sci. U. S. A.*, 99, 16934-16939.
[206] West, S. C. (1997). Processing of recombination intermediates by the RuvABC proteins. *Annu. Rev. Genet.*, 31213-31244.
[207] Kobor, M. S., Venkatasubrahmanyam, S., Meneghini, M. D., Gin, J. W., Jennings, J. L., Link, A. J., Madhani, H. D. and Rine, J. (2004). A protein complex containing the conserved Swi2/Snf2-related ATPase Swr1p deposits histone variant H2A.Z into euchromatin. *PLoS Biol.*, 2, 0587-0599.
[208] Morrison, A. J., Highland, J., Krogan, N. J., Arbel-Eden, A., Greenblatt, J. F., Haber, J. E. and Shen, X. (2004). INO80 and γ-H2AX interaction links ATP-dependent chromatin remodeling to DNA damage repair. *Cell*, 119, 767-775.
[209] van Attikum, H., Fritsch, O., Hohn, B. and Gasser, S. M. (2004). Recruitment of the INO80 complex by H2A phosphorylation links ATP-dependent chromatin remodeling with DNA double-strand break repair. *Cell*, 119, 777-788.
[210] d'Adda di Fagagna, F., Weller, G. R., Doherty, A. J. and Jackson, S. P. (2003). The Gam protein of bacteriophage Mu is an orthologue of eukaryotic Ku. *EMBO Rep.*, 4, 47-52.
[211] Akroyd, J. and Symonds, N. (1986). Localization of the gam gene of bacteriophage mu and characterisation of the gene product. *Gene*, 49, 273-282.
[212] Nick McElhinny, S. A. and Ramsden, D. A. (2003). Polymerase mu is a DNA-directed DNA/RNA polymerase. *Mol. Cell. Biol.*, 23, 2309-2315.
[213] Rydberg, B. and Game, J. (2002). Excision of misincorporated ribonucleotides in DNA by RNase H (type 2) and FEN-1 in cell-free extracts. *Proc. Natl. Acad. Sci. U. S. A.*, 99, 16654-16659.
[214] Cairns, J. and Overbaugh, J. and Miller, S. (1988). The origin of mutants. *Nature*, 335, 142-145.
[215] Hall, B. G. (1990). Spontaneous point mutations that occur more often when advantageous than when neutral. *Genetics*, 126, 5-16.
[216] Prival, M. J. and Cebula, T. A. (1992). Sequence analysis of mutations arising during prolonged starvation of *Salmonella typhimurium*. *Genetics*, 132, 303-310.
[217] Hall, B. G. (1992). Selection-induced mutations occur in yeast. *Proc. Natl. Acad. Sci. U. S. A.*, 89, 4300-4303.

[218] Steele, D. F. and Jinks-Robertson, S. (1992). An examination of adaptive reversion in *Saccharomyces cerevisiae*. *Genetics*, 132, 9-21.

[219] Baranowska, H., Policinska, Z. and Jachymczyk, W. J. (1995). Effects of the *CDC2* gene on adaptive mutation in the yeast *Saccharomyces cerevisiae*. *Curr. Genet.*, 28, 521-525.

[220] Heidenreich, E. and Wintersberger, U. (1997). Starvation for a specific amino acid induces high frequencies of rho$^-$ mutants in *Saccharomyces cerevisiae*. *Curr. Genet.*, 31, 408-413.

[221] Heidenreich, E. and Wintersberger, U. (1998). Replication-dependent and selection-induced mutations in respiration-competent and respiration-deficient strains of *Saccharomyces cerevisiae*. *Mol. Gen. Genet.*, 260, 395-400.

[222] Heidenreich, E. and Wintersberger, U. (2001). Adaptive reversions of a frameshift mutation in arrested *Saccharomyces cerevisiae* cells by simple deletions in mononucleotide repeats. *Mutat. Res.*, 473, 101-107.

[223] Hiom, K. (2003). DNA repair: Bacteria join in. *Curr. Biol.*, 13, R28-R30.

[224] MacNeill, S. A. (2005). Genome stability: a self-sufficient DNA repair machine. *Curr. Biol.*, 15, R21-R23.

[225] Kysela, B., Chovanec, M. and Jeggo, P. A. (2005). Phosphorylation of linker histones by DNA-dependent protein kinase is required for DNA ligase IV-dependent ligation in the presence of histone H1. *Proc. Natl. Acad. Sci. U. S. A.*, 102, 1877-1882.

[226] Downs, J. A., Kosmidou, E., Morgan, A. and Jackson, S. P. (2003). Suppression of homologous recombination by the *Saccharomyces cerevisiae* linker histone. *Mol. Cell*, 11, 1685-1692.

[227] Harvey, A. C. and Downs, J. A. (2004). What functions do linker histones provide? *Mol. Microbiol.*, 53, 771-775.

In: New Developments in Mutation Research
Editor: Charles L. Valon, pp. 189-215

ISBN 1-59454-664-9

Chapter 8

THE DOMINANT LETHAL TEST IN THE SNAIL *BIOMPHALARIA GLABRATA*: AN ASSAY TO DETECTION OF GERM CELL MUTAGENS IN FRESHWATER FOR ENVIRONMENTAL MONITORING

Eliana Nakano* and Lenita de Freitas Tallarico
Laboratório de Parasitologia/Malacologia, Instituto Butantan, São Paulo, Brazil

ABSTRACT

Studies on mutagenic agents have mostly focused on potential effects in humans. Despite the presence of mutagens in most of the pollution sources, the impact of mutagens on natural populations is still poorly understood. Mutagens in the environment may represent a long-term risk for ecosystems. Although exposure of aquatic organisms to genotoxic contaminants can pose a risk to human health via the food chain, there is also an ecological risk that may lead to heritable mutations and loss in the total genetic diversity. Mutagens can also directly affect the reproductive potential of populations by alterations in the fecundity and offspring viability caused by germ cell mutations. Among the biomarkers of germ cell effects, dominant lethal events have been used as convenient indicators of genetic damage since they are easily detectable. The so-called dominant lethal test with rodents has generally been considered one of the most important methods for the estimation of the mutagenic effects in humans. Freshwater mollusks of the genus *Biomphalaria* are good indicators for biomonitoring studies. The genus shows wide geographical distribution, has low dispersion and is easily collected. Other aspects make this system a good model for laboratory studies. The snails are easy to breed, need little space, can reproduce throughout the year under controlled conditions and have a short life-span: an egg-to-egg monitoring can be done in two months in *B. glabrata*. The dominant lethal assay was established in *B.glabrata* to evaluate mutagenic effects on

* Corresponding author. Tel.: +55-11-3726-7222x2158; Fax.: +55-11-3726-1505. E-mail address: eliananakano@butantan.gov.br

germ cells by using a similar approach to that used in the test in rodents. Albinism was used as genetic marker to adapt the test for *Biomphalaria*, since the animal is a simultaneous hermaphrodite and reproduces by self- and cross-fertilization. Adult wild-type snails are exposed for ten to 30 days to the test agent in aqueous solutions. The range of concentrations is defined in preliminary experiments. After the end of the exposure, animals are crossed with albino snails at different intervals to evaluate the effects in the germ cells at the different stages of spermatogenesis. The analysis of lethal effects in the heterozygous offspring indicates the induction of mutations on germ cells of the exposed wild-type snails. Dominant lethal test in B. glabrata can be an useful tool to predict ecological changes in biomonitoring studies. Germ cell effects are biomarkers of great value in determining the implications for ecosystems of exposure to pollution since they reflect changes that affect the overall reproductive capability.

Environmental Mutagens

Environmental contamination by compounds that pose a risk to human and environmental health is a consequence of economic development and technological advancement. The process of industrialization is one of the main sources of environmental pollution, since considerable amounts of residues are formed in most of processes. Besides, production of energy, necessary for all these processes, is also associated to environmental pollution. The basis for the second cause of environmental contamination is the increasing human population density, which forces an increasing food production. As the area of cultivable land cannot grow proportionally, an increase in food production can only be achieved by intensification of agriculture in available areas, which implies the use of pesticides and fertilizers. Moreover, both industrialization and food production are likely to increase with the human population density [1].

Natural freshwaters are the ultimate recipients of most of substances generated by industrial, agricultural and domestic activities and released into the environment [2]. The most readily identifiable sources of contaminants into the aquatic environment are the point sources, which refer to the direct discharge of industrial and municipal effluents into waterways through pipes and sewers. However, chemicals often enter aquatic systems from multiple and diffuse sources such as landfills, atmospheric deposition, ground water, agricultural runoff and storm sewers [3].

Many factors influence the susceptibility of an ecosystem to hazardous chemicals. Among them, the hydrologic regime is critical because water is the medium in which chemicals can be dispersed most easily and quickly through an ecosystem [4]. Therefore, aquatic ecosystems can be considered particularly vulnerable to the presence of contaminants.

There are highly toxic pollutants that cause evident impact; however, certain compounds are capable of reducing long-term survival of exposed populations. Chemical mutagens can affect the overall reproductive capability of populations through induction of genetic damage at non-toxic levels.

Despite of the complex chemical nature of polluted waters, mutagenic compounds can be found in most of pollution sources [5,6].

Industrial effluents and wastes are an important source of chemical mutagens. The presence, potency and potential hazard of genotoxins in a wide variety of industrial effluents, representing a wide range of industrial processes have been assessed in short-term genetic bioassays [5, 6, 7, 8]. From the different industrial categories evaluated, wastewaters of organic chemical manufacturing, pulp and paper facilities, metal refining and found operations and petroleum facilities exhibited noteworthy genotoxic potency; the industrial organic chemicals showed the highest potencies [5, 8]. Treatment and remediation of some wastes, such as pulp and paper mill effluents, have been shown to reduce or eliminate genotoxicity. However, in other cases, treatment and remediation have been shown to enhance genotoxicity, such as for fungal treatment of oils [6].

Despite the noteworthy genotoxic potency of many industrial wastewaters, the genotoxic hazard posed to the downstream ecosystem and its associated biota will be determined by genotoxic loading, which may be viewed as the product of genotoxic potency and discharge rate. Municipal wastewaters, although ranking low in potency, can achieve loading values that are several orders of magnitude greater than those of most industries. Sources of genotoxic substances in domestic wastewaters remain largely unknown, but the genotoxicity of sanitary wastes, including urine and feces wastes was demonstrated [9].

Drinking water can be viewed as another source of chemical mutagens. Chlorination of naturally occurring humic substances during the chlorine disinfection step in raw water treatment is known to produce mutagenic substances [10, 11, 12]. Among the chlorination by-products of raw waters, acidic mutagenic compounds were suggested to play a role in the etiology of cancers of the urinary and gastrointestinal tract [13]. Further epidemiological studies showed an association between the increase on the incidence of kidney, bladder and stomach cancer and the past exposure to chlorinated drinking water containing mutagenic compounds [14].

Genotoxic Effects in Natural Populations

Adverse effects of genetic damage are dependent upon the target cell type. Somatic mutations are associated to the induction of cancer; mutations occurring in germ cells may lead to abortions, reduced fertility, birth defects and malformations. If a new mutation is transmitted to offspring and does not drastically reduce reproductive fitness, it may enter the gene pool [15].

Despite the known multistep nature of carcinogenesis, at least one, but possibly several mutagenic events are required for malignant tumors to develop [16, 17]. With the purpose of screening for human carcinogens, genotoxicity tests were developed in the 1970s [18, 19], showing a great potential in predicting carcinogenic activity of chemicals [20]. Nevertheless, the initial high estimates of 90% correlations between mutagenicity and carcinogenicity have been revised to the 65% range with the emergence of carcinogens with non-genotoxic mechanisms of action [21, 22], and led to concern that short-term tests based on DNA interactions would not detect some compounds that might be carcinogenic to humans [23]. However, further data showed that most of these non-genotoxic carcinogens have been tested in long-term animal bioassays at maximum tolerated doses (MTD) or near MTD, which can cause mitogenic or regenerative cell proliferation at specific target sites [24]. These types of mechanisms may not be operative at lower doses that are more relevant to human exposures, and therefore, may be of less concern in human carcinogenicity [17].

Environmental effects of genotoxins were extensively discussed by Wurgler and Kramer [25]. Although genotoxic chemicals can damage the genetic material of any living species in the environment, mutagenicity studies have almost exclusively focused on potential effects in humans. From the point of view of risk management, the potential effects of an increased frequency of mutations have to be considered differently in humans and in other organisms. In human toxicology, the potential impairment of health in any individual resulting from inherited or somatically acquired mutations is of concern. In ecotoxicology, the concern is rather the maintenance of the integrity of ecosystems, which depends on surviving and reproducing populations. The induction of somatic mutations in animal and plants may, as in humans, give rise to neoplasia. If this occurs after the main reproductive period and does not otherwise interfere with reproduction, this will be not critical for the maintenance of natural populations. In contrast, genotoxic effects in germ cells can be passed on the future generations. Depledge [26] observed that phenotypic effects of genotoxin-induced changes in DNA may thereby become temporally decoupled from chemical and radiation exposures so that transgenerational changes ensue. This is a unique aspect of genotoxicity for it is worth identifying genotoxicity. Thus, genotoxicity can result in rapid alterations in gene frequencies (relative to normal evolutionary rates) in natural populations, the ecological consequences of which are poorly understood, but are likely to be serious.

Wurgler and Kramer [25] discussed the fate of mutations induced by environmental mutagens in natural populations. Mutations can be transmitted to liveborn offspring providing genetic variation. If one consider that most genetic changes are random, it is clear that the great majority will be harmful or at best neutral and only a few will present an improvement given the actual environmental condition. That is, the process of providing the necessary genetic variation is not without cost, which is the proportion of individuals carrying harmful mutations that are thus eliminated rapidly from the population. If the mutation does not drastically reduce reproductive fitness, it may enter the gene pool. Neutral mutations may persist and provide a source of variation useful for adaptations needed in the future. Only when mutations confer a selective advantage in connection with a particular environmental condition will they spread in the population. Nevertheless, the authors pointed that these may be advantages for the species in question but not for its neighbors, e.g., the induction of pesticide resistance, the increased virulence of pathogens, alterations of host ranges of pathogenic forms or the appearance of new virus types and subtle changes in parasite-host or predator-prey relationships, in other words, increased instability in ecosystems. Mutations in a polygenic system affecting a fitness trait such as mating efficiency or reproductive capacity are harmful at the population level. This type of mutation may be induced at high enough frequencies to be no longer 'a rare event' and at the same time it may affect vital population functions [27]. Also, mutations affecting complex behavioral patterns may be included in this category [28]. An important point in fitness traits, in comparison with other quantitative characteristics, is that they are usually optimized, i.e., they show little additive genetic variability [29]. Therefore, a new mutation in a fitness trait is more likely to have a harmful effect [25].

The biochemical and physiological consequences of genotoxicity in individuals, which are of most concern to ecotoxicologists, are those that affect reproductive health [26]. Genotoxic effects on germ cell mutations include gamete loss, reduced fertility, embryo mortality and malformations. Altered fertility, development, and embryonic survival are environmentally significant, because they can reduce reproductive success and thus alter population size or structure [30].

An increasing interest in the preservation of biodiversity and advances in molecular techniques for detecting DNA damage has heightened concerns on genotoxic effects in organisms and the associated detrimental outcomes for exposed populations [31]. The status of Genetic Ecotoxicology was, in this way, assessed and defined as: "The study of chemical- or radiation-induced changes in the genetic material of natural biota. Changes may be direct alterations in genes and gene expression or selective effects of pollutants on gene frequencies." This definition attempts to encompass direct DNA damage, epigenetic effects, and changes in gene pools attributable to toxicant exposure.

Environmental Biomonitoring of Chemical Mutagens

Potent genotoxic compounds as polycyclic aromatic hydrocarbons, polychlorinated biphenyls, organics, metals and pesticides have entered aquatic environments from a wide range of industrial, domestic and agricultural sources [5, 7, 9]. In recent years, some regulatory developments have recommended the genotoxicity testing of chemicals and effluents prior to their discharge into the marine environment. For example, a requirement for evaluating the mutagenic, carcinogenic and teratogenic properties of substances entering the aquatic environment was recommended by the 1995 North Sea Conference (OSPAR Commission, www.ospar.org). Similarly, the US Environmental Protection Agency has recommended adding mutagenicity testing protocols to the current suite of chemical methods used to detect the presence of specific chemicals or effluents (Houk, 1992). The International Standards Organization (ISO) is now recommending a standard bacterial method for testing the genotoxicity of wastewater.

Bacterial tests have been used extensively to evaluate wastewaters from industrial [5] and domestic [9] sources. In this context, it is worth mentioning that while in vitro or bacterial test systems could be used as screening tools to define the 'intrinsic' genotoxicity of a substance, definitive ecotoxicological risk assessments should also consider the 'expressed' genotoxic activity in ecologically relevant organisms. This approach will thus take into account environmentally realistic routes of exposure, the effects of metabolism and DNA repair efficiency [32].

Many of the standard tests already employed for human hazard assessment have been used in eco-genotoxicology in a broad range of invertebrate and vertebrate aquatic species (Table 1). Biomonitoring studies have employed quantification of DNA adducts, DNA strand breaks, micronucleus test and analysis sister chromatid exchanges in several species of marine and freshwater vertebrates and invertebrates. The use and status of short-term tests in genotoxicological studies has been discussed [5, 60, 61, 62].

DNA-adducts in aquatic organisms are effective molecular dosimeters of genotoxic contaminant exposure, and the ^{32}P-postlabeling assay has been used to measure covalent DNA-xenobiotic adducts. In the ^{32}P-postlabeling assay, DNA is hydrolyzed enzymatically to 3'-monophosphates and DNA adducts are enriched by the selective removal of normal nucleotide. The DNA adducts are then labeled with [^{32}P] phosphate and resulting ^{32}P-labeled DNA adducts are usually separated by thin-layer chromatography or high performance liquid chromatography. Radioactivity of DNA adducts are detected by autoradiography and liquid scintillation counting, imaging analysis or a liquid scintillation analyzer.

Table 1. In situ environmental genotoxicology studies with aquatic species

Principal contaminants	Animal	Endpoint	Reference
Marine			
hydrocarbons	mussel	SCE	Brunetti et al [33]
PCBs – PAHs	fish	micronucleus	Hose et al [34]
industrial, urban and agricultural wastes	mussel	micronucleus	Brunetti et al [35]
PAH, PCB, metals, organics	mussel; oyster	DNA strand breaks	Nacci et al [36]
urban and industrial wastes	mussel	DNA strand breaks	Vukmirovic et al [37]
PCB, HCH	oyster	micronucleus	Burgeot et al [38]
metals	mussel	DNA strand breaks	Steinert et al [39]
oil	fish	anaphase aberrations	Hose e Brown [40]
oil	fishes, sponge, mussel	DNA adducts	Harvey et al [41]
Industrial and municipal wastes	mussel	DNA strand breaks	Rank & Jensen [42]
PAHs, nitro-polycyclic aromatic Compounds, polycyclic aromatic ketones	fish	DNA adducts	Lyons et al. [43]
PAHs, PCBs	dab	DNA adducts, DNA strand breaks	Akcha et al.[44]
PAHs	dab	DNA adducts, DNA strand breaks	Akcha et al. [45]
PAHs, PCBs, OCPs, heavy metals	mussel	DNA adducts	Pisoni et al. [46]
Freshwater			
Organic pollutants	fish	SCE	Alink et al [47]
urban and industrial wastes	fish	micronucleus	De Flora et al [48]
PAHs – PCBs	fish	DNA strand breaks	Pandrangi et al [49]
urban wastes	fish	micronucleus	Minissi et al [50]
Pesticides	tadpoles	DNA strand breaks	Ralph e Petras [51]
Anthropic influence	fish	micronucleus	Sánchez-Galán [52]
PAHs	fish	DNA adducts	Lyons et al [53]
Industrial effluents, domestic sewage	mussel	DNA strand breaks, micronucleus	Klobucar et al [54]
PAHs, PCBs, OCPs, heavy metals	chub	DNA strand breaks, DNA adducts	Winter et al. [55]
disinfectants: sodium hypochlorite, peracetic acid and chloride dioxide	fish	DNA strand breaks, micronucleus	Buschini et al. [56]
detergents, pesticides, chrome	fish	DNA strand breaks, micronucleus	Russo et al. [57]
PAHs, PCBs	amphibian larvae	micronucleus	Gauthier et al. [58]
mercury	fish	micronucleus	Porto et al. [59]

List of abbreviations: PCB = polychlorinated biphenyl; PAH = polycyclic aromatic hydrocarbon; HCH = lindane; OCP = organochlorine pesticide.

DNA strand breaks are potential pre-mutagenic lesions and are sensitive markers of genotoxic damage. In the DNA unwinding assay, detection of DNA strand breaks is based on an established quantitative relationship between an increased rate of DNA strand separation, referred to as unwinding, an increased DNA strand breaks. The most commonly used technique for DNA strand break detection is the alkaline single cell gel electrophoresis (comet) assay. This technique permits the efficient visualization of DNA damage in individual cells and any cells that have a nucleus can be used. Nuclear DNA is unwound and electrophoresed under alkaline (>pH 13) conditions, and DNA fragments migrate from the nucleus towards the anode. The distance and/or amount of DNA migration from individual nuclei indicate the extent of DNA damage. Using this high pH level not only helps in the detection of DNA single-strand breaks, but may also reveal other classes of DNA damage (e.g. DNA protein cross-linking, alkali labile sites) and incomplete DNA repair.

The application of the comet assay has revolutionized the field of genetic ecotoxicology. This assay has provided the opportunity to study DNA damage, repair and cell death (apoptosis) in different cell types of aquatic organisms without prior knowledge of karyotype and cell turnover rate. Notwithstanding, Jha [62] pointed out that, despite the high sensitivity of these assays on detecting exposure to genotoxins, it must be remained that both DNA strand breaks and adduct formation are repairable damages and biomarkers of exposure rather than effect; thus, prediction of the potential impact of environmental exposure to genotoxins on ecosystems can only be achieved by using mutagenicity assays.

Clastogenesis and aneuploidy are two major classes of genetic damage and refer to alterations in chromosome structure or number respectively. Effects of clastogenesis include chromosome aberrations, the formation of micronuclei (MN) and sister chromatid exchanges (SCE). Chromosome aberrations include large deletions, translocations, or other rearrangements of DNA. The micronucleus assay is a widely used cytogenetic assay for the assessment of in vivo or in vitro chromosomal damage. Micronuclei are small DNA-containing bodies, which can be present near the cell nucleus during interphase as a consequence of both chromosome breakage and spindle dysfunction. In general, the chromosomes of fish and other aquatic organisms are relatively small in size and/or high in number. Therefore, the metaphase analysis of chromosomal aberrations using these organisms is difficult. However, small size and large chromosome number does not affect the performance of the micronucleus assay, so it can be easily applied to fish or other aquatic organisms. In the sister-chromatid exchange assay (SCE), there is assessment of the exchange of chromosome fragments between sister-chromatids after DNA strand damage. In the anaphase aberration test, lagging chromosomes, multipolar anaphases, bridges and fragments are scored.

A battery of short-term tests is usually employed to establish the mutagenic potential of an agent, since no single assay can detect all types of genetic damage. In mammalian genotoxicology, a tiered screening approach has been employed and universally accepted by the regulatory bodies. The inherent genotoxicity of a substance is evaluated in a suite of in vitro tests followed by expressed genotoxicity under in vivo conditions. If the substance is a somatic mutagen to mammals in vivo, tests are conducted to measure the interaction of the substance with germ cell DNA or other heritable mutations in vivo conditions. For future ecogenotoxicity testing, the adoption of standardized test procedures leading to a tiered approach is recommended [60].

DOMINANT LETHAL TEST IN *BIOMPHALARIA GLABRATA*

Dominant Lethal Mutations

Dominant lethal events are convenient indicators of gross genetic damage. First studies employed the hatchability of eggs as a measure of dominant lethality in insects [63]. The dominant lethal effects of x-rays and chemical mutagens were evaluated in mammals [64, 65]. Studies with radiation in fishes were performed by Newcombe and McGregor [66] and Schröder [67]. The so-called "dominant lethal test" in rodents was proposed by Bateman [68] in screening for mutagens to determine the potential transmissibility of mutagenic effects. Egami et al. [69] proposed the dominant lethal test in the freshwater fish *Oryzias latipes* for detecting mutagens in the aquatic environment. Guidelines for mutagenicity testing recommend the dominant lethal test in rodents for monitoring germ cell effects of chemicals [70, 71, 72].

A dominant lethal mutation is a genetic lesion that occurs in a gamete and, although not interfering with the gamete's ability to fertilize or be fertilized, is lethal to the resulting embryo. Chromosomal examinations have revealed that the death of the embryo is primarily due to chromosome breaks which generally produce break and reunion in the early cleavage stage resulting in bridges and micronuclei [73, 74]. The broken chromosomes are normally lost resulting in monosomy and in the death of the embryo. In studies using known mutagens as methyl methanesulfonate, ethyl methanesulfonate and triethylenemelamine, the chromosomes of the resulting embryos have been analyzed, and a good correlation between chromosomal aberrations in early cleavage divisions and dominant lethality has been demonstrated [73].

Dominant lethal mutations are self-limiting by definition, because they are usually eliminated in the first generation and are not inherited. By themselves, these mutations are believed not to contribute to the genetic load. However, many mutagens produce both lethal and nonlethal chromosomal breakage. Reciprocal translocations and inversions are examples of the latter type break. Hence, the production of dominant lethal mutations could indicate the occurrence of other mutagenic events [73].

There are two principle approaches to conduct dominant lethal tests. In one, adult males are treated acutely or subacutely with sublethal concentrations of the test agent and then mated sequentially with untreated virgin females. A weekly crossing schedule allows sampling the effect of treatment on different stages of spermatogenesis; the timing for these has been well characterized in rats, mice and some other species. In the second approach, males are treated for the entire duration of spermatogenesis and then subsequently mated with untreated virgin females to assess dominant lethality. Only one crossing is usually required, although this may be repeated to confirm any findings; this modified method however does not allow determination of the stage of spermatogenesis affected by the treatment. Females are killed at mid-time of pregnancy and the uterine contents examined for evidence of dead implantations [73, 74].

Dominant Lethal Test in *Biomphalaria Glabrata*

Test Organism

The choice of a freshwater snail as a model for environmental biomonitoring of chemical mutagens took into account that invertebrates represent more than 90% of the aquatic species and among mollusks, 80% correspond to gastropods [75, 76]. Although there is few data on metabolism, cell cycle and DNA repair processes in invertebrates, it is essential to detect possible mutagenic effects in these organisms in order to evaluate the impact of pollution in an aquatic ecosystem [77].

Freshwater mollusks of the genus *Biomphalaria* can be considered good indicators for biomonitoring studies [78]. The genus shows wide geographical distribution, it is represented in Brazil by 10 species [79], has low dispersion and is easily collected. Other aspects make this system a good model for laboratory studies. The snails are easy to breed and need little space. Exposure systems require relatively small quantities of test samples for analysis, hence contributing to reduced costs of aquatic in vivo evaluations. Snails can reproduce throughout the year under controlled conditions and have a short life span: an egg-to-egg monitoring can be done in two months in *B. glabrata*. Being a simultaneous hermaphrodite, different reactions in both sexes do not occur. Embryos also constitute a good system for study. In *B. glabrata*, adult snails lay egg-masses with 20-50 eggs almost daily. The embryonic development from egg to hatching takes about seven to nine days at 25°C. Eggs are placed one by one in a single layer forming a transparent plurioval egg capsule, which allows the visualization of the embryos throughout development. The normal embryonic development has been described in detail [80, 81] and showed well-defined and easily recognizable stages, allowing the detection of morfogenetic effects. This organism has been used in our laboratory to evaluate toxic and mutagenic effects of chemical and physical agents using mortality index, alterations in embryonic development and chromosome aberrations as biomarkers [82, 83, 84, 85, 86].

Embryonic malformations can be used as biomarkers in mutagenicity tests, since they are one of the effects of the induction of mutations in germ cells. In *Biomphalaria*, most of the embryos with unspecific malformations, in which multiple structures are affected (Figure 5), die as the time of hatching approaches [66]: germ cell mutations that cause this type of malformations are, therefore, lethal mutations.

The adaptation of the dominant lethal test in rodents [68] for *Biomphalaria* took into account the fact that the animal is a simultaneous hermaphrodite and reproduces by self- and cross-fertilization; thus, after pairing, two types of embryos are produced. In order to identify the cross-fertilization embryos, exposed homozygous wild-type snails (Figure 1a) are crossed with non-exposed albino snails (Figure 1b); the analysis of the effects on the heterozygous embryos among the offspring of albino snails shows that the observed effects result, indeed, from germ cell mutations. Heterozygous wild-type embryos clearly show, around the fourth day, a black point at the base of each antenna (Figure 2), corresponding to the eyespots; no such points appear in homozygous albino embryos (Figure 2). The early appearance of the eyespots in still intracapsular embryos makes possible the speedy completion of a crossing test. Among the offspring of albino snails, pigmented embryos result from cross-fertilization, whereas albino ones result from self-fertilization [87].

(a)

(b)

Figure 1. B. glabrata – adult specimes. (a) wild-type specimen – arrow indicating the eye; (b) Albino specimen – h: head; f: foot; e: eye (without pigmentation); t: tentacle.

In some species of egg-laying amphibians, fishes and mollusks, losses due to abortions and failures of implantation do not occur, and embryos at all stages are in proximity to their food supplies. Thus, potentially malformed embryos have the nourishment with which to grow, and this circumstance may be regarded as particularly favorable for survival up to a stage at which the potential malformations can be expressed. Furthermore, the embryos can be readily observed before hatching, because of the transparency of the outer coatings of the eggs [66]. All these advantages are displayed by *B. glabrata*, which made it possible to identify the heterozygous malformed embryos among the offspring of albino snails (Figure 2a).

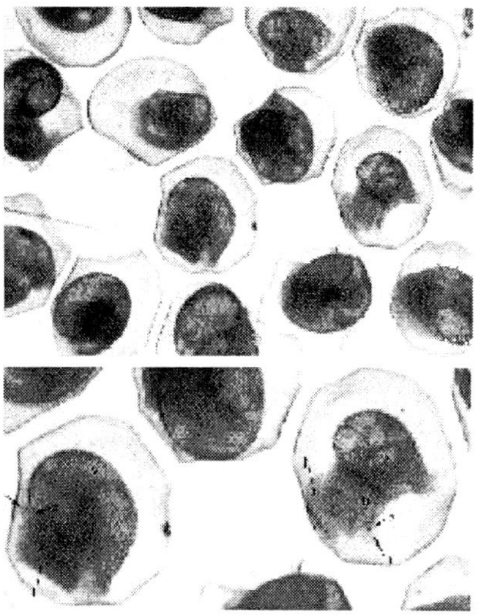

(a)

(b)

Figure 2. B. glabrata – normal wild-type and albino embryos resulting from a crossing between an homozygous wild-type snail and an albino snail; (b) detail of (a) – h: head; f: foot; s: shell; e: eye; t: tentacle.

Test Description

Animals

A pigmented homozygous wild-type strain and a non-pigmented albino strain of *B. glabrata* are used. At least 13 selected wild-type snails are required for each group; a corresponding number of albino snails is required for each pairing.

The colonies are maintained in plastic aquaria with filtered, dechlorinated and aerated water and fed fresh lettuce. For the experiments, animals are isolated for at least 3 weeks before the onset of the experiments in individual aquaria under the same conditions as the colonies. Sexually mature snails at least 2 months old and with a minimal shell diameter of 10 mm are used.

Exposure

Vehicle

Many solvents are not usable because of their toxicity or potential mutagenicity. Dechlorinated filtered water is used for water-soluble substances and dimethylsulfoxide (DMSO) for water-insoluble substances.

Dose Levels – Negative Control

Dose levels are selected on the basis of toxicokinetic data or biological effects reported in the literature, or by conducting a range finding study. A wide dose range is recommended. Low levels are required, markedly in multiple exposures to compounds with cumulative toxicity and high levels ensure that the maximum potential mutagenicity of the agent was explored. At least, two doses should be employed; the highest one should produce some evidence of toxicity. The most unequivocal sign of acute toxicity in *Biomphalaria* is death; the highest dose can be, therefore, the lowest dose which produces lethality.

Dechlorinated filtered water is used in the negative control group for water-soluble substances; for water insoluble substances, DMSO at a maximum concentration of 0,3% is employed. Both experimental and control groups are submitted to the same experimental conditions, including feeding, handling and renewal of solutions.

Treatment Schedule

A single or multiple treatment schedules may be used in the testing. In general, the multiple regimen has the advantage of simulating environmental exposure, as well as involving a smaller number of albino snails for crossing after treatment.

Duration of treatment in the multiple regimen should cover the entire spermatogenic cycle. Single treatment is useful for the determination of stage specificity of an agent in the test with rodents; for this regimen, multiweek crossing is required [73, 74]. For the moment, multiweek crossing in *B. glabrata* can only provide an estimate of the affected stages of spermatogenesis, since the chronology of spermatogenesis is not fully established.

Selection

All snails used in the experiments are selected by previous analysis of background frequencies of embryonic malformations. Three or four egg-masses are collected from albino snails before the crossings and those with malformation frequencies above 5% are discarded. Wild-type snails are evaluated after a crossing with albino snails before the exposure. At least 50 phenotypically wild-type embryos per animal are analyzed among the progeny of the albino snails and those with frequencies of malformations above 3% are discarded.

Crossing

Pairing

Twenty-four hours after the end of the exposure, each wild-type snail is paired with a non-exposed albino snail for 24 hours in individual aquaria. After this period, animals are isolated and egg masses are collected.

Crossing Schedule

Crossing procedure varies depending on the duration of exposure. If wild-type snails are treated with an acute or sub-acute dose of the test agent, they should be mated with albino snails each 5 or 7 days to cover the entire spermatogenic cycle.

The complete chronology of spermatogenesis in *Biomphalaria* has not been established, thus it was not possible to determine all the affected stages. In order to ascertain sampling the effect of treatment on germ cells at all stages, we have estimated some parameters of spermatogenesis of *B. glabrata* based on the results of the dominant lethal effects of ^{60}Co gamma radiation [88]. Total duration of spermatogenesis was estimated to be approximately 36 days, with the following distribution of developmental stages: 1–13 days, spermatogonia; 14–20 days, spermatocytes; 21–36 days, spermatids and spermatozoa. Crossings should cover, hence, at least 36 days, if there is no evidence of delaying effects on the cell cycle (Figure 12).

For multiple treatment, wild-type snails are exposed for the entire duration of spermatogenesis and then subsequently crossed with albino snails; only one crossing is usually required, although this may be repeated to confirm any finding.

Analysis of the Dominant Lethal Effects

Collection

Egg masses are collected daily. Plastic sheets are placed on the water surface to propitiate oviposition and egg masses are then transferred to cell culture plates and maintained in climatic chambers at 25°C until the end of the analysis.

Crossing between Wild-type and Albino Snails

A crossing between homozygous wild-type snails and albino snails produces the following types of embryos (Figure 3):

A. Among the offspring of the albino snails: i) heterozygous wild-type embryos produced by cross-fertilization and ii) albino embryos produced by self-fertilization. Wild-type embryos can be identified among the offspring of albino snails by the presence of pigmentation in the eyes, visible from the fourth day of embryonic development.

B. B. Among the offspring of wild-type snails: i) heterozygous wild-type embryos produced by cross-fertilization and ii) homozygous wild-type embryos produced by self-fertilization; both types are phenotypically indistinguishable.

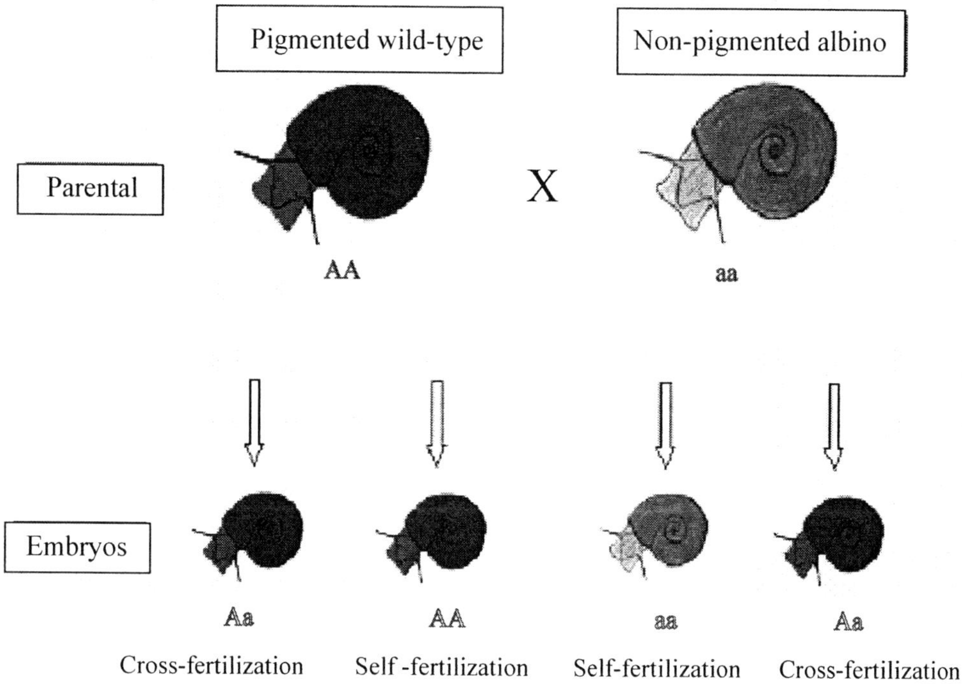

Figure 3. Types of embryos produced after a crossing between homozygous wild-type snails and albino snails. AA, Aa and aa are the genotypes corresponding to the phenotypes wild-type pigmented, wild-type pigmented and albino non-pigmented, respectively

For the dominant lethal test in *Biomphalaria*, only the offspring of the non-exposed albino snails is usually analyzed, considering that the test in rodents is primarily conducted in the treated males. Nevertheless, being *Biomphalaria* a simultaneous hermaphrodite, both effects on male and female lineages can be evaluated in the same exposed animal, since in adult snails, eggs and sperm are produced in a continuous and simultaneous way from a single group of undifferentiated cells [89].

Analysis of the offspring of the exposed wild-type snails shows the effects predominantly on the female lineage. It is known that, in spite of the highly developed ability of *B. glabrata* to self-fertilize, it shows an absolute preference to cross-fertilization [87]. Besides, in a pairing of two snails, both act as male and female without preference to any role [90]. High rates of cross-fertilization are obtained in the pairings if snails are previously isolated. Assuming that the rates of cross-fertilization in the wild-type snails are of the same order of the high rates observed in the albino snails, we may conclude that a minority of self-fertilized embryos are produced.

Scoring of Embryonic Malformations

Embryos are observed for 8 days from the beginning of the development until nearly hatching using a stereomicroscope.

The effects on the embryonic development are classified according to Geilenkirchen [91] in:

- death – this category includes all embryos that die up to the gastrula stage (Figure 4);
- exogastrulation – includes vesicular embryos, in which the invagination of the archenteron is suppressed. In these abnormalities no further development is observed and death usually ensues within a few days (Figure 5);
- unspecific malformations – this group includes all embryos with an abnormal development, but which cannot be described as head or shell malformations. They are, as a rule, teratomorphic or hydropic embryos. In the albino offspring, wild-type embryos with unspecific malformations were gathered in a separate group; in the same progeny, teratomorphic embryos, in which the identification of the eyes was not possible, were classified as embryos with indeterminate malformations (Figure 6).

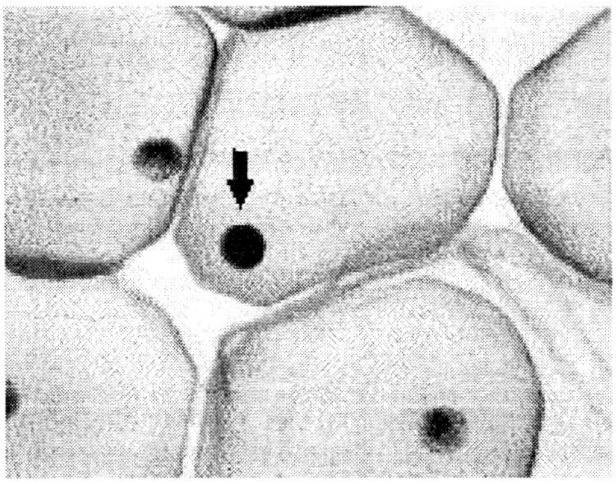

Figure 4. Alterations in the embryonic development of B. glabrata. Dead embryo – blastula stage.

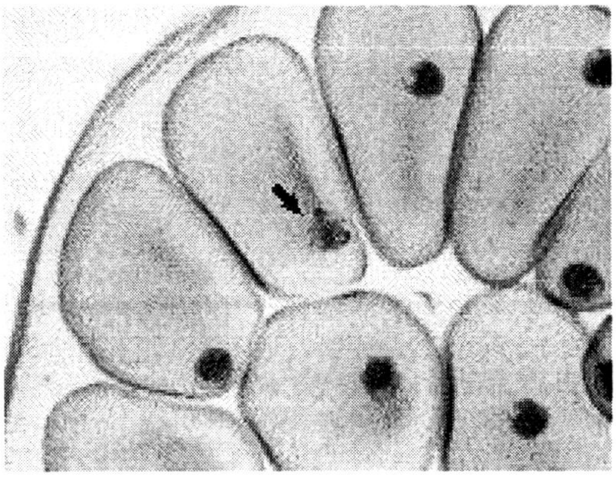

Figure 5. Alterations in the embryonic devepment of B. glabrata. Exogastrule – 24h after gastrulation.

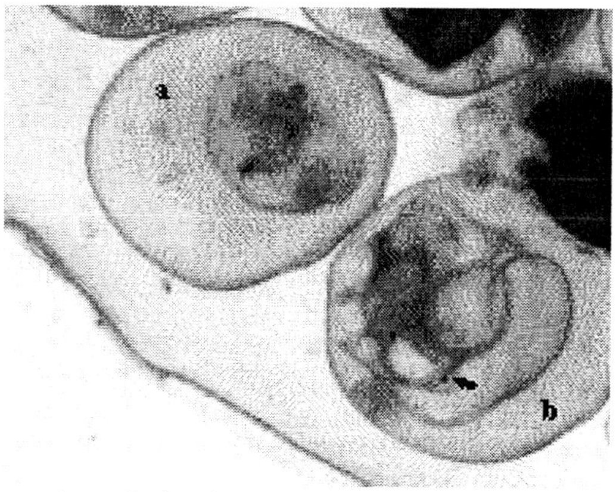

Figure 6. Alterations in the embryonic development of B. glabrata. (a) Embryo with indeterminate malformations; (b) phenotypically wild-type malformed embryo – arrow indicating the eye.

The number of teratomorphic or hydropic wild-type embryos—in which multiple structures are affected — are scored among the offspring of albino snails. For the wild-type snails offspring, all types of effects on the embryonic development are scored.

In this system, the induction of dominant lethal mutations leads to an increase in the frequency of unspecific malformations and cell-killing effects are evidenced by a reduction in the crossing rates. The dominant lethal effect are evaluated by comparing the frequencies of malformations in the heterozygous offspring of the albino snails crossed with the wild-type snails at the different intervals after treatment. To evaluate the effects on crossing rates, numbers of effective crossings, that is, albino snails that produced heterozygous embryos, are compared to the numbers of pairings among the different groups.

All analyses are carried out in coded scoring.

Statistical Analysis

To compare the frequencies of malformations among the different groups, the χ^2-test for Poisson distributions is used. The Poisson model was considered since it has a good adjustment in presence of rare events. To compare the crossing rates among the different groups the $\chi 2$-test for homogeneity is used.

Studies with Reference Mutagens

Mitomycin C and Cyclophosphamide

To develop a dominant lethal test system for environmental biomonitoring using a freshwater mollusk, we first employed two reference chemical mutagens; mitomycin C was employed as a direct mutagen and cyclophosphamide, as an indirect mutagen [92].

Wild-type snails were exposed for 10 days to three concentrations of each agent and crossed with non-exposed albino snails at the end of the treatment. Egg masses were collected from both groups of animals. The total frequencies of malformations were analyzed in the offspring of wild-type snails; among the offspring of albino snails, only the heterozygous wild-type embryos were analyzed for malformations. Both agents induced germ cell mutations.

There was an increase in the frequencies of unspecific malformations in the wild-type offspring of the non-exposed albino snails crossed with the wild-type snails exposed to both agents, demonstrating the induction of dominant lethal mutations (Tables 2, 3, 4 and Figures. 8, 9, 10).

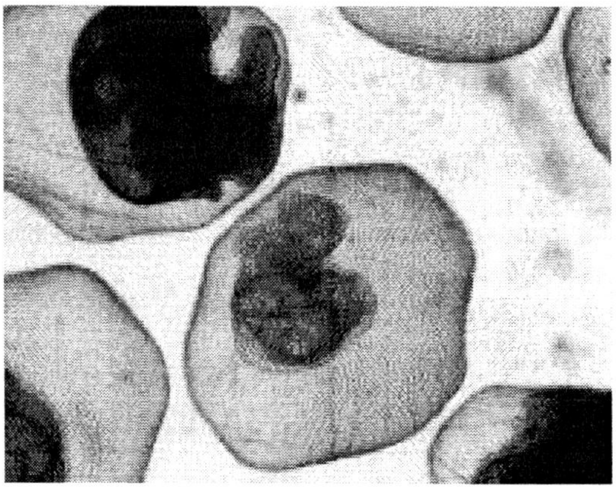

Figure 7. Alterations in the embryonic development of B. glabrata. Albino malformed embryo.

Table 2. Frequencies of malformation in the wild-type offspring of albino snails crossed with the wild-type snails before and 24h after the exposure to cyclophosphamide

Group	No. of animals	No. of embryos		Total of wild-type embryos		No. of malformed (%)	
		before	after	before	after	before	after
C0	11	2919	4564	2317	4417	10 (0.43)	35 (0.79)
C1	13	3556	4372	3435	4215	16 (0.47)	47 (1.12)
C2	10	2851	2982	2729	2625	10 (0.37)	285 (10.86)*
C3	12	3227	5435	2189	933	16 (0.73)	467 (50.05)*

C0 = control group, C1 = $3.6x10^{-5}$M, C2 = $3.6x10^{-4}$M, C3 = $3.6x10^{-3}$M, *$P<0.0001$

Table 3. Frequencies of malformation in the wild-type offspring of albino snails crossed with the wild-type snails before and 5 days after the exposure to cyclophosphamide

Group	No. of animals	No. of embryos		Total of wild-type embryos		No. of malformed (%)	
		before	after	before	after	before	after
C0	9	2348	3162	1866	3041	10 (0.54)	29 (0.95)
C1	11	3158	4012	3065	3891	15 (0.49)	31 (0.79)
C2	9	2487	4393	2412	4193	10 (0.41)	147 (3.51)*
C3	12	3227	5092	2189	2823	16 (0.73)	1554 (55.05)*

C0 = control group, C1 = $3.6x10^{-5}$M, C2 = $3.6x10^{-4}$M, C3 = $3.6x10^{-3}$M, *$P<0.0001$

Table 4. Frequencies of malformation in the wild-type offspring of albino snails crossed with the wild-type snails before and 24h after the exposure to mitomycin C

Group	No. of animals	No. of embryos		Total of wild-type embryos		No. of malformed (%)	
		before	after	before	after	before	after
C0	11	3094	3058	2557	2793	30 (1.17)	25 (0.89)
C1	10	2422	3044	2032	2945	20 (0.98)	37 (1.26)
C2	9	1723	3236	1452	2940	23 (1.58)	68 (2.31)
C3	5	1530	2358	1255	2153	29 (2.31)	302 (14.03)*

C0 = control group, C1 = 10^{-6}M, C2 = 10^{-5}M, C3 = 10^{-4}M, *$P<0.0001$

For the offspring of the exposed wild-type snails, temporal graphs showed that cyclophosphamide affected germ cells of *B. glabrata* at later phases of ovogenesis, probably after meiosis (Figure 8). For mitomycin C, the effects of toxic concentrations of were observed at all stages, but with the lowest concentration, only the early stages of ovogenesis were affected (Figs. 9, 10). These results are in agreement with studies on rodents using the dominant lethal test. Post-meiotic stages of gametogenesis were the most sensitive to the induction of damage by cyclophosphamide [93]. Ehling [94] treated mice with three doses of mitomycin C and obtained different time–response curves for each one. Germ cells at later phases of spermatogenesis were affected only at the highest dose. Comparing mitomycin C with other inducers of dominant lethal mutations, Witt and Bishop [95] emphasized the induction of higher frequencies of lethal mutations in pre-meiotic stages of gametogenesis, a specific effect of mitomycin C, attributed to the inhibition of DNA synthesis.

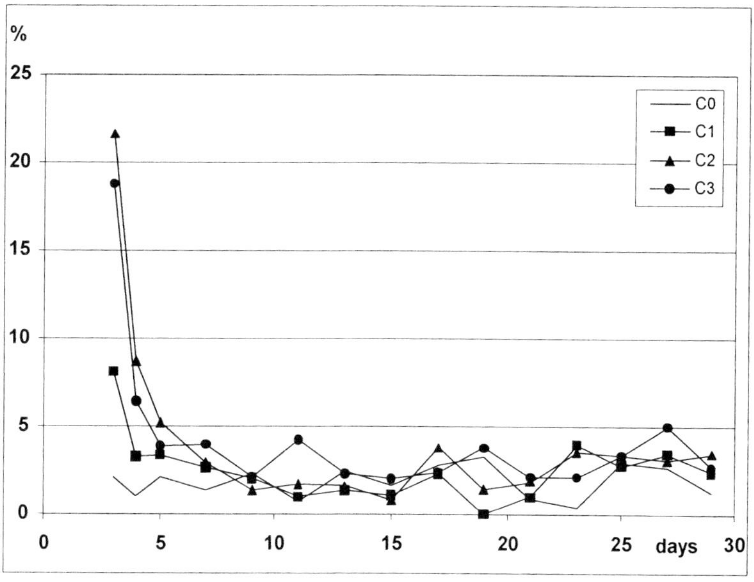

Figure 8. Induction of malformations in *B. glabrata* by cyclophosphamide – offspring of wild-type snails. C0: control group; C1: $3.6x10^{-5}$ M; C2: $3.6x10^{-4}$ M; C3: $3.6x10^{-3}$ M.

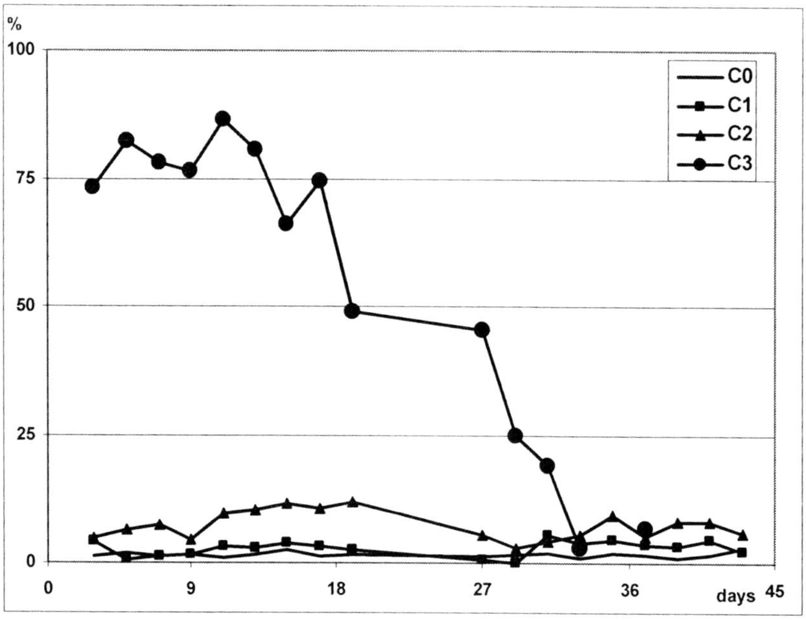

Figure 9. Induction of malformations in *B. glabrata* by mitomycin C – offspring of wild-type snails. C0: control group; C1: 10^{-6} M; C2: 10^{-5} M; C3: 10^{-4} M.

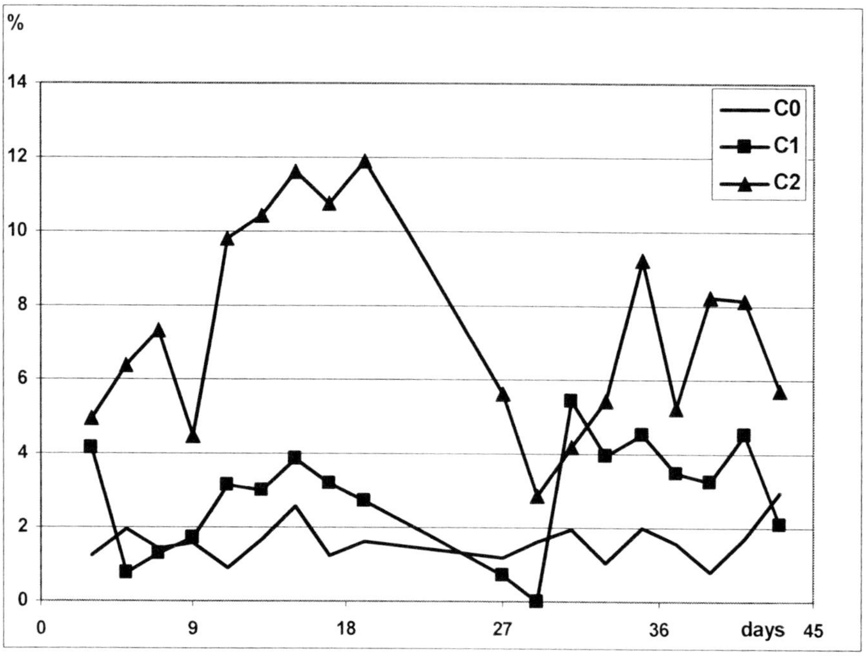

Figure 10. Induction of malformations in B. glabrata by mitomycin C (without C3) – offspring of wild-type snails. C0: control group; C1: 10^{-6} M; C2: 10^{-5} M.

Regarding the sensitivity of the system, results showed that the ability of detection of chemical mutagens by *B. glabrata* was comparable to other aquatic organisms.

Data obtained in this study showed that *B. glabrata* can take up chemical mutagens from the aquatic environment, and the positive response obtained with cyclophosphamide demonstrated that *B. glabrata* can activate indirect mutagens. This allows the use of the model in biomonitoring of environmental mutagens, since many of them require metabolic activation.

^{60}Co gamma Radiation

Ionizing radiation acts as a direct mutagen with minimal latency time of action, since it does not require absorption, distribution or metabolism; furthermore, the time of effective activity is practically zero. Thus, ionizing radiation can be used as reference to infer some kinetic parameters of chemical mutagens, considering that, sometimes, the correlation between the pharmacokinetics of chemicals with their genotoxic activity is complex [96]. ^{60}Co gamma radiation was chosen for this study not only as a reference mutagen, but for a unique aspect of ionizing radiation. A characteristic effect of radiation-induced dominant lethal mutations is the differential response of spermatogenic stages [94]. In general, pre-meiotic germ cells are highly sensitive to the cell-killing effect and less sensitive to the induction of lethal dominant mutations than post-meiotic cells. These effects can be attributed to two factors: i) young gametogenic cells – mitotically active – have greater repair ability from sub-lethal DNA damage and ii) there is a selective elimination of the damaged cells. A striking uniformity in response between germ cells at homologous stages of spermatogenesis in widely different species was observed [69, 97].

Table 5. *B. glabrata* dominant lethal mutation test of ^{60}Co gamma radiation

Crossing schedule (days)	Dose (Gy)	No. of pairings	No. of crossings	Wild-type embryos	
				Total	Malformed (%)
7	0	11	10	2316	12 (0,52)
	2,5	11	11	2108	42 (1,99)†
	10	11	11	1737	377 (21,70)†
	20	13	12	1825	920 (50,41)†
17	0	11	10	2276	2 (0,09)
	2,5	12	11	2207	8 (0,36)
	10	10	10	2176	115 (5,28)†
	20	12	12	1874	284 (15,15)†
23	0	10	10	2267	13 (0,57)
	2,5	12	12	2016	21 (1,04)
	10	11	11	1267	37 (2,92)†
	20	12	7*	809	94 (11,62)†
30	0	13	13	2151	5 (0,23)
	2,5	12	11	2432	5 (0,21)
	10	9	8	877	1 (0,11)
	20	11	0†	-	-
36	0	13	12	1779	4 (0,22)
	2,5	12	12	1838	3 (0,16)
	10	11	8	1302	3 (0,23)
	20	12	4*	578	1 (0,12)

$^{*}P < 0{,}02$; $^{\dagger}P < 0{,}0001$

For this study, three groups of 13 wild-type snails were irradiated with single doses of 2.5; 10 and 20 Gy. Crossings were carried out at intervals of 7, 17, 23, 30 and 36 days after irradiation (Table 5 and Figure 11). The dominant lethal effect was observed only at the first crossing occurring 7 days after irradiation with 2.5 Gy. With 10 and 20 Gy, the induction of lethal mutations was detected at 7, 17 and 23 days after irradiation; a dose-response effect was observed. The effect was stronger 7 days after irradiation, decreasing in the succeeding crossings up to 30 days. Cell-killing effects on germ cells were detected in the crossings at 23 days and 30 days after irradiation with 20 Gy. After 36 days, frequencies of malformations resumed background levels; crossing rates partially recovered. These results show that gamma radiation affected all the stages of spermatogenesis. Germ cells at later phases were more sensitive to the mutagenic effect of radiation and the cell killing effects were observed on the youngest cells. This response was quite similar to the highly homogeneous pattern observed in widely different species and allowed us to estimate some parameters of spermatogenesis in *B. glabrata*.

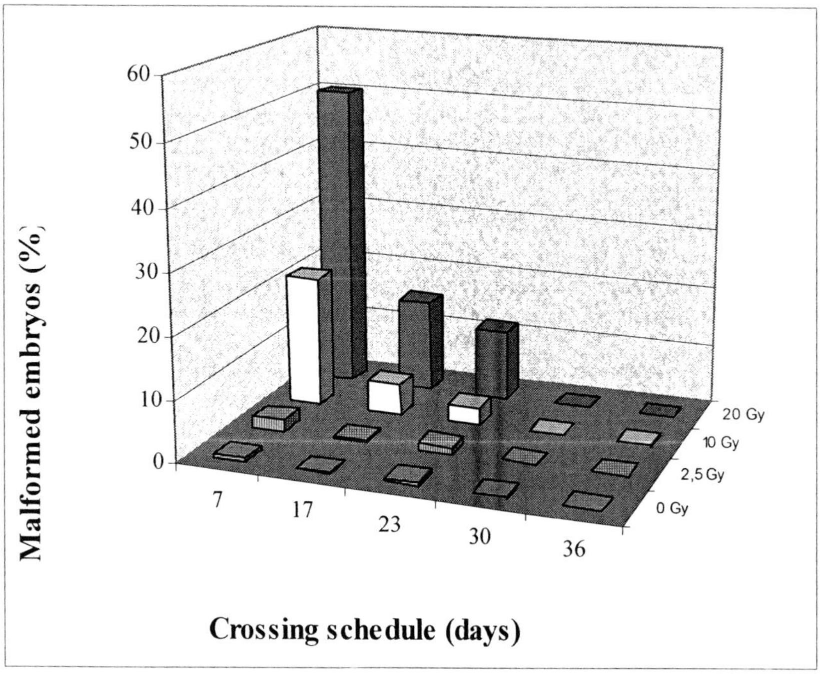

Figure 11. Frequencies of malformation in the wild-type offspring of albino snails crossed with the wild-type snails at different intervals after the exposure to 60Co gamma radiation

Advantages

The dominant lethal test was established in *Biomphalaria glabrata* by using an approach similar to that used in the dominant lethal test in rodents [68]. Despite the lacking data on the timing of spermatogenesis and metabolizing systems in *Biomphalaria,* the test in this freshwater mollusk shows some advantages comparing to the assay in rodents.

Since the snail is a simultaneous hermaphrodite, effects on female germ cell lineages can be obtained from the same exposed animals. The brood size allows the scoring of a greater number of embryos: one adult snail lay egg-masses with 20-50 eggs almost daily and reproduces throughout the year. In the assay with rodents, reduced counts in total implantations can be due to loss of fertilized eggs before implantations or loss of unfertilized eggs; the former reflects gross chromosomal aberrations that result in embryonic death before implantation, and the latter, a toxic effect that reduces the germ cell's ability to fertilize; both are indistinguishable [73]. In *Biomphalaria*, there are no losses due to abortions and failures of implantation, and analysis of crossing rates indicates effects on the fertilization ability; thus cell-killing effects can also be evaluated. At last, the dominant lethal test in *B. glabrata* is the only test system for evaluation of germ cell effects of environmental mutagens in aquatic invertebrates. At present, studies on water samples are being conducted to test for its potential as in vivo model for evaluating the mutagenicity of water-borne contaminants in freshwater environments.

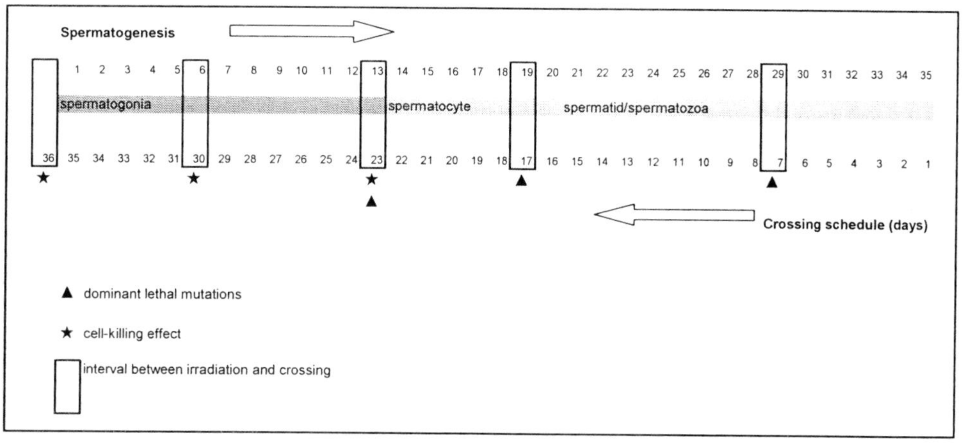

Figure 12. Estimated chronology of spermatogenesis in *Biomphalaria glabrata*.

Acknowledgements

The authors thank Fundação de Amparo à Pesquisa de São Paulo, Conselho Nacional de Desenvolvimento Científico e Tecnológico for financial support, Cristina Antonia de Albuquerque, Patrícia Aoki Miyasato for technical assistance and M.Nazareth Rabello-Gay for critical review.

References

[1] Fellenberg, G. (1980). Introdução aos problemas da poluição ambiental. São Paulo. EPU: Springer: Ed da Universidade de São Paulo, 1980, 196p.

[2] Cooney, J. D. (1995). Freshwater tests. In: RAND, G. M. Ed. *Fundamentals of Aquatic Toxicology*. Washington, D. C., Taylor & Francis, 71-102.

[3] Doust, J.L.; Schmidt, M.; Doust, L.L. (1994). Biological assessment of aquatic pollution: a review, with emphasis on plants as biomonitors. *Biol. Rev.*, 69, 147-186.

[4] Burger, J. (1997). Methods for and approaches to evaluating susceptibility of ecological systems to hazardous chemicals. *Environ. Health Perspect.,* 105, 4, 843-848.

[5] Houk, V. S. (1992). The genotoxicity of industrial wastes and effluents. *Mutation Res.,* 277, 91-138.

[6] Claxton, L. D; Houk, V. S; Hugles, T. J. (1998). Genotoxicity of industrial wastes and effluents. *Mutation Res.,* 410, 237-243.

[7] Stahl, JR. R. G. (1991). The genetic toxicology of organic compounds in natural waters and wastewaters. *Ecotoxicol. Environ. Safety*, 22, 94-125.

[8] White, P. A., Rasmussen, J. B., Blaise, C. (1996). Comparing the presence, potency, and potential hazard of genotoxins extracted from a broad range of industrial effluents. *Environ. Mol. Mutag.*, 27, 116-139.

[9] White, P. A., Rasmussen, J. B. (1998). The genotoxic hazards of domestic wastes in surface waters. *Mutation Res.*, 410, 223-236.

[10] Loper, J. C. (1980). Mutagenic Effects of organic compounds in drinking water. *Mutation Res.*, 76, 241-268.

[11] Meier, J. R., Ringhand, H. P., Coleman, W. E., Munch, J. W., Streicher, R. P., Kaylor, W. H., Schenk, K. M.(1985). Identification of mutagenic compounds formed during chlorination of humic acids. *Mutation Res.*, 157, 111-112.

[12] Liu, Q., Jiao, Q. C., Huang, X. M., Jiang, J. P., Cui, S. Q., Yao, G. H., Jiang, Z. R., Zhao, H. K., Wang, N. Y. (1999). Genotoxicity of drinking water from Chao Lake. *Environ. Res. Section*, 80, 127-131.

[13] Kronberg, L., Vartiainen, T. (1988). Ames mutagenicity and concentration of the strong mutagen 3-chloro-4(dichloromethyl)-5-hydroxy-2(5*H*)-furanone and its geometric isomer *E*-2-chloro-3-(dichloromethyl)-4-oxobutenoic acid in chlorine treated tap water. *Mutation Res.*, 206, 177-182.

[14] Koivusalo, M. T., Jaakkola, J. J. K., Vartiainen, T. (1994). Drinking water mutagenicity in past exposure assessment of the studies on drinking water and cancer: application and evaluation in Finland. *Environ. Res.*, 64, 90-101.

[15] ICPEMC (1983). Commission for protection against environmental mutagens and carcinogens. Committee 4 final report: estimation of genetic risks and increased incidence of genetic disease due to environmental mutagens. *Mutation Res.*, 115, 255-291.

[16] Volgelstein, B., Kinzler, W. (1993). The multistep nature of cancer. *TIG*, 9, 4, 138-141.

[17] Waters, M. D.; Stack H. F.; Jackson, M. A. (1999). Genetic toxicology data in the evaluation of potential human environmental carcinogens. *Mutation Res.*, 437, 1, 21-49.

[18] Ames, B. N., Durston, W. E., Yamasaki, E., Lee, F. D. (1973) Carcinogens are mutagens: A simple test system combining liver homogenates for activation and bacteria for detection. *Proc. Natl. Acad. Sci.*(USA), 70, 2281-2285.

[19] Schmid, W. (1975). The micronucleus test. *Mutation Res.*, 31, 9-15.

[20] McCann, J.; Choi, E.; Yamasaki, E.; Ames, B. N. (1975b). Detection of carcinogens as mutagens in the Salmonella/ microsome test: assay of 300 chemicals. *Proc. Natl. Acad. Sci. USA*, 72, 979-983.

[21] Tennant, R. W.; Margolin, B. H.; Shelby, M. D.; Zeiger, E.; Haseman, J. K.; Spalding, J.; Caspary, W. (1987). Prediction of chemical carcinogenicity in rodents from in vitro genetic toxicity assays. *Science*, 236, 933-941.

[22] Ashby, J.; Tennant, R. W. (1991). Definitive relationships among chemical structure, carcinogenicity and mutagenicity for 301 chemical tested by the US NTP. *Mutation Res.*, 257, 229-306.

[23] Elespuru, R. K. (1996). *Future approaches to genetic toxicology risk assessment*, 365, 191-204.

[24] Ashby, J., Morrod, R. S. (1991) Detection of human carcinogens. *Nature*, 352, 185-186.

[25] Wurgler, F. E.; Kramers, P. G. N. (1992). Environmental effects of genotoxins (eco-genotoxicology). *Mutagenesis*, 7, 321-327.

[26] Depledge, M. H. (1996). Genetic ecotoxicology: an overview. *Journal Exp. Mar. Biol. Ecol.*, 200, 57-66.

[27] Ramel, C.(1983). Polygenic effects and genetic affecting quantitative traits. ICPEMC Publication N°8. *Mutation Res.*, 114, 107-11.

[28] Lowery, M. C.; Au, W. W.; Adams, P. M.; Whorton Jr., E. B.; Legator, M. S. (1990). Male-mediated behavioral abnormalities. *Mutation Res.*, 229, 213-230.

[29] Roff, D. A.; Mousseau, T. A. (1987). Quantitative genetics and fitness: lessons from *Drosophila. Heredity*, 58, 103-118.

[30] Anderson, S. L.; Wild, G. C. (1994). Linking genotoxic responses and reproductive success in ecotoxicology. *Environ. Health Perspect.*, 102 (12), 9-12.

[31] Anderson, S., Sadinski, W., Shugart, L., Brussard, P., Depledge, M., Ford, T., Hose, J., Stegeman, H., Suk, W., Wirgin, I., Wogan, G. (1994). Genetic and molecular ecotoxicology: a research framework. *Environ. Health Perspect.*, 102 (12), 3-8.

[32] Jha, A. N.; Cheung, V. V.; Foulkes, M. E.; Hill, S. J.; Depledge, M. H. (2000). Detection of genotoxins in the marine environment: adoption and evaluation of an integrated approach using the embryo-larval stages of the marine mussel, *Mytilus edulis*. *Mutation Res.*, 464, 213-228.

[33] Brunetti, R.; Gola, I.; Majone, F. (1986). Sister exchange in development eggs *Mytilus galloprovincialis* Lmk. (Bivalvia). *Mutation Res.*, 174, 207-211.

[34] Hose, J. E., Cross, J. N., Smith, S. G., Diehl, D. (1987). Elevated circulating erythrocyte micronuclei in fishes from contaminated sites off southern California. *Mar. Environ. Res.*, 22, 167-176.

[35] Brunetti, R.; Majone, F.; Gola, I.; Beltrame, C. (1988). The micronucleus test: examples of application to marine ecology. *Mar. Ecol. Prog. Ser.*, 44, 65-68.

[36] Nacci, D.; Nelson, S.; Nelson, W.; Jackim, E. (1992). Application of DNA alkaline unwinding assay to detect DNA strand breaks in marine bivalves. *Mar. Environ. Res.*, 33, 83-100.

[37] Vukmirovic, M.; Bihari, N.; Zahn, R.K.; Müller, W.E.G.; Batel, R. (1994). DNA damage in marine mussel *Mytilus galloprovincialis* as a biomarker of environmental contamination. *Mar. Ecol. Prog. Ser.*, 109, 165-171.

[38] Burgeot, T.; His, E.; Galgani, F. (1995). The micronucleus assay in *Crassosstrea gigas* for the detection of seawater genotoxicity. *Mutation Res.*, 342, 125-140.

[39] Steinert, S.A; Streib-Montee, R.; Leather, J.M.; Chadwick , D. B. (1998). DNA damage in mussels at sites in San Diego Bay. *Mutation Res.*, 399, 65-85.

[40] Hose, J.E.; Brown, E. D. (1998). Field applications of the piscine anaphase aberration test: lessons from the Exxon Valdez oil spill. *Mutation Res.*, 399, 167-178.

[41] Harvey, J.S.; Lyons, B.P.; Page, T.S.; Stewart, C.; Parry, J.M. (1999). An assessment of the genotoxic impact of the Sea Empress oil spill by the measurement of DNA adduct levels in selected invertebrate and vertebrate species. *Mutation Res.*, 441, 103-114.

[42] Rank, J.; Jensen, K. (2003). Comet assay on gill cells and hemocytes from the blue mussel *Mytilus edulis*. Ecotox. *Environ. Safety*, 54, 3, 323-329.

[43] Lyons, B. P.; Stentiford, G. D. ; Green, M.; Bignell, J.; Bateman, K.; Feist, S. W.; Goodsir, F.; Reynolds, W. J.; Thain, V. (2004). DNA adduct analysis and histopathological biomarkers in European flounder (*Platichthys flesus*) sampled from UK estuaries. *Mutation Res.*, 552, 1-2, 177-186 .

[44] Akcha, F.; Hubert, F. V.; Pfhol-Leszkowicz, A. (2003). Potential value of the comet assay and DNA adduct measurement in dab (*Limanda limanda*) for assessment of in situ exposure to genotoxic compounds. *Mutation Res.*, 534, 1-2, 21-32.

[45] Akcha, F.; Leday, G.; Pfohl-Leszkowicz, A. (2004). Measurement of DNA adducts and strand breaks in dab (*Limanda limanda*) collected in the field: effects of biotic (age, sex) and abiotic (sampling site and period) factors on the extent of DNA damage. *Mutation Res.*, 552, 1-2, 197-207.

[46] Pisoni, M.; Cogotzi, L. ; Frigeri, A.; Corsi, I.; Bonacci, S.; Iacocca, A.; Lancini, L.; Mastrototaro, F.; Focardi, S.; Svelto, M. (2004). DNA adducts, benzo(*a*)pyrene monooxygenase activity, and lysosomal membrane stability in *Mytilus galloprovincialis* from different areas in Taranto coastal waters (Italy). *Environ. Res.*, 96, 2, 163-175.

[47] Alink, G.M.; Frederix-Wolters, E.M.H.; van der Gaag, M.A; van der Kerkhoff, J.F.J.; Poels, C.L.M. (1980). Induction of sister-chromatid exchanges in fish exposed to Rhine water. *Mutation Res.*, 78, 369-374.

[48] De Flora, S.; Viganò, L.; Agostini, F.D.; Camoirano, A; Bagnasco, M.; Bennicelli , C.; Melodia, F.; Arillo, A (1993). Multiple genotoxicity biomarkers in fish exposed in situ to polluted river water. *Mutation Res.*, 319, 167-177.

[49] Pandrangi, R.; Petras, M.; Ralph ,S.; Vrzoc , M. (1995). Alkaline single cell gel (Comet) assay and genotoxicity monitoring using bullheads and carp. *Environ. Mol. Mutagen.*, 26, 345-356.

[50] Minissi, S.; Ciccotti, E; Rizzoni, M. (1996). MN test in erythrocytes of *Barbu plebejus* (Teleostei, Pisces) from two natural environments: a bioassay for the in situ detection of mutagens in freshwater. *Mutation Res.*, 367, 245-251.

[51] Ralph, S.; Petras, M. (1997). Genotoxicity monitoring of small bodies of water using two species of tadpoles and the alkaline single cell gel (Comet) assay. *Environ Mol Mutagen*, 29, 418-430.

[52] Sánchez-Galán , S.; Linde, A. R.; Izquierdo, J. I.; García-Vázquez, E. (1998). Micronuclei and fluctuating asymmetry in brown trout (*Salmo trutta*): complementary methods to biomonitor freshwater ecosystems. *Mutation Res.*, 412:219-225

[53] Lyons, B. P.; Stewart, C.; Kirby, M. F. (1999). The detection of biomarkers of genotoxin exposure in the European flounder (Platichthys flesus) collected from the River Tyne Estuary. *Mutation Res*, 446, 111-119.

[54] Klobuc ar, G.I. V.; Pavlica, M.; Erben, R.; Pape, D. (2003). Application of the micronucleus and comet assays to mussel *Dreissena polymorpha* haemocytes for genotoxicity monitoring of freshwater environments. *Aquat. Toxicol.*, 64, 1, 15-23.

[55] Winter, M. J.; Day, N.; Hayes, R. A.; Taylor, E. W. ; Butler, P. J.; Chipman, J. K. (2004). DNA strand breaks and adducts determined in feral and caged chub (*Leuciscus cephalus*) exposed to rivers exhibiting variable water quality around Birmingham, UK. *Mutation Res.*, 552, 1-2, 163-175.

[56] Buschini, A.; Martino, A.; Gustavino, B.; Monfrinotti, M.; Poli, P.; Rossi, C.; Santoro, M.; Dörr, A.J.M.; Rizzoni, M. (2004). Comet assay and micronucleus test in circulating erythrocytes of *Cyprinus carpio* specimens exposed in situ to lake waters treated with disinfectants for potabilization. *Mutation Res.*, 557, 119–129.

[57] Russo, C.; Rocco, L.; Morescalchi, M. A.; Stingo, V. (2004). Assessment of environmental stress by the micronucleus test and the Comet assay on the genome of teleost populations from two natural environments .*Ecotoxic. Environ. Safety*, 57, 2, 168-174.

[58] Gauthier, L.; Tardy, E.; Mouchet, F.; Marty, J. (2004) Biomonitoring of the genotoxic potential (micronucleus assay) and detoxifying activity (EROD induction) in the River Dadou (France), using the amphibian *Xenopus laevis*. *Sci. Total Environ.*, 323, 47-61.

[59] Porto, J. I. R.; Araujo, C. S.O.; Feldberg, E. (2005). Mutagenic effects of mercury pollution as revealed by micronucleus test on three Amazonian fish species. *Environ. Res.*, 97, 3, 287-292 .

[60] Dixon, D. R.; Pruski, A. M.; Dixon, L. R. J.; Jha, A. N. (2002). Marine invertebrate eco-genotoxicology: a methodological overview. *Mutagenesis*, 17 (6), 495-507.

[61] Ohe, T.; Watanabe T.; Wakabayashi K. (2004). Mutagens in surface waters: a review. *Mutation Res.*, 567, 2-3, 109-149.

[62] Jha, A. N. (2004). Genotoxicological studies in aquatic organisms: an overview. *Mutation Res.*, 552, 1-2, 1-17.

[63] Muller, H. J. (1927). Artificial transmutation of gene. *Science*, 66, 84-87, 1927.

[64] Bateman, A. J. (1958). Mutagenic sensitivity germ cells in the male mouse. *Heredity*, 12, 213-232.

[65] [65]Bateman, A. J. (1960). The induction of dominant lethal mutations in rats and mice with triethylenemelamine (TEM). *Genet. Res.*, 1, 381-392.

[66] Newcombe, H. B.; McGregor, J. F. (1967). Major congenital malformations from irradiations of sperm and eggs. *Mutation Res.*, 4, 663-673.

[67] Schröder, J. H. (1969). X-Ray-induced mutations in the policiliid fish, *Lebistes reticulatus* Peters. *Mutation Res.*, 7, 75-90.

[68] Bateman, A. J. (1966). Testing chemicals for mutagenicity in a mammal. *Nature*, 5032, 205-206.

[69] Egami, N.; Shimada, A.; Hama-Furukama, A. (1983). Dominant lethal mutation rate after γ-irradiation of the fish *Oryzias latipes*. *Mutation Res.*, 107, 265-277.

[70] Ashby, J.; Waters, M. D.; Preston, J.; Adler, I. D.; Douglas, G. R.; Fielder, R.; Shelby, M. D.; Anderson, D.; Sofuni, T.; Gopalan, H. N. B.; Becking, G.; Sonich-Mullin, C. (1996). IPCS harmonization of methods for the prediction and quantification of human carcinogenic/mutagenic hazard, and for indicating the probable mechanism of action of carcinogens. *Mutation Res.*, 352, 153-157.

[71] Gollapudi, B. B.; Krishna, G. (2000). Practical aspects of mutagenicity testing strategy: an industrial perspective. *Mutation Res.*, 455, 21-28.

[72] MacGregor, J. T.; Casciano, D.; Müller, L. (2000). Strategies and testing methods for identifying mutagenic risks. *Mutation Res.*, 455, 3-20.

[73] Green, S.; Auletta, A.; Fabricant, J.; Kapp, R.; Manandhlar, M.; Sheu, C.; Springer, J.; Whitfield, B. (1985). Current status of bioassays in genetic toxicology – the dominant lethal assay. *Mutation Res.*, 154, 49-67.

[74] Green, S.; Lavappa, K. S.; Manandhar, M.; Sheu, C.; Whorton, E.; Springer, J. A. (1987). A guide for mutagenicity testing using the dominant lethal assay. *Mutation Res.*, 189, 167-174.

[75] Barnes, R. S. K.; Calow, P.; Olive J. W. (1995). Os invertebrados: uma nova síntese. *Ed Atheneu, São Paulo,* 136-155.

[76] Seed, R. (1984). Structural organization, adaptative radiation, and classification of molluscs. In:Tompa, A. S.; Verdonk, N. H.; van der Biggelaar, J. A. M. eds. *The Mollusca,* Academic Press, New York, 3, 1-54.

[77] Jha, A. N. (1998). Use of aquatic invertebrates in genotoxicological studies. *Mutation Res.*, 399, 1-2.

[78] López-Barea, J.; Pueyo, C. (1998). Mutagen content and metabolic activation of promutagens by molluscs as biomarkers of marine pollution. Mutation Res., 399, 3-15.

[79] Carvalho, O. S. (1992). *Intermediate hosts of Schistosoma mansoni in Brazil.* Mem. Inst. Oswaldo Cruz , 87 (IV), 307-309.

[80] Camey, T.; Verdonk, N. H. (1970) The early development of the snail *Biomphalaria glabrata* (Say) and the origin of the head organs. *Neth. J. Zool.*, 20, 1, 93-121.

[81] Kawano, T.; Simões, L. C. S.; Marques, J. (1992). *Biomphalaria glabrata* (Say, 1818): efeitos morfogenéticos do sulfato de nicotina. *Rev. Brasil. Biol.*, 53 (4), 539-547.

[82] Kawano, T; Simões, L. C. G. (1986). Efeito da *Stevia rebaudiana* em *Biomphalaria glabrata*. *Rev. Bras. Biol.*, 46(3), 555-562.

[83] Ré, L.; Kawano, T. (1987). Effects of *Laurus nobilis* (Lauraceae) on *Biomphalaria glabrata* (Say, 1818). Mem. Inst. Oswaldo Cruz, 82 (IV), 315-320.

[84] Okazaki, K.; Kawano, T. (1991). Cytogenetic effects of ^{60}Co gamma radiation on *Biomphalaria glabrata* (Say, 1818) embryos. *Caryologia*, 44(2), 121-136.

[85] Kawano, T.; Okazaki, K.; Ré, L. (1992). Embryonic development of *Biomphalaria glabrata* (Say, 1818) (Mollusca, Gastropoda, Planorbidae): a practical guide to the main stages. *Malacologia*, 34, 1-2, 25-32.

[86] Yamamoto, M. M.; Kawano, T.; Young. M. C. M.; Chu, E. P.; Haraguchi, M.; Hiroki, K. (1996). Molluscicidal activity of three Brazilian plant species. *Fitoterapia*, LXVII(1), 59-62.

[87] Paraense, W. L. (1955). Auto-fecundação e fecundação cruzada em *Australorbis glabratus*. *Men. Inst. Oswaldo Cruz*, 53, 2, 3, 4, 277-284.

[88] Tallarico, L. F. Okazaki, K.; Kawano, T.; Pereira, C. A. B.; Nakano, E. (2004). Dominant lethal effect of ^{60}Co in *Biomphalaria glabrata* (Say, 1818). *Mutation Res.*, 561, 139-145.

[89] Dohmen, M. R. (1983). Gametogenesis. In:Tompa, A.S.; Verdonk, N. H./ Van der Biggelaar, J. A. M. eds. *The Mollusca*. London, Academic Press Inc., 3, 1-47.

[90] Vernon, J. G.; Taylor, J. K. (1996). Patterns of sexual roles adopted by the schistosome-vetor snail *Biomphalaria glabrata* (Planorbidae). *J. Moll. Stud.*, 62, 235-241.

[91] Geilenkirchen, W. L. M. (1966). Cell division and morphogenesis of *Limnaea* eggs after treatment with heat pulses at successive stages in early division cycles, *J. Embryol. Exp. Morphol.* 16 (2) , 321-337.

[92] Nakano, E.; Watanabe, L.C. ; Ohlweiler, F.P.; Pereira, C.A.B.; Kawano, T.(2003). Establishment of the dominant lethal test in the freshwater mollusk *Biomphalaria glabrata* (Say, 1818), *Mutation Res.*, 536, 145-154.

[93] [94]Anderson, D.; Bishop, J. B.; Garner, R. C., Ostroski-Wegman, P. O.; Selby, P. B.; (1995). Cyclophosphamidde: review of its mutagenicity for an assessment of potential germ cell risks. *Mutation Res.*, 330, 115-181.

[94] Ehling, U.H.(1971). Comparison of radiation- and chemically induced dominant lethal mutations in male mice. *Mutation Res.* 11, 35-44.

[95] Witt, K. L.; Bishop, J. B. (1996). Mutagenic of anticancer drugs in mammalian germ cells. *Mutation Res.*, 355, 209-234.

[96] Morales-Ramirez, P.; Vallarino-Kelly, T.; Anguiano-Orozco, G.; Rodríguez-Reyes, R. (1997). Pharmacokinetic parameters of genotoxic activity inferred from the comparison of the kinetics of MN-PCE induced by chemical agents and ionizing radiation, *Mutation Res.*, 391, 127-134.

[97] Mandl, A. M. (1964). The radiosensitivity of germ cells. *Biol. Rev.*, 391, 127-134.

In: New Developments in Mutation Research
Editor: Charles L. Valon, pp. 217-240
ISBN 1-59454-664-9

Chapter 9

DENDRIMERS AS DELIVERY SYSTEMS IN GENE THERAPY

Barbara Klajnert and Maria Bryszewska*
Department of General Biophysics, University of Lodz, Lodz, Poland

ABSTRACT

The dendrimer is the common name of the polymers which are characterized by a specific structure, although their chemical composition can be very different. Dendrimers display several characteristics which distinguish them from linear polymers. All dendrimers are synthesized by attaching branched monomers to the central core molecule. Monomers are attached radially, layer by layer. As a result of such a synthesis, dendrimers possess plenty of terminal, reactive groups on their surface and empty internal cavities. Both of these features can be utilized. Soon after the first synthesis in mid-1980's, dendrimers have become interesting and promising molecules for pharmaceutical aims. First, they turned out to be efficient transfection agents. They are able to transport bigger amount of the genetic material than liposomes and do not cause a strong immunological response as viral vectors do. Second, dendrimers can play an important role as drug-delivery-systems. Drug molecules can be attached to the end groups or encapsulated in the dendrimer's interior.

To continue the studies on the dendrimers' role in medicine, it is important to know more about their toxicity and biocompatibility. The last ten years brought many new facts about dendrimers impact on biological systems. Both, general cytotoxic studies, as well as detailed research devoted to particular cellular compounds, as DNA or lipid bilayer, were carried out.

In this review the development of last ten years studies how dendrimers affect different biological systems is reported. The emphasis is put on basic studies on interactions between dendrimers and DNA or a lipid bilayer to explain dendrimers' high efficiency as transfection agents.

---

* Correspondence: Department of General Biophysics, University of Lodz; 12/16 Banacha St., 90-237 Lodz, Poland; marbrys@biol.uni.lodz.pl

DENDRIMERS' SYNTHESIS, STRUCTURE AND PROPERTIES

Dendrimers were first synthesized in the mid-1980's by Tomalia's group [1]. Their name originates from a Greek word 'dendron' which means a tree, and it well describes their branched architecture.

There are two main ways to obtain dendrimers: a divergent method or a convergent one [2]. In the divergent method, dendrimers grow outwards from a central core molecule which reacts with branched monomers. Monomers are attached layer after layer creating so called generations. The more layers are attached, the higher generation is obtained. In the convergent approach, dendrimers are constructed stepwise starting from building branches called dendrons. When they are large enough they are attached to the core molecule (Figure 1).

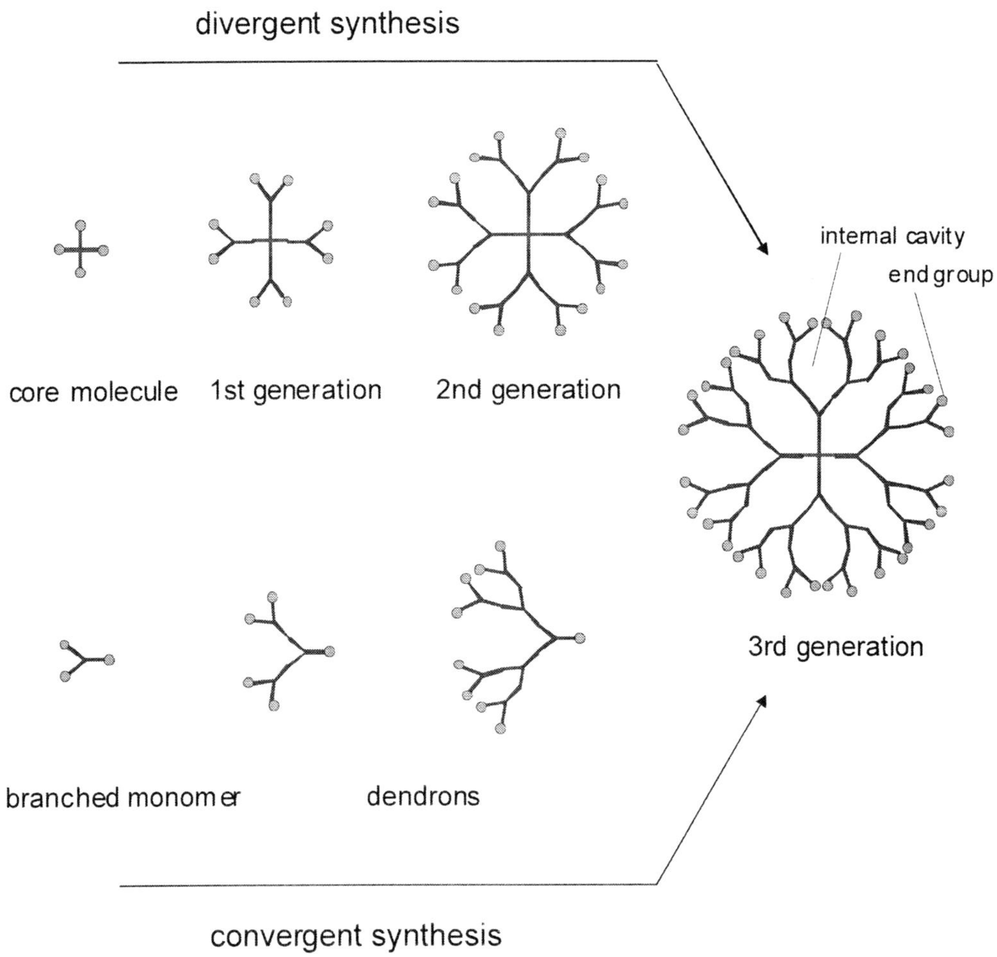

Figure 1. Scheme of dendrimer's synthesis

The specific methods of synthesis result in a specific structure of these polymers. They possess many terminal groups which make them highly reactive and well soluble. Dendrimers' solubility is strongly influenced by the nature of surface groups. Dendrimers terminated in hydrophilic groups are soluble in polar solvents, while dendrimers having hydrophobic end groups are soluble in nonpolar solvents. High generations of dendrimers, which adopt globular shape, have densely packed surface and empty internal cavities. Lower generations have asymmetric shape and possess a more open structure. One of the most important dendrimers' features is their monodispersity which distinguish them from classical polymers, which are usually of different sizes.

Now, there are many families of dendrimers, each synthesized from different monomers and possessing unique properties, since the surface and interior can be tailored to specific types of applications. Nevertheless, there are two kinds of dendrimers which are most often used in biomedical applications and which are commercially available. The first group is polyamidoamine dendrimers, known also as PAMAM dendrimers or Starburst dendrimers. The term Starburst is a trademark [3]. Ethylenediamine is used as a core molecule. In the presence of methanol, it reacts with methyl acrylate and then ethylenediamine is added. At the end of each branch there are two free amino groups that can react again with ethylenediamine molecules, so the process is repeated. Each completed reaction sequence results in a next dendrimer generation. The second group is polypropylenimine dendrimers (PPI dendrimers) available under the commercial name Astramol. A core molecule is butylenediamine and propylenimine branches are added to it by an iterative procedure, which includes the Michael addition reaction followed by reduction of nitryle end groups to amine groups [4].

DENDRIMERS AS DRUG CARRIERS

Two specific features make dendrimers especially suitable as drug delivery systems (Figure 2).

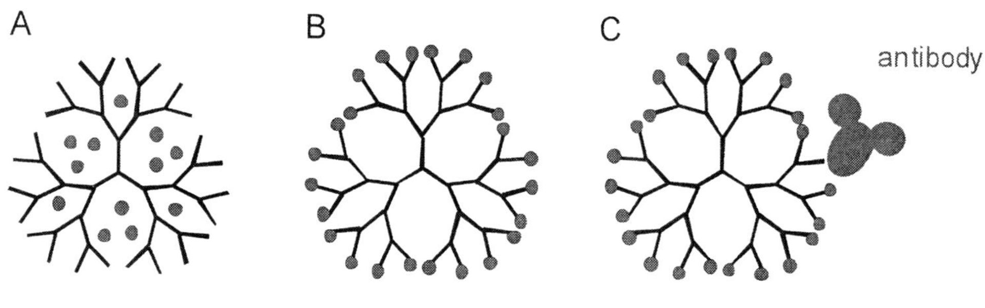

Figure 2. Different strategies of drug delivery: A – drug molecules encapsulated inside a dendrimer; B – drug molecules attached to a dendrimer's surface; C – dendrimer in a targeted therapy with an attached antibody

The first one is the presence of internal empty cavities. It enables to encapsulate guest molecules inside e.g. it was possible to trap rose bengal, p-nitrobenzoic acid or eosin Y inside so called 'dendritic box' of polypropylenimine dendrimer [5-7]. Soon afterwards first attempts to utilize dendrimers as drug carriers appeared. It was shown that dendrimers with a hydrophobic interior and a hydrophilic surface were able to solubilize hydrophobic drugs in aqueous solutions, so they acted analogously to micelles [8]. That is why they were called unimolecular micelles.

Drugs entrapped inside the dendrimer can be released slower. The slow release is important for antitumor drugs because it reduces their toxicity. Adriamycin and methotrexate were encapsulated inside polyamidoamine dendrimers [9]. The ability to encapsulate these drugs enhanced with an increasing dendrimer generation.

The second important dendrimers' characteristic is the presence of a large number of terminal groups to which drug molecules can be conjugated. Such an use of dendrimers is based on multiplication of drug activity, because one molecule of dendrimer is capable of carrying many molecules of a drug, e.g. attaching α-sialinic acid moieties to the dendrimer surface enhanced their therapeutic effect and allowed to attain a higher activity in inhibiting influenza infection [10, 11]. The inhibiting effect increased proportionally to the number of sialinic acid groups.

Dendrimers can play an important role in a targeted anticancer therapy which is based on using antibodies modified by attaching radioisotopes or cytotoxic drugs. Unfortunately sometimes it is necessary to carry more than one molecule of radioisotope or drug to achieve a therapeutic effect. Such a modification of an antibody molecule often eliminates its targeting potential. An attractive approach is to use dendrimers as linker molecules. Dendrimer surface can bear many isotopes or drugs and then the dendrimer can be attached to the antibody. Thus, only a single site on the surface of the antibody is changed [12]. A good example of such application is using dendrimers in a boron neutron capture therapy. Dendrimers with covalently attached boron atoms were prepared and first tests on these compounds gave positive results [13].

Dendrimers in Gene Therapy

A gene therapy is an exciting approach to treat many diseases. It is based on a simple assumption: if there is a lack or insufficient expression of natural proteins, it can be compensated by a delivery of exogenous gene into a cell in order to express the encoded protein. Special systems, called vectors, must be employed to achieve a successful delivery of DNA into cells. The vectors can be divided into two categories: viral and non-viral ones. Most gene therapy experiments and clinical trials have utilized viral vectors for a gene therapy [14]. Genetically engineered viruses are the most efficient, but a concern about immunological aspects as well as potential oncogenicity limit their use *in vivo* [15]. Therefore, non-viral delivery systems are increasingly investigated as an alternative. These systems mainly include DNA complexes with liposomes and with different cationic polymers. Complexes with liposomes are called lipoplexes, and with polymers – polyplexes. Polyplexes are created due to self-assembling based on electrostatic binding between the positively charged polymers and negatively charged phosphate groups of DNA. Generally these systems are less efficient than viral ones. It encourages to look for other synthetic carriers to overcome this problem. Dendrimers with positively charged terminal groups can also bind to DNA and form complexes, which by analogy to lipoplexes and polyplexes are called dendriplexes. This class of polymers has structural advantages for gene transfer. First, dendrimers possess many terminal groups that can be ionized, it means that they can efficiently bind large amount of genetic material. Second, dendrimers of large generations can mimic, both in size and shape, histone proteins which condense and store DNA within an eukaryotic cell nuclei.

During last ten years many reports appeared about successful uses of dendrimers as transfection agents, both *in vitro* and *in vivo*.

Haensler and Szoka first published that luciferase and β-galactosidase containing plasmids were transfected into diverse mammalian cultured cells using polyamidoamine dendrimers [16]. The ability of different generations of polyamidoamine dendrimers to transport complexed DNA into cells was evaluated. High-efficiency transfection was achieved for many cell lines including primary human fibroblasts that are difficult to transfect using other techniques [17].

Studies on polyamidoamine dendrimers as transfection agents were not narrowed to *in vitro* experiments. In some cases studies were broaden to *ex vivo* or *in vivo* level. In *in vivo* systems the gene is directly introduced into the body, whereas in *ex vivo* systems cells are first harvested from the body and after genetic manipulation they return to the body.

Hudde and co-workers extended studies on dendrimers as gene transfer vectors to *ex vivo* culture [18]. They investigated the efficiency of activated polyamidoamine dendrimers to transfect corneas. On the inner surface of the cornea there is a cell monolayer that is critical in the maintenance of normal visual acuity. Damage to these cells results in blindness and is an indication for a transplantation of a cornea. A corneal transplantation is generally successful in most cases, however in some cases endothelial graft rejection occurs. There are attempts to use gene-based approaches to modulate an immune response against the corneal allograft. Viral vectors are not suitable in this case due to immunogenic problems, especially in transplant setting. Dendrimers did not cause these problems. Moreover, serum and antibiotics did not interfere with transfection efficiency so cells and tissues did not need to be deprived of any important component of their culture medium.

Wang and co-workers transferred DNA into murine cardiac transplants *ex vivo* [19]. They used the 5th generation of polyamidoamine dendrimers. Gene transfer into heart is important for treatment of several cardiovascular diseases like coronary disease, heart failure and thrombosis. It also holds therapeutic potential in decreasing graft rejections by modulating immune response. To achieve a therapeutic aim, efficient delivery of genetic material into myocytes and sustaining high levels of expression of the transferred gene are required. It was shown that DNA could be transferred to cardiac myocytes when dendriplexes were administered by coronary infusion during the donor heart harvest. Pretreatment with serotonin and prolonged incubation time enhanced gene transfer and gene expression in cardiac transplants.

This is not the only article which shows dendrimer's potency in gene transfer into cardiac systems. The same group demonstrated that polyamidoamine dendrimers (generation 5th) significantly increased plasmid-mediated gene transfer efficiency and expression in a murine, nonvascularized cardiac transplantation models [20]. Dendrimers prolonged allograft survival.

Polyamidoamine dendrimers (generation 5th, 7th and 9th) were successfully used for DNA delivery in solid phase transfection *in vitro* and *in vivo* [21]. These studies have big therapeutic relevance, as a direct topical transfection of DNA into skin cells seems to be an attractive strategy for gene therapy of many human diseases [22]. Such approach requires a carrier which enables prolonged contact and enhances skin permeability. In contrast to other carriers, dendriplexes retained the ability to transfect even after drying. It was checked *in vitro* by transfection of cultured cells with dendriplexes coated on the surface of membranes of poly(DL-lactide-co-glycolide). Desiccated complexes dissociated from the membranes and transfected surrounding cells. The presence of exogenous collagenase was necessary to prolong release of dendriplexes *in vitro.* Dendrimers ability to serve as carriers for skin cell transfection was confirmed by *in vivo* experiments. Topical application of the dendrimer/DNA containing membranes onto denuded skin of hairless mice resulted in detectable expression of the transgenic chloramphenicol transacetylase.

Dendrimers seem to be a promising strategy for gene delivery in anticancer therapy. Gene transfer is an attractive approach to fight cancer, especially because chemotherapy and radiation have a limited effect. Tumor invasion can be inhibited by controlling of angiogenesis [23].

Polyamidoamine dendrimers were associated with 36-mer anionic oligomers for delivering angiostatin and tissue inhibitor of metalloproteinase (TIMP-2) genes [24]. First, dendriplexes capacity to promote gene transfer into breast cancer and endothelial cells was checked *in vitro* using plasmid coding for a green fluorescence protein. It allowed to establish the best conditions for a gene therapy. Next, *in vitro* gene transfer efficiency of angiostatin and TIMP-2 to endothelial and cancer cells was analyzed. Gene transfer markedly reduced proliferation of endothelial cells, wound healing reparation for endothelial and cancer cells and formation of capillary tube. Finally, gene transfer was tested *in vivo* on mice. The results were very encouraging because primary tumor growth was inhibited dramatically and decreased vascularization within tumors was observed.

The use of dendrimers as gene delivery agents has been largely focused on polyamidoamine dendrimers but there are several studies on polypropylenimine dendrimers too.

It was shown that large molecules (generation 5th) had poor transfection efficiency [25]. Moreover, high generations were too toxic to consider them for *in vivo* use [26]. However, Zinselmeyer and co-workers proved that low generations could be considered for using them as transfection agents [27]. An ethidium bromide intercalation assay was employed to study the level of DNA condensation. DNA was fully condensed by higher generations of polypropylenimine dendrimers (generation 3rd – generation 5th) and only partially condensed by lower generations (1st and 2nd). It is interesting that complete DNA condensation was not necessary for efficient gene transfer. Surprisingly, lower generations were more efficient and what is even more important they were non-toxic. It is a big difference between polyamidoamine dendrimers, for which optimum gene transfer was obtained for molecules with molecular weight around 20000 Da, whereas for polypropylenimine dendrimers the same effect was achieved for small compounds (around 800 Da).

Recently it has been shown that gene therapy offers not only a possibility to insert DNA into a cell to express a protein, but also allows disrupting the expression of disease-related genes. There are two approaches to the disruption: on the translational level or on the transcriptional level. In the first case, designed antisense oligonucleotides (ODNs) specifically bind to short complementary sequence of target mRNA to prevent the translation of the target gene [28]. In the second case, transcription is disrupted by the binding of a triplex-forming oligonucleotide at the promoter region of a target gene [29]. ODNs require an effective delivery system, as naked ODNs are poorly transported across cell membranes by endocyto mechanistic and are rapidly destroyed by cellular nucleases. Dendrimers have been utilized as carriers for ODNs in both strategies.

Studies performed by Bielinska and co-workers showed that polyamidoamine dendrimers could be used as transfection agents for the delivery of antisense nucleotides and plasmid expression vectors coding antisense mRNA [30]. They achieved inhibition of targeted gene expression with picomolar concentrations of specific oligonucleotides. Dendrimers were not cytotoxic in concentrations required for gene transfer. Moreover, oligonucleotides appeared to have the increased stability when complexed to dendrimers.

Hollins and co-workers evaluated the potential of low-generation (2nd and 3rd) polypropylenimine dendrimers to cellular delivery of antisense oligonucleotides targeted to the epidermal growth factor receptor in epidermoid carcinoma cells [31]. The receptor plays a central role in the initiation and development of breast, brain and lung tumors [32]. The tested dendrimers turned out to be good delivery systems for antisense ODNs. Targeted gene expression was inhibited, and consequently cell growth was inhibited. For both generations of dendrimers, antisense oligonucleotide uptake increased about ten times. Not surprisingly, it was found that the 2nd generation was less toxic than the 3rd one.

Santhakumaram and co-workers investigated the efficiency of five generations of polypropylenimine dendrimers (1st-5th) to deliver a 31 nt triplex-forming oligonucleotide targeted to the c-*myc* oncogene in breast, prostate and ovarian cancer cell lines [33]. C-*myc* is involved in cell proliferation. Their results showed that dendrimers were capable of facilitating the uptake of the ODNs in several cancer cell lines. In the case of breast cancer cells dendrimers increased the uptake by 14-fold, compared to control ODNs. Dendrimers had not significant effect on cell viability in a used concentration range. Dendrimers' efficiency depended on the generation number and was the highest for the 4th generation. It is interesting that in case of Zinselmeyer and co workers' studies, who worked with the same type of dendrimers, the optimal generation for complexing a long DNA molecule was the 2nd one [27]. It means that achieving the sufficient binding sites depends on the size of transfected material.

Studies to optimize gene transfer by dendrimers were carried out. Transfection efficiency of polyamidoamine dendrimers was significantly increased by heat treatment of dendrimers in solvents such as water or butanol. As a result of solvolysis some branches of the molecule are trimmed away. This loss is accompanied by an increase in flexibility but the size and shape of the molecule remain the same. Such activated molecules are termed fractured or degraded ones. Tang and co-workers observed that fractured dendrimers, compared to intact ones, exhibited much higher transfection activity [34]. However, there was an optimum level of degradation beyond which dendrimers started to be less efficient due to the lack of charge density to complex DNA. Nowadays, fractured polyamidoamine dendrimers are commercially available and known as SuperFect. Fractured dendrimers were employed in several studies.

Efficient arterial gene transfer can be a beneficial tool for a treatment of vascular diseases. Turunen and co-workers checked efficiency of different liposomes and cationic polymers for *in vitro* gene transfer to smooth muscle cells and endothelial cells to evaluate the most promising vesicles for *in vivo* gene delivery to rabbit carotid artery [35]. Fractured polyamidoamine dendrimers (generation 6th) were chosen for *in vivo* studies and in *in vivo* experiments they showed the highest gene transfer efficiency.

Ohashi and co-workers used degraded polyamidoamine dendrimers (SuperFect) for gene delivery into chondrocytes [36]. Gene delivery into chondrocytes seems to be an interesting approach to cartilage repair and prevent cartilage degeneration, especially because it is difficult to achieve a satisfactory concentration of therapeutic proteins in joints when they are administered in a conventional way. When genes are delivered to chrondrocytes, it results in in-situ expression of the protein. SuperFect was an efficient system for this purpose and it did not induce cytotoxic effects.

There were also attempts to use fractured polyamidoamine dendrimers to mediate gene transfer to the lung. It is very important in a context of treatment of chronic lung diseases such as cystic fibrosis. Unfortunately, in the last case dendrimers turned out not to be efficient enough and branched polyethylenimine was found to be better for pulmonary gene delivery *in vivo* [37].

The main disadvantage of using dendrimers as nonviral transfection agents is a long multi-step synthesis which is required to obtain a dendrimer of a size comparable to histone proteins. However, recently an interesting new method was developed that allowed converting hyperbranched polyglycerols into fully branched dendritic polymers in only two synthetic steps [38]. The only drawback was that the product was not monodisperse, but this problem could be reduced by using membrane-filtered hyperbranched polyglycerols. Simple synthetic route allowed to study a broad range of dendritic polymers and to estimate the influence of a size and a degree of branching on efficiency of DNA transfection in several adherent cell lines. It was found that maximum transfection efficiency was reached when a degree of branching was around 60% and a molecular weight was in the range of 6000-25000 Da. It is interesting that using higher degree of branching resulted in a decrease of gene transfection efficiency. This is the similar situation as for fractured dendrimers i.e. the flexibility is a very important factor.

Luo and co-workers synthesized polyamidoamine dendrimers (generation 5th) modified with 3400 molecular weight polyethylene glycol chains (PEG) [39]. Their aim was to create a cheaper dendrimer that could mimic some essential properties, like flexibility, of a fractured high generation PAMAM dendrimer. The novel conjugate produced a 20-fold increase of transfection efficiency when compared with partially degraded dendrimers. Probably PEG chains not only added flexibility but also accommodated DNA helices without creating strong electrostatic interactions, thus intracellular release of DNA molecules was more efficient. Moreover, the cytotoxicity of PEGylated dendrimers was lower due to the reduction of dendrimer generation and addition of biocompatible PEG branches. PEG chains have been previously coupled to polycationic polymers to improve their biocompatibility [40].

To improve the activity of gene transfer, Choi and co-workers synthesized L-arginine-grafted polyamidoamine dendrimers, which were composed of backbones of polyamidoamine dendrimers and their surfaces were covered with basic L-arginine residues [41]. Gene delivery potency greatly increased in comparison with that of unmodified polyamidoamine dendrimers. The transfection efficiency was higher than for polyamidoamine dendrimers modified with L-lysine. The level of the cytotoxicity was relatively low, and much lower than for polypropylenimine dendrimers.

Dendrimers can be not only modified by attaching a layer of amino acid residues, they can also be synthesized from branched amino acids such as lysines. Water-soluble, amphipathic polypeptide molecules were synthesized. These polypeptides lack the symmetrical architecture of dendrimers, but due to high-branching and positive charge they were also considered as transfection agents [42, 43]. Their efficiency of transfection suggests that they can be the next suitable systems for gene delivery *in vivo*. It is important that when polypeptides were complexed with DNA their cytotoxicity was low.

In case of polylysine dendrimers, not all studies show that branched dendritic polylysines are better than their linear analogues, e.g. in some cases the dendritic shape seemed to be unfavorable for the cell uptake, which reduced the transfection efficiency [44].

The efficiency of polyamidoamine dendrimer-mediated DNA transfer can be improved by the addition of substituted β-cyclodextrins [45]. Cyclodextrins are cyclic oligosaccharides containing a hydrophobic interior and hydrophilic outer surface that have been used to enhance stability, solubility and bioavailability of drugs. They have previously been applied to oligonucleotides delivery [46], that is why, one could expect the synergistic effect of dendrimers and cyclodextrins. The addition of β-cyclodextrins to dendrimer-DNA complexes improved a distribution of complexes within aqueous solutions. Moreover, *in vitro* chloramphenicol transacetylase expression increased approximately 200-fold, when multipolymer formulations including polyamidoamine dendrimers, β-cyclodextrins and plasmid DNA were applied on the surface of collagen membranes. It means that the system can be suitable for solid support gene delivery.

Also Arima and co-workers prepared polyamidoamine dendrimers (2nd generation) conjugated with cyclodextrins [47]. It was shown that modified dendrimers created complexes with plasmid DNA, which were protected from degradation by DNase I. Transfection activity increased 100 times when compared to an unmodified dendrimer. Although the conjugation improved gene transfer activity comparing to using only a dendrimer, still it was not a sufficient level and the ability to reach a target was too low. To improve these defects they conjugated cyclodextrins with mannosylated dendrimers with various degrees of substitution of mannose [48]. It was shown that conjugates formed complexes with plasmid DNA, but not in a case when many mannose residues were covalently bound to a dendrimer. It means that some number of positively charged primary amino groups on the surface of a dendrimer is necessary to create a complex with DNA. The improvement of gene transfer was observed for cells which recognized the mannose moiety. It can be an interesting solution for cell-specific gene transfer to human lung adenocarcinoma cells and to immunological cells which possess mannose receptors.

Interactions between Dendrimers and Gene Material and Intracellular Pathway of Dendriplexes

Although there are many articles which describe successful transfections with dendrimers as synthetic carriers, some issues related to understanding the delivery process still remain unresolved. In order to understand DNA-dendrimer interactions it is necessary to study them on a molecular level.

Ottaviani and co-workers focused on the initial step in the gene transfection process, i.e., the interactions of earlier (2nd) and later (6th) generations of polyamidoamine dendrimers with polynucleotides [49]. Several types of polynucleotides were employed e.g. calf thymus DNA and a short sequence of 12 base pairs. Interactions were studied by CW EPR spectroscopy by labeling dendrimers with nitroxide radicals. They expected that dendrimers would interact electrostatically with DNA bases, thus the extent of interactions should depend on the degree of protonation of surface amino groups. When pH decreases, the percent of protonated amino groups increases, that is why experiments were done in pH 8 and 5.5. From these studies two main conclusions could be drawn: DNA interacted stronger with larger dendrimers because it was able to 'wrap' around them. Interactions were stronger for more protonated dendrimers due to the electrostatic interactions with DNA base pairs.

When DNA is mixed with dendrimers, complexes are formed immediately by self-assembling. Disruption of the complexes requires strong ionic detergents, such as SDS [17]. Primary and tertiary amino groups of dendrimers, which are partially ionized at physiological pH, must provide electrostatic attraction to negatively charged phosphate groups of DNA. Generally, the efficiency of transfection is dependent on the ratio N/P, which is the ratio of concentrations of the total nitrogen atoms of dendrimer (N) to phosphate groups of DNA (P) [25]. Complexes of dendrimers with polyanions containing equal amount of cationic and anionic groups, in which almost all cationic and anionic groups are ion-paired, are insoluble in water. Water soluble nonstoichiometric complexes can be obtained if a dendrimer is a deficient component of the complex [50]. Different groups of researchers prepared complexes characterized by different N/P ratios, which can partially explain why obtained results are not always coherent. Usually dendrimers are mixed with DNA at a proportion providing excessive amino nitrogens to phosphates. It has been reported that for correct N/P ratio, the size of the complex is small and aggregation between complexes is rare [51].

To determine if any permanent alteration in DNA occurred as a result of interactions with dendrimers, SDS was added to formed complexes [52]. This treatment disrupted them and electrophoretical mobility of DNA was analyzed. It revealed that DNA was intact and behaved identically as never complexed plasmid.

There are parallels between dendrimers and naturally occurred cationic materials such as spermine and spermidine in bacteria [53] and particularly histones, proteins encountered in eukaryotic cell nuclei, which are known to complex and condense DNA. X-ray crystallography has shown that DNA is wrapped around the histone octamers like a thread around a spool [54]. The net charge of the complex is negative because negative charge of the wrapped DNA is larger than the total positive charge of the histone protein octamer [55]. Undoubtedly the fact that polyamidoamine dendrimers (generations 7th-10th) have similar shape and size as histones explains why they create very stable complexes with DNA. These histone-mimic compounds are more efficient transfection agents than lower generations (1st-5th) [56]. Kabanov and coworkers studied interactions between two kinds of native DNA (from salmon sperm and from bacteriophage T4dC) and polypropylenimine dendrimers (generations 1st-5th) [57]. Using fluorescence microscopy they observed that for all generations dendrimers induced DNA transitions from an extended coil to a compact conformation. They distinguished two types of interactions dependent on dendrimers' generation. Complexes for generations 1st and 2nd were electroneutral, whereas generations 4th and 5th created positively charged complexes. It means that part of amino groups were unpaired. CD spectroscopic measurements performed for DNA in the presence of dendrimers (generations 4th and 5th) allowed to obtain spectra characteristic of a highly wound DNA double helix. Consistent results were obtained when dendronized polystyrene polymers (generations 2nd, 3rd and 4th) were adsorbed on negatively charged mica, then DNA was added and complexes were visualized by scanning force microscopy [58]. The structural model for the complex was proposed, according to which DNA was wrapped around charged polymers.

The overall charge of the complex was dependent on the generation creating the complex. Within the complex, it was possible to distinguish the rigid cylinder with dendronized polymer and semiflexible chain of DNA. Similar model was proposed by Chen and co-workers [59]. They postulated that there were two regions of DNA within the complex: a tightly bound region and a linker region which was more flexible (Figure 3). Their model was based on experiments with ethidium bromide (EB). EB is a planar aromatic dye which easily intercalates between a base pair of DNA and emits fluorescence. It is a very convenient fluorescent probe of DNA conformation changes and it is widely used to study binding properties of different compounds to DNA through competition studies. Using EB as a fluorescent probe, Chen and co-workers characterized binding between DNA and dendrimers (generations 2nd, 4th and 7th). Interactions between dendrimers and DNA and creating complexes reduced the number of ethidium bromide binding sites, thus fluorescence decreased. Complexes with the 7th generation had the largest fraction of tightly bound DNA regions. It means that highest generations created more compact complexes. It explains why these dendrimers are more effective as gene carriers.

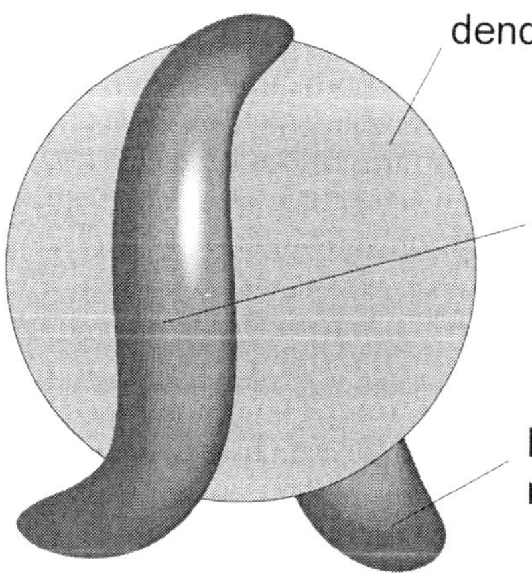

Figure 3. The structure of a DNA/dendrimer complex [59]

Evans and coworkers studied the structure of DNA/dendrimer complexes by synchrotron X-ray diffraction methods [60]. They used polypropylenimine dendrimers (generations 4th and 5th, with hydrodynamic radius 1.6 and 2 nm, respectively). It turned out that both dendrimers were too small to act as model histones. Created complexes were columnar mesophares consisting of arrays of DNA rods intercalated with dendrimers.

Tang and Szoka compared complexes between DNA and two types of polyamidoamine dendrimers: intact and fractured ones, using electron microscopy [61]. They observed that fractured dendrimers created single, non-aggregated complexes with DNA. Their diameters were measured by dynamic light scattering and they ranged from 90 to 130 nm. On a contrary, DNA complexes with intact dendrimers had a form of aggregated clusters with diameters larger than 1000 nm. Creating aggregates can be a reason why intact dendrimers are less effective in gene transfer, as there is a correlation between the level of transfection and the size of complexes [62].

Not only creating stable complexes between DNA and dendrimers, but also the protection of DNA from *in vivo* degradation is an important issue in gene delivery. Abdelhady and co-workers applied atomic force microscopy to visualize at the molecular scale and in real time, the effect of DNase I on dendriplexes built from polyamidoamine dendrimers (generation 4th) [63]. DNase I is a pancreatic endonuclease, which catalyses the hydrolysis of double-stranded DNA [64]. The presence of dendrimers provided a degree of protection to DNA. This protection is related to the structural morphology of the formed complex. It is important that DNA/dendrimer complexes remained highly soluble which indicates that nuclease resistance can be achieved without forming insoluble complexes. The transfection efficiency of nonviral systems is often reduced in the presence of serum [65], but dendrimers are efficient in delivering oligonucleotides even in the presence of high concentrations of serum proteins [20]. This is another proof that dendrimers protect oligonucleotides from degradation by exonuclease.

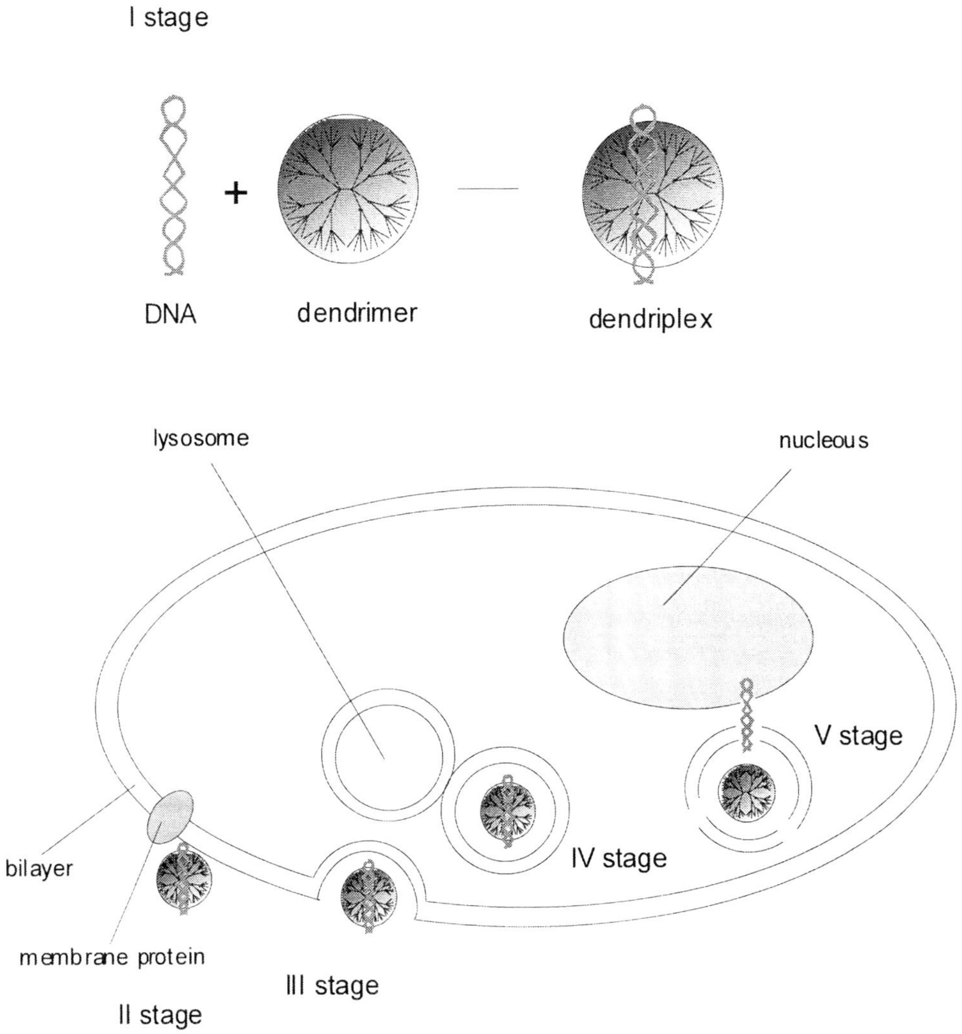

Figure 4. Intracellular pathway of a dendriplex. Stage I – creating electrostatic complexes between a dendrimer and DNA; stage II – interactions between a dendriplex and lipid bilayer and its receptors; stage III – endocytotic uptake; stage IV – fusion between an endosome and a lysosome; stage V – release of DNA into cytosol.

Usually, gene transfer by polyplexes (among them also dendriplexes) is less efficient *in vivo* than *in vitro* due to interactions with different compounds. Ruponen and co-workers postulated that in case of polyplexes it can be a consequence of extracellular interactions between complexes and anionic glycosaminoglycans [66]. Negatively charged glycosaminoglycans (e.g. hyaluronic acid) are long, linear polysacharides and important constituents in extracellular matrices [67]. Since every gene carrier has to move through extracellular space, glycosaminoglycans can bind to positively charged polyplexes. It can result in a release of DNA from complexes or affect the cellular uptake and distribution in tissues. Using ethidium bromide intercalation assay allowed for revealing that hyaluronic acid did not release or relax DNA in any complexes (also in dendriplexes); however, the transfection was partially inhibited [68].

After dendriplexes are created, still multiple steps are required for a successful transfection, among them there are: cellular uptake of the complex, release of DNA, and DNA transfer into nucleus. There are attempts to figure out the pathway of cellular uptake and intracellular trafficking of dendrimer/DNA complexes (Figure 4).

Generally, it is believed that positively charged polyplexes bind to negative charges of proteoglycans on a cell surface and then they are endocytosed by a cell [69]. Dendriplexes also cannot penetrate through lipid membranes and require active transport mechanisms for uptake into the cell. The uptake occurs through energy-dependent endocytosis. It was proved by preincubation of cells with inhibitors of cellular metabolism (5 mM sodium azide) or with specific inhibitors of endocytosis (deoxyglucose and cytocholasin B). In both cases the uptake of DNA/dendrimer complexes was inhibited [17].

Usually after endocytosis endocytosed material ends up in lysosomes. Lysosomes contain about forty types of hydrolytic enzymes. For optimal activity they require an acid environment and lysosomes provide this by maintaining pH around 5 in their interior [70]. Dendrimers buffer the lysosome after it is fused with the endosome, leading to pH inhibition of lysosomal nucleases and providing stability of dendriplexes. It protects DNA from lysosomal degradation and allows the transport of intact DNA to the nucleus [18].

After uptake, the genetic material is still separated from the cytoplasm. Once within the cell, the dendriplex or at least DNA has to escape into the cytosol before trafficking to the nucleus. Therefore, the last step requires the transport of genetic material across endosomal membrane. Two processes might be considered: transfer through membrane fusion or transfer across vesicular membrane by vesicle disruption. It was found that gene transfer with dendrimers as carriers can be 10-fold enhanced by endosomolytic agents, such as peptides or glycerol [71]. It is not big enhancement when comparing with 1000-fold for polylysine-mediated transfer. Thus, it is the next dendrimers' advantage that they provide efficient gene transfer without the need for agents facilitating endosomal escape. It is a consequence of the fact that the electrostatic bonding between oligonucleotides and dendrimers is weakened in an acidic environment in the lysosome, which allows migration of intact DNA to the nucleus.

The first who proposed that dendrimers with surface and internal amino groups could act as endosomal buffering agents and cause the vesicle rapture and the release of the complex to the cytoplasm were Haensler and Szoka [16]. They postulated that dendrimers might act as a 'proton sponge' leading to osmotic lysis of endosomes. Previously it was reported that lysosomotropic agents induced osmotic swelling and lysis in response to the pH drop in endosomal compartment [72].

Moreover, processes which occur in endosomes can also facilitate disruption of dendriplexes (Figure 5). Dendrimers at neutral pH adopt a fully extended conformation due to electrostatic repulsion between protonated primary amino groups presented on the surface. When these terminal groups are charge neutralized by complexation with DNA, a dendrimer collapses into a compact form. In an endosomal compartment pH drops, thus tertiary amines presented in the dendrimer's interior become protonated. There is no enough DNA to neutralize dendrimer's positive charge. It means that less amount of a dendrimer is required to maintain charge neutralization of DNA, thus excess dendrimer is released from the complex. No longer charge-neutralized by DNA dendrimer swells again and causes endosomal rapture which allows the release of DNA into the cytoplasm [34].

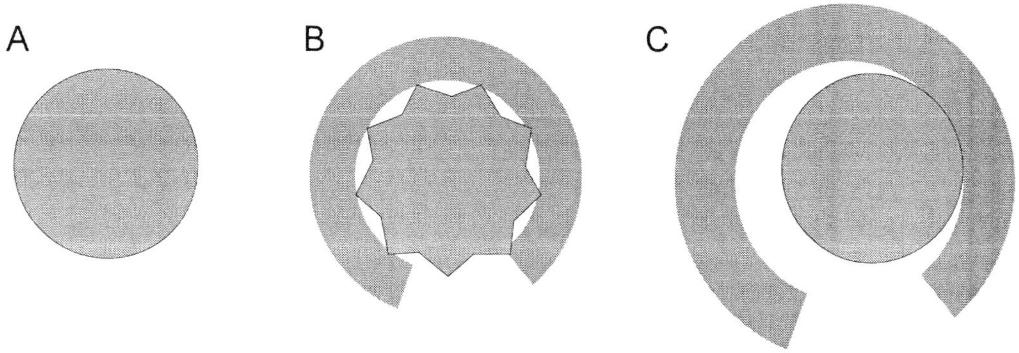

Figure 5. Illustration of 'proton sponge' hypothesis. A – at neutral pH a dendrimer adopts fully extended conformation; B – upon interactions with DNA a dendrimer collapses into a compact form; C – as pH drops (in endosomes) a dendrimer swells and DNA is released.

According to the described above hypothesis, it seems that although content of amino groups, net charge, size and molecular weight of dendrimers are important characteristics; a key parameter influencing transfection efficiency is dendrimer's flexibility. It can explain why fractured dendrimers are more efficient. Intact dendrimers cannot swell or collapse in the same extent as degraded dendrimers because these processes are sterically restricted. Removing some of dendrimer branches diminishes the restriction. This conclusion is supported by Gebhart and Kabanov studies. They evaluated the broad group of polyplexes with the respect to their effectiveness as transfection agents [25]. Dendriplexes were represented by fractured polyamidoamine dendrimer (SuperFect) and by polypropylenimine dendrimer (Astramol). The luciferase encoding plasmid was used as a model DNA in these studies. It turned out that SuperFect was one of the most active transfection agents in practically all cell lines. In contrast, Astramol belonged to the group which was significantly less active. Both types of dendrimers are weak bases having ionizable nitrogen atoms that can be protonated. The difference between them is their structure and the level of flexibility.

INTERACTIONS BETWEEN DENDRIMERS AND A LIPID BILAYER

Reports that dendrimers are efficient gene transfer vectors in a wide range of mammalian cells raised motivation for basic scientific investigations of interactions between dendrimers and model lipid bilayer. Membranes are important components on the pathway of the transfection process. First, these studies help to rationalize the way dendrimers enter a cell; second, at some point in transfection process an endosomal membrane must be breached to release DNA material. Interactions between dendrimers and membranes hold a great interest for the field of dendrimer-mediated drug delivery, as dendrimers assist in the passage of a drug through cell membrane. Curved bilayers and multilayers, termed vesicles or liposomes, are widely employed as models of biological membranes.

Ottaviani and co-workers studied interactions between polyamidoamine dendrimers and dimyristoylphosphatidylcholine (DMPC) multilamellar vesicles (MLVs) [73]. They selected for studies 2nd and 6th generations as representatives of small and large molecules. They also investigated the interactions as a function of the level of protonation of amino groups on a dendrimer's surface, by changing pH. To characterize liposome-dendrimer mixtures and size of compounds they employed dynamic light scattering and negative-staining transmission microscopy. EPR techniques with dendrimers which were labeled with nitroxide radical (TEMPO) were used to achieve information on the structures existing in the solution. It allowed monitoring the localization of dendrimers. It was found that dendrimers interacted with membrane surface but did not significantly perturbed membrane properties. In fact, dendrimers were not permanently attached to the liposome surface and they did not form any stable assembly. However, protonated dendrimers showed more effective dendrimer-vesicle interactions and the larger generation created more disturbances. The studies were continued and they were focused on large polyamidoamine dendrimers (generation 7th) [74]. Similarly like previously, by changing pH from 8.5 to 4.5 unprotonated and protonated dendrimers were obtained. This time spin probes to monitor membrane fluidity at different depths (5DXSA, 12DXSA and 16DXSA) were used. Generally, the obtained results were consistent with the previous ones. The bilayer structure was modified, but only partially perturbed by addition of dendrimers. In the presence of dendrimers, the bilayer mobility increased in the region close to the polar heads and remained essentially unchanged in the rest of the chain. Protonated dendrimers showed more effective interactions and they caused swelling of the vesicle structure. It is known that cell membranes are characterized by heterogeneity of their composition. To mimic better these conditions, mixed vesicles were produced [75]. They consisted of DMPC and its phosphatidylcholate salt – DMPA-Na. DMPA-Na is a negatively charged phospholipid. Two membrane spin probes were inserted into vesicles to monitor changes within bilayer. One of them located near the negatively charged headgroup layer of the vesicles (CAT16) and another in the lipid hydrophobic layer (5DSA). It was shown that dendrimers bound to the liposome surface due to interactions between phospholipid headgroups with dendrimers' amino groups. Not surprisingly, the biggest changes were observed for protonated dendrimers. These studies allowed to formulate a model of interactions between vesicles and dendrimers. According to the model, the interactions are different with respect to the generation number, for large dendrimers a lipid bilayer bends to wrap around the dendrimer in such a way that direct interactions between the charged groups of the dendrimer and the vesicle are prevented, whereas small and flexible dendrimers directly interact with the vesicle, not causing such disturbances in the bilayer shape (Figure 6).

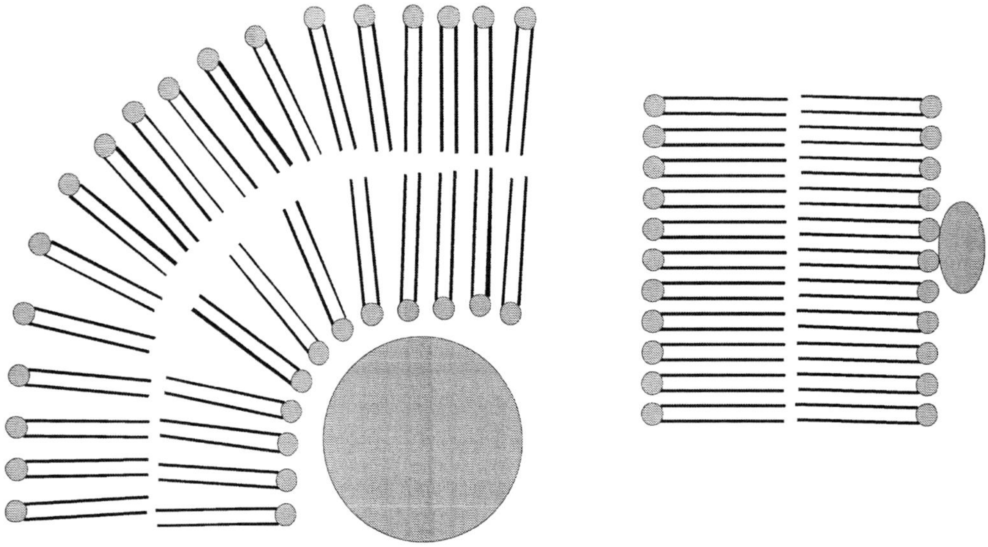

Figure 6. Model of interactions of large and small dendrimers with a lipid bilayer [75]

Zhang and Smith also chose anionic lipid vesicles as a model of a lipid bilayer, but they worked with large unilamellar vesicles (LUVs) [76]. They checked the impact of several generations of polyamidoamine dendrimers (from 4th to 7th) on LUVs stability. Several different experimental approaches were employed, such as: leakage, lipid mixing and content mixing assays. Additionally ^{31}P NMR spectroscopy was used to monitor changes in membrane morphology, and membrane fluidity was evaluated by using fluorescent probe – pyrene. It was shown that all dendrimers were able to disrupt anionic vesicle membranes. The destruction ability increased with an increasing generation and was strongly dependent on membrane composition. Phosphatidyletanolamine (PE) vesicles were disrupted, but phosphatidylcholine (PC) vesicles and even mixed vesicles with a high fraction of PC were protected from disruption. It can explain some of the differences in transfection efficiency and cytotoxicity observed for different cells. Two dendrimers features seem to be crucial: a positive charge and a globular shape. ^{31}P NMR experiments showed that dendrimers of generation 7th acted like a near-spherical polycationic protein – cytochrome c. Moreover, also cationic but linear polylysine did not have ability to disrupt the membrane. Based on the experiments, Zhang and Smith formulated the hypothesis how dendrimers act. They excluded the option that dendrimers disrupt membrane due to a simple detergent effect. The outer surface of large dendrimers is too densely packed to be penetrated by phospholipids. They proposed that dendrimers induce membrane biding and create a local region of inverse curvature. Dendrimers facilitate lipid mixing by electrostatically stabilizing these inversely curved surfaces. It means that dendrimers can promote vesicle-cell fusion. It is not surprisingly that it happens for PE-containing liposomes as PE is a phospholipid that has a tendency to adopt an inverted hexagonal phase, that is why it has been commonly used to study membrane fusion in artificial membranes [77].

Karoonuthaisiri and co-workers reported on interactions of polyamidoamine dendrimers (generations from 1st to 8th) with PE vesicles by leakage assay [78]. They showed that lipid bilayers containing PE were susceptible to disruption upon dendrimers. Providing electrostatic interactions was not enough to cause membrane disruption, because phospatidylserine-containing liposomes did not leak when treated with dendrimers. Generally, their results are consistent with Zhang and Smith studies as both groups showed that dendrimers mediated both the contact between vesicles, and the creation of local hexagonal phase. Membranes containing PE are characterized by flexibility which allows them to wrap partially around the dendrimer surface. It is interesting that the disrupting activity decreased for high dendrimer concentrations probably due to a steric barrier to vesicle aggregation created by too many dendrimer molecules in the system.

To study interactions between dendrimers and a model bilayer in detail Mecke and co-workers used atomic force microscopy to directly image supported DMPC lipid bilayers upon exposure to polyamidoamine dendrimers [79]. The pictures showed that the 7th generation caused the formation of small (15-40 nm) holes in bilayer. The dendrimers of generation 5th did not create holes but they removed lipids from edges of existing holes. These results were confirmed by checking a release of cytosolic enzymes lactate dehydrogenase (LDH) and luciferase (Luc) from KB and Rat2 cells [80]. The observed increased permeability was consistent with the hole formation. It is important that used dendrimer concentrations did not cause cell death and membrane permeability returned to normal level after dendrimers were removed. Basing on these results Hong and co-workers built alternative to endocytosis hypothesis of dendrimer uptake, according to which dendrimers can enter the cell through the holes.

Confocal fluorescence microscopy was used to measure the binding of fluorescein-labelled polyamidoamine dendrimers to individual fibroblast cell surfaces [81]. As expected, the higher generations bound more tightly than the lower ones. There were more unbound dendrimers of generation 4th than unbound dendrimers of generation 6th in the system. It was suggested that larger dendrimers were able to interact at the same time with larger amount of negatively charged membrane compounds, whereas smaller dendrimers, because of size restrictions, could not. It is interesting that the binding varied significantly from cell to cell and the variation could not be explained on the basis of cell size. Thus, the variation must reflect nanometer-scale differences in membrane topology.

To obtain information relevant to a transfection process, interactions between membranes and dendriplexes were studied.

It was found out that the internalization of dendriplexes took place at preferential sites of the cell surface such as membrane rafts. Membrane rafts are microdomains of cholesterol and sphingolipids in the exoplasmic leaflet of the plasma membrane. It was already proved that these dynamic structures play a role in cellular uptake and endocytosis of other non-viral vectors, such as liposomes [82]. Recently it has been shown that when cholesterol is removed from the cells, the gene delivery by dendrimers is significantly reduced. The presence of cholesterol is especially important at the level of two important processes: binding and internalization of dendriplexes. Cholesterol may have a direct role in the binding due to interactions between its polar head and charged amino groups of the dendrimer, or dendrimers can associate with rafts because rafts are surrounded by the annular anionic phospholipids. It is interesting that the presence of cholesterol is also crucial for internalization of dendriplexes. It suggests that cholesterol is also involved in a signaling cascade that initiates endocytosis [83]. The presence of cholesterol in the endosomes can influence the endosomal escape by altering the lipid-dendrimer interaction or by changing membrane characteristics such as strength, rigidity or permeability. However, cholesterol removal after transfection did not inhibit expression of a delivered gene, indicating that the results are not due to a reduction in transcription or translation processes.

The aspect of interactions between dendriplexes and proteins is also important for the successful gene delivery. It was found that ODN/dendrimer uptake was mediated, at least in part, by binding to cell surface proteins. To assess the extent of protein-bound fraction, trypsin was used to strip off cell surface-bound ODN/dendrimer by digesting cell surface protein-binding sites [84]. It turned out that 70% of complexes were removed. It means that a large fraction of dendriplexes associates via trypsin-sensitive cell surface proteins.

General Studies on Dendrimers' Toxicity

During last ten years general studies on dendrimers cytotoxicity were done for several groups of polymers. Roberts and co-workers checked an influence of polyamidoamine dendrimers (generations 3rd, 5th and 7th) in *in vitro* (V79 cell line) and *in vivo* (Swiss-Webster mice) systems [85]. Dendrimers did not show properties that could preclude their use as therapeutic agents. The cytotoxicity increased with both concentration and the generation number. Potential biological complications were observed only for the highest concentration of polyamidoamine dendrimers of generation 7th. No evidence of immunogenicity was seen. Broader range of dendrimers was tested by Malik and co-workers [26]. To estimate the effect of the dendrimer generation and the type of surface end groups, they used polyamidoamine dendrimers terminated with amino groups (generations 1st, 2nd, 3rd, 4th), polyamidoamine dendrimers terminated with carboxylate groups (so called half-generations: 1.5, 2.5, 3.5, 5.5, 7.5, 9.5), polypropyleneimine dendrimers and polyethylene oxide (PEO) grafted carbosilane dendrimers. Dendrimers with carboxylate or PEO groups on the surface were neither hemolytic nor cytotoxic towards a panel of cell lines *in vitro*. The most toxic were dendrimers possessing amino groups. For high concentrations changes in red cell morphology were observed which led to hemolysis [26, 86]. The reduction of dendrimers' cytotoxicity can be obtained by a modification of their surface, which decreases the positive charge. It was done by conjugating lauroyl chains or polyethylene glycol (PEG) chains onto the surface of polyamidoamine dendrimers (generations 2nd, 3rd and 4th) [87, 88]. Although general toxic studies revealed some level of undesirable effects for cationic polymers, it is important to remember that in most therapeutic uses, it can be compensate by low concentrations of dendrimers necessary to use. It is very encouraging that even cationic dendrimers showed lower level of cytotoxicity than other polymers like polyethylenimine, poly L-lysine or dextran [89].

To summarize, dendrimers appear to be non-cytotoxic in the concentrations relevant for gene transfer. These polymers are not immunogenic or carcinogenic. Even though they are less efficient than viral vectors, they have a potential for use in a gene therapy due to their high biocompatibility. They are also promising materials for other therapeutic applications like drug delivery.

REFERENCES

[1] Tomalia, D. A., Baker, H., Dewald, J. R., Hall, M., Kallos, G., Martin, S., Roeck, J., Ryder, J. & Smith, P. (1985) A new class of polymers: starburst dendritic macromolecules. *Polym. J.*, 17, 117-132.

[2] Hodge, P. (1993) Polymer science branches out. *Nature*, 362, 18-19 .

[3] Tomalia, D. A., Naylor, A. M. & Goddard III, W. A. (1990) Starburst dendrimers: molecular-level control of size, shape, surface chemistry, topology, and flexibility from atoms to macroscopic matter. Angew. *Chem. Int. Ed.*, 29, 138-175.

[4] Weener, J.-W., van Dongen, J. L. J. & Meijer, E. W. (1999) Electrospray mass spectrometry studies of poly(propylene imine) dendrimers: Probing reactivity in the gas phase. *J. Am. Chem. Soc.*, 121, 10346-10355.

[5] Jansen, J. F. G. A., de Brabander van den Berg, E. M. M. & Meijer, E. W. (1994) Encapsulation of guest molecules into a dendritic box. *Science* 266, 1226-1229.

[6] Jansen, J .F. G. A. & Meijer, E. W. (1995) The dendritic box: Shape-selective liberation of encapsulated guests. *J. Am. Chem. Soc.* 117, 4417-4418.

[7] Archut, A., Azzellini, G. C., Balzani, V., Cola, L. D. & Vögtle, F. (1998) Toward photoswitchable dendritic hosts. Interaction between azobenzene-functionalized dendrimers and eosin. *J. Am. Chem. Soc.* 120, 12187-12191.

[8] Twyman, L. J., Beezer, A. E., Esfand, R., Hardy, M. J. & Mitchell, J. C. (1999) The synthesis of water soluble dendrimers, and their application as possible drug delivery systems. *Tetrahedron Lett.*, 40, 1743-1746.

[9] Kojima, C., Kono, K., Maruyama, K. & Takagishi, T. (2000) Synthesis of polyamidoamine dendrimers having poly(ethylene glycol) grafts and their ability to encapsulate anticancer drugs. *Bioconjug. Chem.* 11, 910-917.

[10] Roy, R., Zanini, D., Meunier, S. J., & Romanowska, A. (1993) Solid-phase synthesis of dendritic sialoside inhibitors of influenza A virus haemagglutinin. *J. Chem. Soc., Chem. Commun.* 1869-1872.

[11] Zanini, D. & Roy, R. (1998) Practical synthesis of Starburst PAMAM α-thiosialodendrimers for probing multivalent carbohydrate-lectin binding properties. *J. Org. Chem.* 63, 3486-3491.

[12] Roberts, J. C., Adams, Y. E., Tomalia, D., Mercer-Smith, J. A. & Lavallee, D. K. (1990) Using starburst dendrimers as linker molecules to radiolabel antibodies. *Bioconjug. Chem.*, 2, 305-308.

[13] Barth, R. F., Adams, D. M., Soloway, A. H., Alam, F. & Darby, M. V. (1994) Boronated starburst dendrimer-monoclonal antibody immunoconjugates: evaluation as a potential delivery system for neutron capture therapy. *Bioconjug. Chem.* 5, 58-66.

[14] Mulligan, R. C. (1993) The basic science of gene therapy. *Science*, 260, 926-932.

[15] Briand, P. & Kahn, A. (1993) Values and limits of adenoviral vectors for gene therapy *in vivo*. *Pathol. Biol.* (Paris), 41, 663-671.

[16] Haensler, J. & Szoka Jr., F. C. (1993) Polyamidoamine cascade polymers mediate efficient transfection of cells in culture. *Bioconjug. Chem.*, 4, 372-379 .

[17] Kukowska-Latallo, J. F., Bielinska, A. U., Johnson, J., Spindler, R., Tomalia, D. A. & Baker Jr., J. R. (1996) Efficient transfer of genetic material into mammalian cells using Starburst polyamidoamine dendrimers. *Proc. Natl. Acad. Sci*, 93, 4897-4902.

[18] Hudde, T., Rayner, S. A., Comer, R. M., Weber, M., Isaacs, J. D., Waldmann, H., Larkin, D. F. P. & George, A. J. T. (1999) Activated polyamidoamine dendrimers, a non-viral vector for gene transfer to the corneal endothelium. *Gen. Ther.*, 6, 939-943.
[19] Wang, Y., Boros, P., Liu, J., Qin, L., Bai, Y., Bielinska, A. U., Kukowska-Latallo, J. F., Baker Jr., J. R. & Bromberg, J. S. (2000) DNA/dendrimer complexes mediate gene transfer into murine cardiac transplants *ex vivo*. *Mol. Ther.*, 2, 602-608.
[20] Qin, L. H., Pahud, D. R., Ding, Y. Z., Bielinska, A. U., Kukowska-Latallo, J. F., Baker Jr., J. R. & Bromberg, J. S. (1998) Efficient transfer of genes into murine cardiac grafts by Starburst polyamidoamine dendrimers. *Hum. Gene Ther.*, 9, 553-560.
[21] Bielinska, A. U., Yen, A., Liang Wu, H., Zahos, K. M., Sun, R., Weiner, n. D., Baker Jr., J. R. & Roessler, B. J. (2000) Application of membrane-based dendrimer/DNA complexes for solid phase transfection *in vitro* and *in vivo*. *Biomaterials*, 21, 877-887.
[22] Trainer, A. H. & Alexander, M. Y. (1997) Gene delivery to the epidermis. *Hum. Mol. Gen.*, 6, 1761-1767.
[23] Hanahan, D. & Folkman, J. (1996) Patterns and emerging mechanisms of the angiogenic switch during tumorigenesis. *Cell*, 86, 353-364 .
[24] Vincent, L., Varet, J., Pille, J.-Y., Bompais, H., Opolon, P, Maksimenko, A., Malvy, C., Mirshahi, M., Lu, H., Vannier, J.-P., Soria, C. & Li, H. (2003) Efficacy of dendrimer-mediated angiostatin and TIMP-2 gene delivery on inhibition of tumor growth and angiogenesis: *in vitro* and *in vivo*. *Int. J. Cancer*, 105, 419-429.
[25] Gebhart, C. L. & Kabanov, A. V. (2001) Evaluation of polyplexes as gene transfer agents. *J. Control. Release*, 73, 401-406.
[26] Malik, N., Wiwattanapatapee, R., Klopsch, R., Lorenz, K., Frey, H., Weener, J.-W., Meijer, E. W., Paulus, W. & Duncan, R. (2000) Dendrimers: relationship between structure and biocompatibility in vitro, and preliminary studies on the biodistribution of ^{125}I-labelled polyamidoamine dendrimers in vivo. *J. Control. Release*, 65, 133-148.
[27] Zinselmeyer, B. H., Mackay, S. P., Schatzlein, A. G. & Uchegbu, I. F. (2002) The lower-generation polypropylenimine dendrimers are effective gene-transfer agents. *Pharm. Res.*, 19, 960-967.
[28] Braasch, D. A. & Corey, D. R. (2002) Novel antisense and peptide nucleic acid strategies for controlling gene expression. *Biochem.*, 41, 4503-4510 .
[29] Seidmann, M. M. & Glazer, P. M. (2003) The potential for gene repair via triple helix formation. *J. Clin. Invest.*, 112, 487-494.
[30] Bielinska, A., Kukowska-Latallo, J. F., Johnson, J., Tomalia, D. A. & Baker Jr., J. R. (1996) Regulation of *in vitro* gene expression using antisense oligonucleotides or antisense expression plasmids transfected using starburst PAMAM dendrimers. *Nucl. Acid. Res.*, 24, 2176-2182.
[31] Hollins, A. J., Benboubetra, M., Omidi, Y., Zinselmeyer, B. H., Schatzlein, A. G., Uchegbu, I., F. & Akhtar, S. (2004) Evaluation of generation 2 and 3 poly(propylenimine) dendrimers for potential cellular delivery of antisense oligonucleotides targeting the epidermal growth factor receptor. *Pharm. Res.*, 21, 458-466.
[32] Arteaga, C. L. (2002) Epidermal growth factor receptor dependence in human tumors: more than just expression? *Oncologist*, 7, 31-39 .

[33] Santhakumaram, L. M., Thomas, T. & Thomas, T. J. (2004) Enhanced cellular uptake of a triplex-forming oligonucleotide by nanoparticle formation in the presence of polypropylenimine dendrimers. *Nucleic Acids Res.*, 32, 2102-2112.

[34] Tang, M. X., Redemann, C. T. & Szoka Jr., F. C. (1996) *In vitro* gene delivery by degraded polyamidoamine dendrimers. *Bioconjug. Chem.*, 7, 703-714.

[35] Turunen, M. P., Hiltunen, M. O., Ruponen, M., Virkamäki, L, Szoka Jr., F. C., Urtii, A. & Ylä-Herttuala, S. (1999) Efficient adventitial gene delivery to rabbit carotid artery with cationic polymer-plasmid complexes. *Gen. Ther.*, 6, 6-11.

[36] Ohashi, S., Kubo, T., Ikeda, T., Arau, Y., Takahashi, K., Hirasawa, Y., Takigawa, M., Satoh, E., Imanishi, J & Mazda, O. (2001) Cationic polymer-mediated genetic transduction into cultured human chondrosarcoma-derived HCS-2/8 cells. *J. Orthop. Sci.*, 6, 75-81.

[37] Rudolph, C., Lausier, J., Nandorf, S., Müller, R. H. & Rosenecker, J. (2000) *In vivo* gene delivery to the lung using polyethylenimine and fractured polyamidoamine dendrimers. *J. Gene Med.*, 2, 269-278.

[38] Krämer, M, Stumbé, J.-F., Grimm, G., Kaufmann, B., Krüger, U., Weber, M. & Haag, R. (2004) Dendritic polyamines: Simple access to new materials with defined treelike structures for application in nonviral gene therapy. *ChemBioChem*, 5, 1081-1087.

[39] Luo, D., Haverstick, K., Belcheva, N., Han, E. & Saltzman, W. M. (2002) Poly(ethylene glycol)-conjugated PAMAM dendrimer for biocompatible high-efficiency DNA delivery. *Macromolecules*, 35, 3456-3462.

[40] Kataoka, K., Kwon, G. S., Yokoyama, M., Okano, T. & Sakurai, Y. (1993) Block copolymer micelles as vehicles for drug delivery. *J. Control. Release*, 24, 119-132.

[41] Choi, J. S., Nam, K., Park, J.-Y., Kim, J.-B., Lee, J.-K. & Park, J.-S. (2004) Enhanced transfection efficiency of PAMAM dendrimer by surface modification with L-arginine. *J. Controlled Release*, 99, 445-456.

[42] Sakthivel, T., Toth, I. & Florence, A. T. (1998) Synthesis and physicochemical properties of some lipidic polyamide dendrimers. *Pharm. Res.*, 15, 776-781.

[43] Shah, D. S., Sakthivel, T, Toth, I., Florence, A. T. & Wilderspin, A. F. (2000) DNA transfection and transfected cell viability using amphipathic asymmetric dendrimers. *Int. J. Pharm.*, 208, 41-48.

[44] Männistö, M., Vanderkerken, S., Toncheva, V., Elomaa, M., Ruponen, M., Schacht, E. & Urtti, A. (2002) Structure-activity relationships of poly(L-lysines): effect of pegylation and molecular shape on physicochemical and biological properties in gene delivery. *J. Control. Release*, 83, 169-182.

[45] Roessler, B. J., Bielinska, A. U., Janczak, K., Lee, I & Baker Jr., j. R. (2001) Substituted β-cyclodextrins interact with PAMAM dendrimer-DNA complexes and modify transfection efficiency. *Biochim. Biophys. Res. Commun.*, 283, 124-129.

[46] Addou, S., Collomb, J., Sallas, F., Marsura, A. & Finance, C. (1997) β-cyclodextrin derivatives as carriers to enhance the antiviral activity of an antisense oligonucleotide directed toward a coronavirus intergenetic consensus sequence. *Arch. Virol.*, 142, 1585-1602.

[47] Arima, H., Kihara, F., Hirayama, F. & Uekama, K. (2001) Enhancement of gene expression by polyamidoamine dendrimer conjugates with α-, β-, and γ-cyclodextrins. Bioconjug. *Chem.* 12, 476-484.

[48] Arima, H., Wada, K., Kihara, F., Tsutsumi, T., Hirayama, F. & Uekama, K. (2002) Cell-specific gene transfer by α-cyclodextrin conjugates with mannosylated polyamidoamine dendrimers. *J. Inclusion Phenomena Macrocyclic Chem.*, 44, 361-364.
[49] Ottaviani, M. F., Sacchi, B., Turro, n. J., Chen, W., Jockusch, S. & Tomalia, D. A. (1999) An EPR study of the interactions between starburst dendrimers and polynucleotides. *Macromolecules*, 32, 2275-2282.
[50] Kabanov, V. A., Zezin, A. B., Rogacheva, V. B., Gulyaeva, Zh. G., Zansochova, M. F, Joosten, J. G. H. & Brackman, J. (1999) Interaction of Astramol poly(propyleneimine) dendrimers with linear polyanions. *Macromolecules*, 32, 1904-1909.
[51] Boussif, O., Lezoualch, F., Zanta, M. A., Mergny, N. D., Scherman, D., Demeneix, B. & Behr, J. (1995) A versatile vector for gene and oligonucleotide transfer into cells in culture and *in vivo*: polyethylenimine. *Proc. Natl. Acad. Sci. USA*, 92, 7297-7301.
[52] Bielinska, A. U., Kukowska-Latallo, J. F. & Baker Jr., J. R. (1997) The interaction of plasmid DNA with polyamidoamine dendrimers: mechanism of complex formation and analysis of alterations induced in nuclease sensitivity and transcriptional activity of the complexed DNA. *Biochim. Biophys. Acta*, 1353, 180-190.
[53] Bloomfield, V. A. (1996) DNA condensation. *Curr. Opin. Struct. Biol.* 6, 334-341.
[54] Luger, K., Mäder, A. W., Richmond, R. K., Sargent, D. F., Richmond, T.J. (1997) Crystal structure of the nucleosome core particle at 2.8 Å resolution. *Nature*, 389, 251-260.
[55] Decher, G. (1997) Fuzzy nanoassemblies: toward layered polymeric multicomposites *Science*, 277, 1232-1237.
[56] Esfand, R. & Tomalia D. A. (2001) Poly(amidoamine) (PAMAM) dendrimers: from biomimicry to drug delivery and biomedical applications. *DDT*, 6, 427-435.
[57] Kabanov, V. A., Sergeyev, V. G, Pyshkina, O. A., Zinchenko, A. A., Zezin, A. B., M. F, Joosten, J. G. H., Brackman, J. & Yoshikawa, K. (2000) Interpolyelectrolyte complexes formed by DNA and Astramol poly(propyleneimine) dendrimers. *Macromolecules*, 33, 9587-9593.
[58] Gössl, I., Shu,L., Schlüter, D. & Rabe, J. P. (2002) Molecular structure of single DNA complexes with positively charged dendronized polymers. *J. Am. Chem. Soc.*, 124, 6860-6865.
[59] Chen, W., Turro, N. J. & Tomalia, D. A. (2000) Using ethidium bromide to probe the interactions between DNA and dendrimers. *Langmuir*, 16, 15-19 .
[60] Evans, h. M., Ahmad, A., Ewert, K., Pfohl, T., Martin-Herranz, A., Bruinsma, R. F. & Safinya, C. R. (2003) Structural polymorphism of DNA-dendrimer complexes. *Phys. Rev. Lett.*, 91, 755011-755014.
[61] Tang, M. X. & Szoka, F. C. (1997) The influence of polymer structure on the interactions of cationic polymers with DNA and morphology of the resulting complexes. *Gen. Ther.*, 4, 823-832.
[62] Wagner, E., Cotten, M., Foisner, R. & Birnstiel, M. L. (1991) Transferrin-polycation DNA complexes: the effect of polycations on the structure of the complex and DNA delivery to cells. *Proc. Natl. Acad. Sci. USA*, 88, 4255-4259.
[63] Abdelhadz, H. G., Allen, S., Davies, M. C., Roberts, C. J., Tendler, S. J. B. & Williams, P. M. (2003) Direct real-time molecular scale visualisation of the degradation of condensed DNA complexes exposed to DNase I. *Nucleic Acids Research*, 31, 4001-4005.

[64] Pan, C. Q., Ulmer, J. S., Herzka, A. & Lazarus, R. A. (1998) Mutational analysis of human DNase I at the DNA binding interface: implications for DNA recognition, catalysis and metal ion dependence. *Protein Sci.*, 7, 628-636.

[65] Lewis, J. G., Lin, K. Y., Kothavale, A., Flanagan, W. M, Matteucci, M. D., DePrince, R. B., Mook, R. A., Hendren, R. W. & Wagner, R. W. (1996) A serum-resistant cytofectin for cellular delivery of antisense oligodeoxynucleotides and plasmid DNA. *Proc. Natl. Acad. Sci. USA*, 93, 3176-3181 .

[66] Ruponen, M., Ylä-Herttuala, S. &Urtti, A. (1999) Interactions of polymeric and liposomal gene delivery systems with extracellular glycosaminoglycans: physicochemical and transfection studies. *Biochim. Biophys. Acta*, 1415, 331-341.

[67] Wight, T. N., Heinegård, D. K. & Hascall, V. C. (1991) *Proteoglycans: structure and function.* In E. D. Hay (Ed.), Cell biology of extracellular matrix (45-78) New York, Plenum Press.

[68] Ruponen, M, Honkakoski, P, Rönkkö, S, Pelkonen, J., Tammi, M. & Urtti, A. (2003) Extracellular and intracellular barriers in non-viral gene delivery. *J. Controlled Release*, 93, 213-217.

[69] Mislick, K. A. & Baldeschwieler, J. D. (1996) Evidence for the role of proteoglycans in cation-mediated gene transfer. *Proc. Natl. Acad. Sci. USA*, 93, 12349-12354.

[70] Alberts, B. Bray, D., Lewis, J. Raff, M, Roberts, K. & Watson, J. J (1994) *Molecular biology of the cell* (3rd edition) New York, Garland Publishing Inc.

[71] Wagner, E. (1998) Effects of membrane-active agents in gene delivery. *J. Control. Release*, 53, 1550158.

[72] DeDuve, C., DeBarsy, T., Poole, B., Trouet, A., Tulkens, P. & Van Hoof, F. (1974) Lysosomotropic agents. *Biochem. Pharmacol.*, 23, 2495-2531 .

[73] Ottaviani, M. F., Matteini, P., Brustolon, M., Turro, N. J., Jockusch, S. & Tomalia, D. A. (1998) Characterization of starburst dendrimers and vesicle solutions and their interactions by CW- and Pulsed-EPR, TEM, and dynamic light scattering. *J. Phys. Chem. B*, 102, 6029-6039.

[74] Ottaviani, M. F., Daddi, R., Brustolon, M., Turro, N. J. & Tomalia, D. A. (1999) Structural modifications of DMPC vesicles upon interaction with polyamidoamine dendrimers studied by CW-electron paramagnetic resonance and electron spin-echo techniques. *Langmuir*, 15, 1973-1980.

[75] Ottaviani, M. F., Favuzza, P., Sacchi, B., Turro, N. J., Jockusch, S. & Tomalia, D. A. (2002) Interactions between starburst dendrimers and mixed DMPC/DMPA-Na vesicles studied by spin label and spin probe techniques, supported by transmission electron microscopy. *Langmuir*, 18, 2347-2357.

[76] Zhang, Z.-Y. & Smith, B. D. (2000) High-generation polycationic dendrimers are unusually effective at disrupting anionic vesicles: membrane bending model. *Bioconjug. Chem.*, 11, 805-814.

[77] Düzgünes, N., Straubinger, R., Baldwin, P., Friend, D. & Papahadjopoulos, D. (1985) Proton-induced fusion of oleic acid-phosphatidylethanolamine liposomes. *Biochemistry*, 24, 3091-3098.

[78] Karoonuthaisiri, N., Titiyevskiy, K. & Thomas, J. L. (2003) Destabilization of fatty acid-containing liposomes by polyamidoamine dendrimers. Colloids Surfaces B: *Biointerfaces*, 27, 365-375.

[79] Mecke, A., Lee, I., Baker Jr., J. R., Banaszak Holl, M. M. & Orr, B. G (2004) Deformability of poly(amidoamine) dendrimers. *Eur. Phys. J. E.*, 14, 7-16.
[80] Hong, S., Bielinska, A. U., Mecke, A., Keszler, B., Beals, J., Shi, X., Balogh, L., Orr, b. G., Baker Jr., J. R. & Banaszak Holl, M. M. (2004) Interactions of poly(amidoamine) dendrimers with supported lipid bilayer and cells: Hole formation and the relation to transport. Boconjug. *Chem.*, 15, 774-782.
[81] Lai, J. C., Yuan, C. & Thomas, J. L. (2002) Single-cell measurements of polyamidoamine dendrimer binding. *Ann. Biomed. Eng.*, 30, 409-416.
[82] Zuhorn, I. S., Kalicharan, R. & Hoekstra, D. (2002) Lipoplex-mediated transfection of mammalian cells occurs through the cholesterol-dependent clathrin-mediated pathway of endocytosis. *J. Biol. Chem.*, 277, 18021-18028.
[83] Manunta, M., Hong Tan, P., Sagoo, P., Kashefi, K. & George A. J. T. (2004) Gene delivery by dendrimers operates via a cholesterol dependent pathway. *Nucleic Acids Research*, 32, 2730-2739.
[84] Hussain, M., Shchepinov, M. S., Sohail, M., Benter, I. F., Hollins, A. J., Southern, E. M. & Akhtar, S. (2004) A novel anionic dendrimer for improvement cellular delivery of antisense oligonucleotides. *J. Controlled Release*, 99, 139-155.
[85] Roberts, C. J., Bhalgat, M. K. & Zera, R. T. (1996) Preliminary biological evaluation of polyamindoamine (PAMAM) Starburst dendrimers. *J. Biomed. Mater. Res.*, 30, 53-65.
[86] Domanski, D. M., Klajnert, B. & Bryszewska, M. (2004) Influence of PAMAM dendrimers on human red blood cells. *Bioelectrochem.*, 63, 189-191.
[87] Jevprasesphant, R., Penny, J. Jalal, R., Attwood, D., McKeown, N. B. & D'Emanuele, A. (2003) The influence of surface modification on the cytotoxicity of PAMAM dendrimers. *Int. J. Pharm.*, 252, 263-266.
[88] Jevprasesphant, R., Penny, J., Attwood, D., McKeown, N. B. & D'Emanuele, A. (2003) Engineering of dendrimer surfaces to enhance transepithelial transport and reduce cytotoxicity. *Pharm. Res.*, 20, 1543-1550.
[89] Fischer, D., Li, Y., Ahlemeyer, B., Krieglstein, J. & Kissel, T. (2003) In vitro cytotoxicity testing of polycations: influence of polymer structure on cell viability and hemolysis. *Biomater.*, 24, 1121-1131.

In: New Developments in Mutation Research
Editor: Charles L. Valon, pp. 241-255
ISBN 1-59454-664-9

Chapter 10

DNA OXIDATION AND ALKYLATION IN DIABETES

*Yildiz Dincer**
Istanbul University Cerrahpasa Medical Faculty,
Department of Biochemistry, Istanbul, Turkey

ABSTRACT

Diabetes is a growing threat to public health worldwide. It is clasified into two main types. Type 1 diabetes is characterized by loss of insulin production due to autoimmune destruction of the pancreatic islet β cells. Type 2 diabetes is characterized by a deficiency in the capacity of insulin secretion and insulin resistance in peripheral tissues. Genetic factors seem to play an important role in the pathogenesis of diabetes. Several genes, which are involved in the metabolic process including the synthesis and secretion of insulin in pancreatic β cells and the insulin action on various target cells, have been proposed to be associated with or responsible for the pathogenesis of the disease. Some mutations in genes encoding for insulin, insulin receptor, glucokinase, glycogen synthase and suppressor of cytokine signalling 3 detected diabetic patients. Also mitochondrial DNA (mtDNA) has been suggested to be one of the genetic factors for diabetic development. mtDNA was shown to be required for the regulation of glucose-stimulated insulin secretion in a pancreatic β cell line of the mouse. Further studies which show a somatic mutation in mtDNA from peripheral blood cells in diabetic patients, have provided strong evidence to support the notion that mitochondrial function is crucial for the secretion of insulin by pancreatic cells. Not only mutational but also quantiative changes in mtDNA have been observed in diabetic patients. In diabetes, both hyperglycaemia and hyperinsulinemia can increase oxidative stress by produces ROS through the polyol pathway, glyco-oxidation, impairment of antioxidant system and activation of endothelial NAD(P)H oxidase. Accordingly, diabetes has been suggested to enhance DNA damage by ROS, which in turn, can cause diabetic complications. It has also been reported that point mutations and deletions of mtDNA accumulate due to

* Correspondence: Dere Dok. Umut Ap. 21/44, Sahrayicedid, Erenkoy, Istanbul, Turkey; Tel: 90 212 4143000/ 22407; E-mail: stare63@yahoo.com

intensive exposure of mtDNA to reactive oxygen species (ROS). On the other hand, methylation of chromosomal DNA has genetically important regulatory effects on gene expression , especially when involving cpG-islands located in the promoter regions of many genes. There is considerable indirect evidence that aberrant DNA methylation plays a role in the development of type 2 diabetes. Recently, defective repair of DNA alkylation damage by the DNA repair enzyme, O^6-methylguanine DNA methyltransferase, have shown in streptozotocin-induced experimental diabetes as well in the patients with type 1 and type 2 diabetes in our laboratory.

The purpose of the present chapter is to discuss the mechanisms and role of DNA oxidation and alkylation in the pathogenesis of diabetes.

INTRODUCTION

Diabetes mellitus is a chronic disease marked by elevated blood glucose and urinary glucose excretion. Generally, it is classified into two main types. Type 1, insulin-dependent diabetes mellitus (IDDM), is characterized by a loss of insulin production due to autoimmune destruction of the pancreatic islet β cells. Type 2, non- insulin-dependent diabetes mellitus (NIDDM), is characterized by a deficiency in the capacity of insulin secretion and impared insulin-stimulated glucose disposal in peripheral tissues (as an another word insulin resistance). Recently, a novel subtype of diabetes mellitus, maternally inherited diabetes and deafness (MIDD) has been identified and its clinical features include maternal inheritance, hearing disturbance, insulin deficiency, and a tendency toward progression [1]. Complications of diabetes include systemic vascular disease, microvascular disease of the eye causing bleeding and retinal degeneration, cataract, kidney damage leading to renal failure and damage to peripheral nerves. The development of diabetic complications is positively associated with prolonged hyperglycemia, hence poorly-controlled, long duration diabetic patients are more likely to suffer diabetic complications than well-controlled patients.

Although not all are known, genetic factors seem to play an important role in the pathogenesis of diabetes mellitus. Several genes, which are involved in the metabolic process including the synthesis and secretion of insulin in pancreatic β cells and the insulin action on various target cells, have been proposed to be associated with or responsible for the pathogenesis of the disease [2-4]. Regulation of glucose-induced insulin secretion is crucially dependent on glucokinase function in pancreatic β cells. Some mutations on the gene encoding glucokinase have been identified in patients suffering from maturity-onset diabetes of young (MODY) [5,6]. Mitochondrial oxidative phosphorylation and the ATP production in pancreatic β cells are generally accepted to play significant roles in insulin secretion in response to glucose. An impairment of the oxidative phosphorylation system in mitochondria could cause diabetes mellitus as a result of defective insulin secretion. As a matter of fact, mitochondrial gene mutations in diabetes have been shown [7,8]. On the other hand, insulin receptor gene is an attractive target for in vivo studies of insulin resistance, as it has been shown to be the master switch of the metabolic and growth-promoting actions of insulin [9].

Lets have a look for the mechanisms of DNA damage leading genetic alterations which are related to diabetes mellitus.

DNA DAMAGE

DNA molecule including all genetic specialities of individuals is exposured to damage as spontaneously and by environmental factors. Unavoidable hydrolytic decay of DNA by depurination, depyrimidination and deamination accounts for the introduction of ~10^4 non-coding or mis-coding lesions per day in the DNA of each human cell. Furthermore, similar levels of damage appear to be caused by base methylation due to non-enzymatic DNA alkylation by the methyl group donor S-adenosylmethionine, and subsequent loss of purines, and by degradation of base residues by free radicals, spontaneously [10]. In addition the spontaneous DNA damage, ionising radiation, UV and chemicals (alkylating agents, cross-linking agents, agents which can be metabolized to electrophilic reactives in the body) are the environmental factors which lead similar DNA lesions. Minor lesions are generally repaired by DNA repair systems. Major lesions induce programmed death, therefore organism conserves itself. Accumulation of moderate lesions on DNA causes mutations and genetic abnormalities.

OXIDATIVE STRESS

Atoms or molecules with an unpaired electron in an outher orbital are termed free radicals. Depending on its reactivity, the free radical is able to oxidize other molecules. It removes an electron from a neighboring molecule and, by becoming stabilized itself, generates another free radical. This process, if perpetuated, induces a so-called chain reaction of redox-processes. Antioxidants, in this scenario, are compounds that donate electrons to free radicals. Thereby, the antioxidant itself becomes a free radical that needs to be detoxified by enzymatic or non-enzymatic mechanisms. At its simplest, antioxidant defence takes the form of low molecular weight compounds, antioxidants such as vitamin C, and vitamin E, which intercept free radicals. Other, more complex approaches involve enzymes, such as superoxide dismutase, catalase and glutathione peroxidase, which have evolved to limit the levels of free radicals. The imbalance between oxidants and antioxidants, caused either by increased generation of oxidants, reduced availability of antioxidants, or both, is frequently called oxidative stress [11].

The living cell is constantly exposed to potentially damaging free radicals, whose origin may be intracellular, such as those arising from normal cellular metabolism (the mitochondrial respiratory chain, neutrophil oxidative burst reaction, the tight regulation of enzyme activities), or extracellular, produced as a consequence, for example, of exposure to ultraviolet radiation, or ionising radiations. Of particular interest are the free radicals, which include the highly reactive hydroxyl radical ($OH^{\cdot}$), superoxide radical ($O_2^{\cdot-}$), non-radical hydrogen peroxide (H_2O_2), and nitric oxide ($NO^{\cdot}$). They are generally termed as reactive oxygen species (ROS). Irrespective of their origin, ROS may interact with cellular biomolecules, such as proteins, lipids, and DNA, leading to modification and potentially serious consequences for the cell. While modified lipids and proteins may be removed via normal turnover of the molecules, damage to DNA is required to be repaired [12]. DNA, unlike proteins and lipids, once modified, cannot be replaced, and as a result has become a focus of much research interest, and its involvement in the pathogenesis of many diseases has been revealed. Recently, increased DNA damage in patients with diabetes mellitus has been reported by various investigators. [13-18].

Mechanisms of Oxidative Damage to DNA

While neither $O_2^{\cdot -}$ nor H_2O_2 reacts directly with DNA, in vivo transition metal ions such as Fe^{2+} or Cu^{1+} are believed to catalyse their conversion into the highly reactive $OH^{\cdot}$ radical, which in turn provokes a broad spectrum of DNA lesions [19]. $NO^{\cdot}$ is an important bioregulatory agent involved in a variety of biological processes including the immune response, cytotoxicity, neurotransmission, and vasodilation. $NO^{\cdot}$ is synthesized in vivo from L-arginine by isoforms of nitric oxide synthase. $NO^{\cdot}$ and $O_2^{\cdot -}$ can react to form peroxynitrite anion. Peroxynitrite anion is stable, but its protonated form, peroxynitrous acid, is a powerful oxidant with properties close to the $OH^{\cdot}$. High concentrations of $NO^{\cdot}$ have been shown to cause deamination of deoxynucleotides and bases within intact DNA in vitro and are mutagenic in vivo [20].

DNA damage derived from ROS include base damage, sugar damage and DNA-protein cross-links.

DNA Base Damage

Hydroxyl radical adds to double bonds of heterocyclic DNA bases and abstracts an hydrogen atom from the base. A base OH-adduct radical occurs. The base OH-adduct radicals are oxidised or reduced depending on their redox properties, redox environment and reaction partners, yielding a variety of products. Addition rate of hydroxyl radical to the double bond of pyrimidines and purines is diffusion-controlled. The distribution of hydroxyl radical attack depends on the electron density of the site of the attacked molecule. Because of its electrophilic nature, hydroxyl radical preferentially adds to the site with the highest electron density [12].

Most of modified bases were identified in DNA in vitro and in mammalian cells upon exposure to free radical-generating systems. Among these, 8-hydroxydeoxyguanosine (8-oxo-dG) is one of the major DNA base modifications and has attracted special attention. 8-oxo-dG is strongly mutagenic, having the propensity to mispair with A residues, leading to an increased frequency of spontaneous G:C $\rightarrow$ T:A transversion mutations unless repaired prior to DNA replication [21].

Sugar Damage

Hydroxyl radical abstracts an hydrogen atom from each of the carbon atoms of the sugar moiety in DNA. The resulting C-centred radicals undergo further reactions to yield a variety of sugar modifications. On the basis of product analyses, the mechanisms of sugar damage by $OH^{\cdot}$ in DNA have been elucidated including that of DNA strand breakage [22,23]. Some sugar products are released from DNA as free modified sugars, whereas others remain within DNA or constitute end groups of broken DNA strands.

DNA–protein Cross-links

As well as base and sugar damages, nuclear proteins can be attacked by radicals.The resulting protein-derived radicals can cross-link to base derived radicals if the two meet in chromatin , giving DNA-protein cross-links that interfere with chromatin unfolding, DNA repair, replication and transcription [24].

Cellular Impacts of Oxidative DNA Damage

Mutation

The presence of oxidized base lesions in DNA is well established and the number of lesions identified is growing. Not surprisingly, 8-oxo-dG is by far the most frequently studied and is often used here as an example of how oxidative lesions may exert their effect. Many oxidative base lesions are mutagenic, irrespective of whether they are formed in situ or arise by misincorporation from the deoxynucleotide pool.

For lesions that can induce conformational changes in DNA in addition to the structural alterations to the native base itself, the potential for enhancing mutagenicity exists. When deoxyguanosine (dG) is oxidized to 8-oxo-dG the native *anti* conformation of dG is maintained; however, when the DNA is made single-stranded, whether it occurs during the replication or transcription, 8-oxo-dG can then adopt the energetically more favorable *syn* conformation. This prevents pairing with deoxycytosine and results in mispairing with deoxyadenosine [25]. However, it seems that oxidative DNA lesions are best described as weakly mutagenic; for example, 8-oxo-dG has mutation frequencies of 2.5–4.8% in mammalian cells although lesion formation, persistence, and accumulation in vivo could give this value greater significance [25]. A further factor affecting the mutagenicity or otherwise of a lesion is the ease with which the lesion is repaired. In mammalian cells, the 8-oxo-dG damage is effectively repaired [26]. 8-OH-Guanine glycosylase 1 (OGG1) liberates 8-oxo-dG via a glycolytic mechanism from double-stranded substrates, relying on an internal Lys residue [27]. Another 8-OH-Guanine glycosylase 2 (OGG2) also repairs this lesion, it predominantly removes 8-oxo-dG from 8-oxo-dG:A pairs that may be formed by misincorporation of 8-oxo-dG into nascent DNA [28]. This is one route by which misincorporated 8-oxo-dG may be addressed. Two other enzymes, MutY homologue (MYH) and MutT homologue 1 (MTH1), may also be involved in the repair of 8-oxo-dG. The former enzyme removes adenine, which mispaired opposite 8-oxo-dG. This mispair may arise from either misincorporation of 8-oxo-dGTP opposite adenine in the template strand or misincorporation of dATP opposite unrepaired 8-oxo-dG in the template strand, directed during DNA synthesis [29,30].

Indeed, oxidative events are reported to be largely responsible for spontaneous mutagenesis [31]. However, mutations are not the only effect of oxidative DNA damage.

Replicative Block

An other DNA oxidation product, thymidine glycol (Tg) occurs mainly as the *cis* isomer, and the mutational specificity of this lesion has been much studied in bacterial and mammalian cells. Even though some mutations have been identified associated with Tg, the general agreement appears to be that Tg does not have a significant mutagenic potential. More recent studies have suggested that rather than being mutagenic, Tg blocks replication one nucleotide before and after the lesion [31]. On the other hand, formamidopyrimidine derived from 3-methyladenine or 8-oxo-dG by opening imidazole ring has been shown to block DNA replication [32].

Deletions

Some oxidative DNA lesions may result deletions. Chaung and Boorstein [33] have reported that, rather than inducing point mutations, the presence of 5-OHMetil Uracyl (OHMUra) leads to large/intermediate deletions in mammalian cells. However, these deletions do not arise as a result of mispairing or misincorporation; instead, it seems that base excision repair by OHMUra-DNA glycosylase may lead to the deletions [33].

Microsatellite Instability/loss of Heterozygosity

In normal cells, the length of repetitive sequences of DNA, so-called microsatellites, is constant, but the length of these repeats can be variable in tumor cells. This microsatellite instability (MI) derives from DNA damage. The association between oxidative events and MI appears to be increasing, with oxidative DNA damage shown to increase the frequency of microsatellite instability through induction of mutations in the repeat sequences [34], and discontinuous loss of heterozygosity being described as a signature mutational pattern of oxidative DNA damage [35].

Epigenetic Effects of Oxidative DNA Damage

While the mutagenic effects of oxidative DNA damage are largely well recognized, epigenetic effects have also gained interest recently. When exposed to oxidants, mammalian cells express stress-induced genes or genes encoding antioxidant defenses. Such adaptive responses to oxidative insults are not surprising; however, it remains to be established whether DNA damage itself has any involvement in gene expression or whether this is merely a by-product of ROS generated during stress. Many transport proteins, receptors, signal transduction systems contain essential –SH groups. Attack upon these by ROS may oxidize them and alter their functions. Indeed, ROS are themselves intracellular signaling molecules [36], although whether sufficient levels of ROS can be induced by an oxidative insult, such that gene expression is affected. It has been demonstrated that the presence of 8-oxo-dG in promoter elements can affect transcription factor binding [37]. The regulation of transcription occurs through promoter recognition by general transcription factors. The binding of site-specific DNA binding proteins regulates levels of transcription. The transcription factors AP-1 and NF-κB are both regulated by the intracellular redox state and share GC-rich consensus binding sites, prime targets for 8-oxo-dG formation. Whilst ROS can induce gene expression through AP-1, Ghosh and Mitchell [37]demonstrated that a single 8-oxo-dG moiety in the promoter region of AP-1 could completely prevent transcription factor binding, and hence gene expression. However, this was not the case for NF-κB which, the authors suggested, was due to the role ROS have in its dissociation from the inhibitory protein IκB.

Induction of mutation does not necessarily derive solely from oxidative modification of DNA, modification of proteins does occur and modification of DNA polymerases has been reported to decrease fidelity [38].

All data imply that events at the DNA level, other than mutation, are also involved in pathogenesis.

MITOCHONDRIAL DNA DAMAGE AND DIABETES

Mitochondrial DNA (mtDNA), the only organelle DNA in animal cells, encodes 13 subunits of the respiratory chain enzyme complexes, 22 transfer RNAs and two ribosomal RNAs while nuclear DNA encodes for the majority of respiratory chain subunits and the machinery for mitochondrial translation via mitochondrial biogenesis. Mitochondria produce reactive oxygen species (ROS) under physiological conditions in association with activity of the respiratory chain in aerobic adenosine triphosphate (ATP) production. The production of ROS is essentially a function of O_2 consumption. Hence, increased mitochondrial activity per se can be an oxidative stress to cells [39]. It has been observed that mtDNA contains a higher steady-state amount of oxidative DNA damage than nuclear DNA [40], because (1) mtDNA is located near the respiratory chain, potentially generating ROS; (2) mitochondria have less efficient DNA repair systems compared with the nucleus; (3) mtDNA is not protected by histone like nuclear DNA; (4) mtDNA polymerase-gamma which is responsible for mtDNA replication has a relatively high insertion error-rate of about 1 per 7000 bases; (5) mutation in a single mtDNA molecule can be propagated within cells due to the continuous turnover of mtDNA in all cells of an organism [41]. The damage of mtDNA by ROS could result in errors in the synthesis of respiratory chain proteins, leading to dysfunction in ATP production and increasing a leak of ROS to cause further mtDNA mutations. This vicious cycle can decrease cell viability and cause the ageing process. Somatic mtDNA mutations not only accumulate during normal ageing but also in various diseases associated with the enhancement of ROS, such as cancer [42], coronary atherosclerotic heart disease [43], neurodegenerative diseases [44] and diabetes [7,8]. Since mitochondria are organelles producing ATP, and insulin secretion from pancreatic cells is dependent on ATP generation, abnormality in mtDNA may result defective insulin secretion in response to glucose and other nutrients [45]. Recently, mtDNA was shown to be required for the regulation of glucose-stimulated insulin secretion in a pancreatic β cell line of the mouse [46]. Moreover, it was also observed that ethidium bromide-induced inhibition of mitochondrial gene transcription leads to the suppression of glucose-stimulated insulin secretion in another mouse pancreatic β cell line [47].

The mtDNA 3243 nucleotide position is one of the vulnerable spots for somatic mutation, because this point has remained single stranded for a relatively long time during the duplication of mtDNA. A point mutation of mtDNA at nucleotide point 3243 A to G (A3243G) has been suggested as a major genetic cause of diabetes [1,7]. The mutation in the gene for tRNA-Leu at position 3243 of mtDNA is associated with maternally inherited diabetes mellitus with deafness [7,48]. Stimulation of pancreatic islet β cells by glucose normally induces a series of responses, including NADH generation by glycolysis, enhanced oxygen consumption, a rise in the ATP/ADP ratio, closure of ATP-sensitive K^+ channels, membrane depolarization, elevation of cytoplasmic free Ca^{2+}, and finally insulin secretion [45]. Thus a dramatic decrease in oxygen consumption rate and the ATP/ADP ratio elicited by the A3243G mtDNA mutation of pancreatic β cells may lead to an impairment of insulin secretion in response to glucose stimulation. Furthermore, the activity of ATP-dependent hexokinase, which is one of the most important regulatory enzymes in glycolysis, may be affected by the decrease of energy charge elicited by the A3243G mutation of mtDNA and thereby lead to impared glucose uptake by muscle and pancreatic β cells [49].

On the other hand, depletion of mtDNA could impair mitochondrial and pancreatic β cell function. There is also considerable evidence that point mutations in other sites of mtDNA and deletions of mtDNA are related to diabetes mellitus[50]. Lee et al [51] have demonstrated decreased mtDNA in peripheral blood of patients with NIDDM as well in offspring of diabetic patients [52].

All these studies have provided strong evidence to support the notion that mitochondrial function is crucial for the glucose-stimulated secretion of insulin by pancreatic β cells. Mitochondrial transcription factor A (mtTFA), which provides the replication primer, plays a key role for the regulation of mtDNA replication. The gene expression of mtTFA is altered depending on the cellular conditions and directly regulated by several transcription factors such as nuclear respiratory factor 1 (NRF-1) and NRF-2. When the factors affecting the transcriptional activity of the mtTFA promoter investigated, it was found that the mtTFA promoter may be regulated by methylation of NFR-1 site and by exogenous stimuli such as hydrogen peroxide and glucose [53]. There are also descriptive epidemiological data suggesting that alkylating agents may contribute to an increased prevalence of IDDM [54-56] as well NIDDM [57]. Therefore, in respect with those data, it is of inerest the role of DNA methylation (as a general term, alkylation) in diabetes.

DNA Methylation and Diabetes

The location of a gene in eukaryotic DNA is important; if the promoter is burried in highly condensed chromatin the gene is unlikely to be transcribed. Cytosine can be methylated to give 5-methylcytosine and about 70% of cytosine bases in C-G sequences in mammalian DNA are methylated. Methylcytosine deters the transcription. Therefore, DNA methylation is a major regulator of transcriptional activity and is involved in the regulation of expression of a broad variety of genes. DNA methylation represents a stable cellular memory for the maintenance of gene expression patterns. There is evidence that DNA methylation memory is involved in maintaining gene expression patterns associated with insulin resistance in NIDDM [58]. Several genes involved in glucose metabolism have been shown to exhibit differential DNA methylation in their promoters, e.g., facilitative glucose transporter4 (GLUT4), the major glucose transporter in adipose and muscle tissues and uncoupling protein 2, a major candidate gene for development of NIDDM [58].

DNA methylation changes have been to shown to occur during aging of organisms in several tissue types [59-61]. Accumulating age-related DNA methylation changes are involved in a number of different diseases such as diabetes, atherosclerosis and cancer [58-60]. NIDDM is strongly age-related: not only is its incidence increased in older populations, but also the metabolic situation of individual patients deteriorates over time. DNA methylation errors that accumulate with increasing age could provide an explanation of the relation between DNA methylation and diabetes. On the countrary, recent insights into the pathogenesis of transient neonatal diabetes, a rare subtype of diabetes that is characterized by transient hyperglycemia in the neonatal period and a predisposition for diabetes in adult life, provide a link between methylation, gene dosage effects and diabetes [62]. These contradictory data for the relation between age-dependent DNA methyltaion status and diabetes , imply the necessity of further studies on this subject.

A general defect in DNA methylation in diabetes is suggested by the recent observation that S-adenosylmethionine (SAM), the main phsiologic donor of methyl groups, is decreased in erythrocytes of diabetic patients. DNA methylation errors could add to the development of NIDDM by reducing gene expression that needs to be adjusted to fast changing glucose levels. In addition, decreased erythrocyte concentrations of SAM, SAM/S-adenosyl homocystein ratios, and lymphocyte methylenetetrahydrofolate reductase activity were found to be associated with disease progression [63].

Environmental alkylating agents, as a contributory factor for development of diabetes have been gained considerable interest for a long time. There are descriptive epidemiological data suggesting that *N*-nitroso compounds from smoke-cured mutton may contribute to an increased prevalence of Type 1 diabetes in Iceland [54]. A case–control study in Sweden has shown a dose response relationship between the risk of Type 1 diabetes and intake of food containing nitrosamines [55]. Another study from the USA showed a correlation between the content of nitrate, a precursor of alkylating agent, in drinking water and the incidence of Type 1 diabetes [56]. Moreover, it has been suggested that *N*-nitroso compounds from the Betel-nut are related to the increased prevalence of Type 2 diabetes in Asians [57].

The role of DNA methylation in the etiology of diabetes has also been examined in animal studies. Streptozotocin (SZ) is an alkylating agent and widely used for the induction of diabetes in experimental animals due to damaging action on the pancreatic β cells which causes cell death [64]. Intracellular action of SZ results in changes of DNA in pancreatic β cells comprising its fragmentation [65]. Although determined a few mechanisms are present, recent experiments have proved that the main way for the SZ-induced β cell death is alkylation of DNA [66]. The alkylating activity of SZ is related to its nitrosourea moiety, especially at the O^6 position of guanine [67]. After SZ injections to rats, different methylated purines were found in tissues of these animals [68]. O^6–methyl guanine (O^6 -MG) adducts on DNA lead to the formation of stable cytotoxic DNA crosslinks. Therefore methylation of guanines by SZ induces apoptosis in cells [69]. Moreover, O^6-MG pairs with thymine, in addition to cytidine, during DNA replication, causing G≡C → A= T transition mutation [70]. O^6-MG is repaired in mammalian cells by O^6-methylguanine DNA methyltransferase O^6–MGMT (E.C 2.1.1.63), a DNA repair enzyme. The methyl group at the O-6 position of guanine is transferred irreversibly to a cysteine residue within O^6–MGMT, thus inactivating the enzyme. O^6 –MGMT activity was found to be lower in pancreas of SZ-induced diabetic rats [71] as well in leukocytes of patients with diabetes mellitus [72]. These data imply that O^6 –MGMTdepletion may be one of the mechanisms which are contribute the development of diabetes.

CONCLUSION

The role of ROS on the pathophysiology of diabetes has been known for a long time. However, mtDNA has attracted special attention recently. Through electron transport chain mitochondria generates ROS, therefore mtDNA exposure much more oxidation than nuclear DNA. The damage of mtDNA by ROS could result in errors in the synthesis of respiratory chain proteins, leading to dysfunction in ATP production and to cause further mtDNA mutations. Insulin secretion from pancreatic cells is dependent on ATP generation and abnormality in mtDNA may result defective insulin secretion. As a matter of fact, an A to G mutation in the tRNA-Leu gene at nucleotide position 3243 of mtDNA has been determined in the some diabetic patients. Besides, deletions and decreased mtDNA in peripheral blood of diabetic patients and in offspring of diabetics have been determined. All those data emphasize the importance of diet-derived antioxidants in health. The potential therapeutic role of antioxidant molecules might be appreciated by clinicans.

Although the role of environmental alkylating agents in the develpment of diabetes has been implied, the mechanism has not been explained until recently. Recent molecular studies revealed that methylation of the promoter region of a gene regulates the expression. It has been found that several genes involved in glucose metabolism exhibit differential DNA methylation in their promoters. Antioxidants just can be protective against oxidative damage, not protective for abnormal methylation. Methylation errors can be repaired by DNA repair enzymes.Recently, O^6 –MGMT depletion has been suggested as a contributory factor for development of diabetes. More detailed studies on O^6 –MGMT and on other DNA repair enzymes which are reponsible for repair of alkylated bases may provide different therapeutic approaches for clinicans.

REFERENCES

[1] Kadowaki, T; Kadowaki, Y; Mori, Y; Tobe, K; Sakuta, R; Suzuki, Y; Tanabe, Y; Sakura, H; Awata, T; Goto, Y; Hayakawa, T; Matsuoka, K; Kawamori, R; Kamada, T; et al. A subtype of diabetes mellitus associated with a mutation in the mitochondrial gene. *New England Journal of Medicine*, 1994, 330, 962-968.

[2] Steiner, DF; Tager, HS; Chan, SJ; Nanjo, K; Sanke, T; Rubenstein, AH. Lessons learned from molecular biology of insulin-gene mutations. *Diabetes Care*, 1990, 13 600-609.

[3] Kadowaki, T; Bevins, CL; Cama, A; Ojamaa, K; Marcus-Samuels, B; Kadowaki, H; Beitz, L; McKeon, C; Taylor, SI. Two mutant alleles of the insulin receptor gene in a patient with extreme insulin resistance. *Science* 1988, 240, 787-790.

[4] Vionnet, N; Stoffel, M; Takeda, J; Yasuda, K; Bell, GI; Zouali, H; Lesage, S; Velho, G; Iris, F; Passa, PH; Cohen, F&D. Nonsense mutation in the glucokinase gene causes early-onset non-insulin-dependent diabetes mellitus. *Nature*, 1992, 356, 721-722.

[5] Velho, G; Froguel, P; Clement, K; Pueyo, ME; Rakotoambinina, B; Zouali, H; Passa, P; Cohen, D; Robert, JJ. Primary pancreatic beta-cell secretory defect caused by mutations in glucokinase gene in kindreds of maturity onset diabetes of the young. *Lancet*, 1992, 340, 444-448.

[6] Permut, MA; Chiu, KC; Tanizawa, Y. Glucokinase and NIDDM. A candidate gene that paid off. *Diabetes*, 1992, 41, 1367-1372.
[7] Chen, Y; Liao, WX; Roy, AC; Loganath, A; Ng, SC. Mitochondrial gene mutations in gestational diabetes mellitus. *Diabetes Research and Clinical Practise*, 2000, 48, 29-35.
[8] Nomiyama, T; Tanaka, Y; Hattori, N; Nishimaki, K; Nagasaka, K; Kawamori, R; Ohta, S. Accumulation of somatic mutation in mitochondrial DNA extracted from peripheral blood cells in diabetic patients. *Diabetologia*, 2002, 45, 1577-1583.
[9] Okamoto, H. & Accili, D. (2003) In vivo mutagenesis of the insulin receptor. *Journal of Biological Chemistry*, 278(31), 8359-8362.
[10] Lindahl, T. Repair Of intrinsic DNA lesions. *Mutation Research*, 1990, 238, 305-311.
[11] Gutteridge, J.M. & Mitchell, J. (1999) Redox imbalance in the critically ill. *British Medical Bulletin*, 55, 49-75.
[12] Evans, MD; Dizdaroglu, M; Cooke, MS. Oxidative DNA damage and disease: induction, repair and significance. *Mutation Research*, 2004, 567(1), 1-61.
[13] Dandona, P; Tnusu, K; Cook, S; Synder, B; Makowski, J; Armstrong, D. Oxidative damage to DNA in diabetes mellitus. *Lancet*, 1996, 347 (17), 444-445.
[14] Collins, AR; Raslova, K; Somorovska, M; Petrovska, H; Ondrusova, A; Vohnout, B; Fabry, R; Dusinska, M. DNA damage in diabetes: correlation with a clinical marker. *Free Radical Biology and Medicine*, 1998, 25(3), 373-377.
[15] Hanon-Fletcher, MPA; O'Kane, MJ; Moles, KW; Weatherup, C; Barnett,CR; Barnett, YA. Levels of peripheral blood cell DNA damage in insulin dependent diabetes mellitus human subjects. *Mutation Research*, 2000, 460, 53-60.
[16] Şardaş, S; Yılmaz, M; Öztok, U; Çakır, N; Karakaya, AE. Assessment of DNA strand breakage by comet assay in diabetic patients and the role of antioxidant supplementation. *Mutation Research*, 2001, 490, 123-129.
[17] Dinçer, Y; Akçay, T; Alademir, Z; İlkova, H. Assesment of DNA Base Oxidation and Glutathione Level in Patients with Type 2 Diabetes. *Mutation Research*, 2002, 505(1-2), 75-81.
[18] Dinçer, Y; Akçay, T; Ilkova, H; Alademir, Z; Ozbay, G. DNA Damage and Antioxidant Defense in Peripheral Leukocytes of Patients with Type I Diabetes Mellitus. *Mutation Research*, 2003, 527, 49-55.
[19] Hartwig, A; Dally, H; Schlepegrell, R. Sensitive analysis of oxidative DNA damage in mammalian cells: use of the bacterial Fpg protein in combination with alkaline unwinding. *Toxicology Letters*, 1996, 88, 85-90.
[20] Thomas, S; Lowe, JE; Knowles, RG; Green, IC; Green, MHL. Factors affecting the DNA damaging activity of superoxide and nitric oxide. *Mutation Research*, 1998, 402, 77-84.
[21] Cheng, KC; Cahill, DS; Kasai, H; Nishimura, S; Loeb, LA. 8-Hydroxyguanine, an abundant form of oxidative DNA damage, causes G→ T and A → C substitutions. *Journal of Biological Chemistry*, 1992, 267 (1),166-172.
[22] Dizdaroglu, M; Schulte-Frohlinde D, von Sonntag, C. Isolation of 2-deoxy-erythro-pentonic acid from an alkali-labile site in gamma-irradiated DNA. *International Journal of Radiation Biology & Related Studies in Physics, Chemistry & Medicine*, 1977, 32, 481–483.

[23] Beesk, F; Dizdaroglu, M; Schulte-Frohlinde D; von Sonntag, C. Radiation-induced DNA strand breaks in deoxygenated aqueous solutions. The formation of altered sugars as end groups. *International Journal of Radiation Biology & Related Studies in Physics, Chemistry & Medicine,* 1979, 36, 565–576.

[24] Altman, SA; Zastawny, TH; Randers-Eichhorn, L; Cacciuttolo, MA; Akman, SA; Dizdaroglu, M; Rao, G. Formation of DNA-protein cross-links in cultured mammalian cells upon treatment with iron ions. *Free Radical Biology and Medicine*, 1995, 19 (6), 897-902.

[25] Cooke, MS; Evans, MD; Dizdaroglu, M; Lunec, J. Oxidative DNA damage: mechanisms, mutation, and disease. *FASEB Journal*, 2003, 17, 1195 - 1214.

[26] Boiteux, S. & Radicella, J. P. (2000) The human OGG1 gene: structure, functions, and its implication in the process of carcinogenesis. *Archieves of Biochemistry & Biophysics,* 377,1-8

[27] Bruner, SD; Norman, DP; Verdine, GL. Structural basis for recognition and repair of the endogenous mutagen 8-oxoguanine in DNA. *Nature (London),* 2000, 403, 859-866.

[28] Hazra, TK; Izumi, T; Maidt, L; Floyd, RA; Mitra, S. The presence of two distinct 8-oxoguanine repair enzymes in human cells: their potential complementary roles in preventing mutation. *Nucleic Acids Research,* 1998, 26, 5116-5122.

[29] McGoldrick, JP; Yeh, YC; Solomon, M; Essigmann, JM; Lu, AL. Characterization of a mammalian homolog of the Escherichia coli MutY mismatch repair protein. *Molecular and Cellular Biology,* 1995, 15, 989-996.

[30] Slupska, MM; Baikalov, C; Luther, WM; Chiang, JH; Wei, YF; Miller, JH. Cloning and sequencing a human homolog (hMYH) of the *Escherichia coli* mutY gene whose function is required for the repair of oxidative DNA damage. *Journal of Bacteriology* 1996, 178, 3885-3892.

[31] Rossman, T.G.& Goncharova, E.I. (1998) Spontaneous mutagenesis in mammalian cells is caused mainly by oxidative events and can be blocked by antioxidants and metallothionein. *Mutation Research,* 402, 103-110.

[32] Boiteux, S; O'Connor, TR; Lederer, F; Gouyette, A; Laval, J. Homogenous Escherichia coli FPG protein. *Journal of Biological Chemistry*, 1990, 265(7), 3916-3922.

[33] Chaung, W.& Boorstein, R.J.(1997) Molecular spectrum of mutations induced by 5-hydroxymethyl-2'-deoxyuridine in (CHO)-PL61 cells. *Mutation Research,* 373,125-137.

[34] Jackson, AL; Chen, R; Loeb, LA. Induction of microsatellite instability by oxidative DNA damage. *Proceedings Of The National Academy Of Sciences USA,* 1998, 95,12468-12473.

[35] Turker, MS; Gage, BM; Rose, JA; Elroy, D; Ponomareva, ON; Stambrook, PJ; Tischfield, JA. A novel signature mutation for oxidative damage resembles a mutational pattern found commonly in human cancers. *Cancer Research,* 1999, 59,1837-1839.

[36] Halliwell, B. & Gutteridge, J.M.C. (1996) *Free Radicals in Biology and Medicine: Oxidative stress: adaptation, damage, repair and death (*3.edition). Oxford, Oxford University Press.

[37] Ghosh, R. & Mitchell, D.L. (1999) Effect of oxidative DNA damage in promoter elements on transcription factor binding. *Nucleic Acids Research.* 27, 3213–3218.

[38] Chevion, M; Jiang, Y; Har-El, R; Berenshtein, E; Uretzky, G; Kitrossky, N. Copper and iron are mobilized following myocardial ischemia: possible predictive criteria for tissue injury. *Proceedings Of The National Academy Of Sciences USA,* 1993, 90 1102–1106.

[39] Kang, D. & Hamasaki, N. (2003). Mitochondrial oxidative stress and mitochondrial DNA. *Clinical Chemistry and Laboratory Medicine* , 41(10), 1281-1288.

[40] Richter, C. (1995). Oxidative damage to mitochondrial DNA and its relationship to ageing. *International Journal of Biochemistry & Cell Biology*. **27**, 647-653.

[41] Kadenbach, B; Muncher, C; Frank, V; Muller-Hocker, J; Napiwotzki, J. Human aging is associated with stochastic somatic mutations of mitochondrial DNA. *Mutation Research*, 1995, 338, 161-172.

[42] Fliss, MS; Usadel, H; Cabarello, OL; Wu, L; Buta, MR; Eleff, SM. Facile detection of mitochondrial DNA mutations in tumors and bodily fluids. *Science*, 2000, 287, 2017-2019.

[43] Corral-Debrinski, M; Shoffner, JM; Lott, MT; Wallace, DC. Association of mitochondrial DNA damage with aging and coronary atherosclerotic heart disease. *Mutation Research*, 1992, 275, 169-180.

[44] Corral-Debrinski, M; Horton, T; Lott, MT; Shoffner, JM; McKee, AC; Beal, MF; Graham, BH; Wallace, DC. Marked changes in mitochondrial DNA deletion levels in Alzheimer brains. *Genomics*, 1994, 23, 471-476.

[45] Newgard, C.B. & McGarry, J.D. (1995). Metabolic coupling factors in pancreatic beta-cell signal transduction. *Annual Review of Biochemistry*, 64, 689-719.

[46] Soejima, A; Inoue, K; Takai, D; Kaneko, M; Ishihara, H; Oka, Y; Hayashi, JI. Mitochondrial DNA is required for regulation of glucose-stimulated insulin secretion in a mouse pancreatic beta cell line MIN 6. *Journal of Biological Chemistry,* 1996, 271, 26194–26199.

[47] Hayakawa, T; Noda, M; Yasudo, K; Yorifuji, H; Taniguchi, S; Miwa, I; Sakura, H; Terauchi, Y; Hayashi, J; Sharp, GW; Kanazawa, Y; Akanuma, Y; Yazaki, Y; Kadowaki T. Ethidium bromide-induced inhibition of mitochondrial gene transcription suppresses glucose-stimulated insulin release in the mouse pancreatic β-cell line β HC9. *Journal of Biological Chemistry,* 1998, 273(32), 20300–20307.

[48] van den Ouweland, JM; Lemkes, HH; Trembath, RC; Ross, R; Velho, G; Cohen, D; Froguel, P; Maassen, JA. Maternally inherited diabetes and deafness is a distinct subtype of diabetes and associates with a single point mutation in the mitochondrial tRNA (Leu(UUR)) gene. *Diabetes*, 1994, 43, 746-751.

[49] van den Ouweland, JMW; Lemkes, HHPJ; Gerbitz, KD; Maassen, JA. Maternally inherited diabetes and deafness (MIDD): a distinct subtype of diabetes associated with a mitochondrial $tRNA^{Leu(UUR)}$ gene point mutation. *Muscle Nerve,* 1995, (3 Suppl), S124–S130.

[50] Liang, P; Hughes, V; Fukagawa, NK. Increased prevalence of mitochondrial DNA deletions in skeletal muscle of older individuals with impaired glucose tolerance. *Diabetes*, 1997, 46, 920-923.

[51] Lee, HK; Song, JH; Shin, CS; Park, DJ; Park, KS; Lee, KU; Koh, CS. Decreased mitochondrial DNA content in peripheral blood precedes the development of non-insulin-dependent diabetes mellitus. *Diabetes Research and Clinical Practise*, 1998, 42, 161-167.

[52] Song, J; Oh, JY; Sung, Y-A; Pak, YK; Park, KS; Lee, HK. Peripheral blood mitochondrial DNA content is related to insulin sensitivity in offspring of type 2 diabetic patients. *Diabetes Care*, 2001, 24, 865-869.

[53] Choi, YS; Kim, S; Pak, YK. Mitochondrial transcription factor A (mtTFA) and diabetes. *Diabetes Research and Clinical Practise*, 2001, 54 (2 Suppl), S3-S9.

[54] Helgason, T; Ewen, SWB; Ross IS; Stowers, JM. Diabetes produced in mice by smoked/cured mutton. *Lancet,* 1982, 2(8306), 1017–1022.

[55] Dahlquist, G; Blom, L; Persson, LA; Sandström A; Wall, S. Dietary factors and the risk of developing insulin-dependent diabetes mellitus in childhood. *British Medical Journal,* 1990, 300, 1302–1306.

[56] Kastraba, JN; Gay, EC; Rewers M; Hamman, RF. Nitrate levels in community drinking waters and risk of IDDM. *Diabetes Care* , 1992,15, 1505–1508.

[57] Boucher, BJ; Ewen, SWB; Stowers, JM. Betel-nut (*Areca catechu*) consumption and the induction of glucose intolerance in adult CD1 mice and in their F1 and F2 offspring. *Diabetologia*, 1994, 37, 49–55.

[58] Maier, S. & Olek, A. (2002). Diabetes: a candidate disease for efficient DNA methylation profiling. *Journal of Nutrition*, 2002, 132(8 Suppl), 2440S-2443S.

[59] Post, WS; Goldschmidt-Clermont, PJ; Wilhide, CC; Heldman, AW; Sussman, M S; Ouyang, P; Milliken, EE; Issa, JP. Methylation of the estrogen receptor gene is associated with aging and atherosclerosis in the cardiovascular system. *Cardiovascular Research*, 1999, 43, 985-991.

[60] Bornman, DM; Mathew, S; Alsruhe, J; Herman, JG; Gabrielson, E. Methylation of the E-cadherin gene in bladder neoplasia and in normal urothelial epithelium from elderly individuals. *American Journal of Pathology*, 2001, 159, 831-835.

[61] Issa, JP; Ahuja, N; Toyota, M; Bronner, MP; Brentnall, TA. Accelerated age-related CpG island methylation in ulcerative colitis. *Cancer Research*, 2001, 61, 3573-3577.

[62] Temple, IK; Gardner, RJ; Mackay, DJ; Barber, JC; Robinson, DO; Shield, J P. Transient neonatal diabetes: widening the understanding of the etiopathogenesis of diabetes. *Diabetes*, 2000, 49, 1359-1366.

[63] Poirier, LA; Brown, AT; Fink, LM; Wise, CK; Randolph, CJ; Delongchamp, RR; Fonseca, VA. Blood S-adenosylmethionine concentrations and lymphocyte methylenetetrahydrofolate reductase activity in diabetes mellitus and diabetic nephropathy. *Metabolism*, 2001, 50, 1014-1018.

[64] Sandler, S. & Swenne, I.(1985) DNA repair synthesis in the pancreatic islets of streptozotocin treated mice. *Diabetes Research*, 2, 255-258.

[65] Morgan, NG; Cabble, HC; Newcombe, NR; Williams, GT. Treatment of cultured pancreatic β cells with streptozotocin induces cell death by apoptosis. *Bioscience Reports,* 1994, 14, 243-250.

[66] Elsner, M; Guldbakke, B; Tiedge, M; Munday, R; Lenzen, S. Relative importance of transport and alkylation for pancreatic beta-cell toxicity of streptozotocin. *Diabetologia*, 2000, 43(12), 1528-1533.

[67] Szkudelski T. (2001). The mechanism of alloxan and streptozotocin action in B cells of the rat pancreas. *Physiological Research*, 50(6), 537-546.

[68] Bennett, R.A. & Pegg, A.E. (1981) Alkylation of DNA in rat tissues following administration of streptozotocin. *Cancer Research,* 41, 2786-2790.

[69] Murata, M; Takahashi, A; Saito, I; Kawanishi, S. Site-specific DNA methylation and apoptosis induction by diabetogenic streptozotocin. *Biochemical Pharmacology*, 1999, 57, 881-887.

[70] Iwakuma, T; Sakumi, K; Nakatsuru, Y; Kawate, H; Igarashi, H; Shiraishi, A; Tsuzuki, T; Ishikawa, T; Sekiguchi, M. High incidence of nitrosamine-induced tumorigenesis in mice lacking DNA repair methyltransferase. *Carcinogenesis* , 1997, 18, 1631-1635.

[71] Dinçer, Y; Çakatay, U; Kayalı, R; Yılmaz, İA; Telci, A; Çelebi, N; Hatemi, H; Bağrıaçık, N; Akçay, T. Pancreatic O^6-Methylguanine DNA methyltransferase level in streptozotocin-induced diabetic rats. *Biomedical Research*, 2002, 23(5), 203-207.

[72] Akçay, T; Dinçer, Y; Çelebi, N; İlkova, H. O^6-Methylguanine DNA methyltransferase activity in diabetic patients. *Diabetes. Research Clinical Practise*, 2003, 61, 1-6.

INDEX

D

E

F

G

H

I

N

O

P

R

S

T

U

V

W

X

Y

Z